Revit Architecture 2023 for Electrical Workers

Elise Moss

SDC
PUBLICATIONS

SDC Publications
P.O. Box 1334
Mission, KS 66222
913-262-2664
www.SDCpublications.com
Publisher: Stephen Schroff

ISBN-13: 978-1-63057-529-8
ISBN-10: 1-63057-529-1

Printed and bound in the United States of America.

Preface

This book began with a call to Nancy Tremblay at Autodesk's Training Program. I am an Autodesk Certified Instructor. This means I am "on call" to provide training at Autodesk Training Centers and also to provide in-house corporate training on Autodesk software. Susan Bowron is an instructor for the IBEW Local 35 in San Leandro, CA. The union provides training to local electricians to provide them with the necessary skills to remain competitive in the job market. Susan wanted to provide a training class in Revit specific to electricians, so they could provide the necessary construction documentation to their clients. She needed an instructor and Nancy recommended me.

I was intrigued by the idea for a lot of reasons. Firstly, I am an avid Revit user. I started using Revit before it was an Autodesk software. I advocated Autodesk to acquire Revit to add to their offerings because I felt strongly that it was a major player in the BIM market. Secondly, I am a strong advocate for users and training. Knowledge is power. Learning the right skills is so very important in today's job market. It can make the difference between being able to feed your family and being homeless. I have seen the results in my own classrooms. Thirdly, I have always felt that the existing MEP training books are insufficient for many students like the members of the IBEW. They need step-by-step instructions and explanations behind why they are doing each step. This method has been the foundation for the types of books I write. Trying to make complex software and ideas accessible to users has been my primary goal.

Susan and some of her colleagues have reviewed the content in the text and provided valuable feedback.

I also am struck that in an era where people are calling for more women in STEM: Nancy, Susan, and I are all female. There are plenty of women in STEM, we just tend to work quietly to make things happen…like this textbook.

Files used in this text can be downloaded from
www.SDCpublications.com/downloads/978-1-63057-529-8.

Acknowledgements

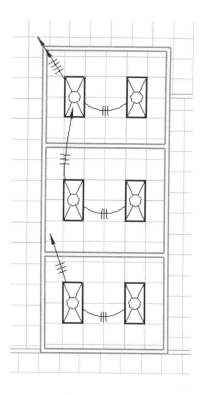

Feel free to email me if you have any questions, comments, or problems with any of the exercises in this text. I get email from all over the world and usually respond within twenty-four hours. My email address is elise_moss@mossdesigns.com.

I have been teaching for more than a decade. I started using AutoCAD in 1982 while I was still attending college. Even then, there was no doubt in my mind that Computer Aided Drafting, as it was called then, was the future of design. Revit continues to evolve, and its power continues to grow in the marketplace. My students amaze and inspire me every day. When I write my textbooks, I imagine one of my students sitting in front of me struggling, trying to figure out how to use a mouse, how to locate the right icon, and feeling frustrated with the effort, but not giving up. This text is intended for classroom use and for beginner learners.

This is the second edition of the text. The content has been expanded and adjusted based on feedback from electricians who attended training at the JATC in San Leandro, CA. Thanks to Kevin Meyer, Doug Rose, Tom Shimabukuro, Brett Hoffman, Christine Sigel, Anthony Barrera, Jack Waller, Emily Chen, Jeffrey Basque, and Susan Bowron. This text has improved thanks to these students.

Infinite thanks to Ari for his encouragement and his faith.

Elise Moss
San Jose, CA

Table of Contents

Lesson Three
Revit Systems

Lesson Four
Wiring

Lesson Seven
Views

Lesson Eight
Projects

Lesson Nine
Annotations, Dimensions, and Symbols

The Revit Interface

Go to Start→Programs→Autodesk → Revit 2023.

You can also type Revit in the search field under Programs and it will be listed.

When you first start Revit, you will see this screen:

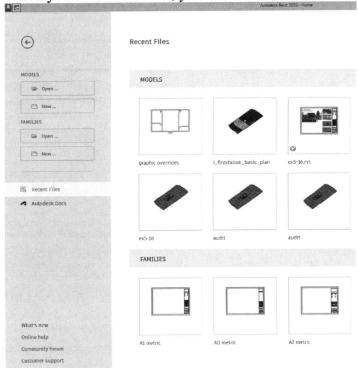

It is divided into three sections:

The Left section has two panels. The top panel is to Open or Start a new project. Revit calls the 3D building model a project. Some students find this confusing.

The bottom panel in the left section is used to open, create, or manage Revit families. Revit buildings are created using Revit families. Doors, windows, walls, floors, etc., are all families.

Recent Files show recent files which have been opened or modified as well as sample files.

There are links for Online help and to connect to the Community forum, where you can post questions to other users.

There is also a link to Autodesk Docs. Autodesk Docs is a cloud-based server where users can collaborate with other team members. This requires a paid subscription additional to the Revit subscription.

Some users have had difficulty locating the Revit Content that is required for some of the exercises. The Revit Content is free and should automatically be installed with Revit, but that sometimes doesn't happen. In the event you do not see the Revit libraries or Template directories, go to Autodesk's website and search for Revit Content. Then download the desired links.

You should be able to identify the different areas of the user interface in order to easily navigate around the software.

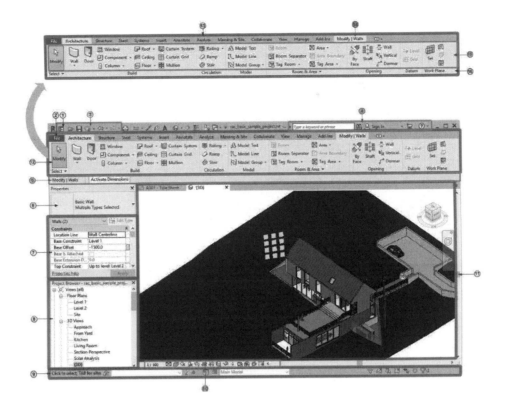

1	Revit Home	9	Status Bar
2	File tab	10	View Control Bar
3	Quick Access Toolbar	11	Drawing Area/Window
4	Info Center	12	Ribbon
5	Options Bar	13	Tabs on the ribbon
6	Type Selector	14	Contextual Ribbon – appears based on current selection
7	Properties Palette	15	Tools on the current tab of the ribbon
8	Project Browser	16	Ribbon Panels – used to organize Ribbon Tools

The Revit Ribbon

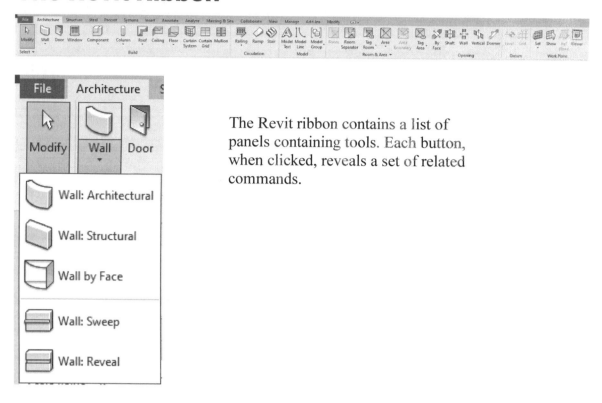

The Revit ribbon contains a list of panels containing tools. Each button, when clicked, reveals a set of related commands.

The Quick Access Toolbar (QAT)

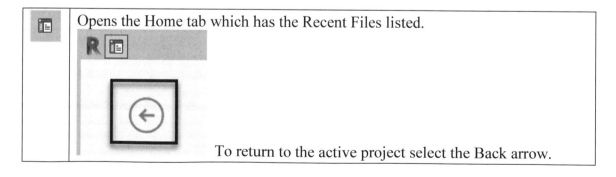

Most Windows users are familiar with the standard tools: New, Open, Save, Undo, and Redo.

🔲	Opens the Home tab which has the Recent Files listed. To return to the active project select the Back arrow.

	Synchronize to Central is used in team environments where users check in and check out worksets on a shared project. The Central location should be a shared drive or server that all team members can access. The **Synchronize to Central** tool is greyed out unless you have set up your project as a shared project with a central location.
	Create a PDF using sheets and views of the active model
	Enable visibility of pins, temporary dimensions, and constraints when elements are selected.
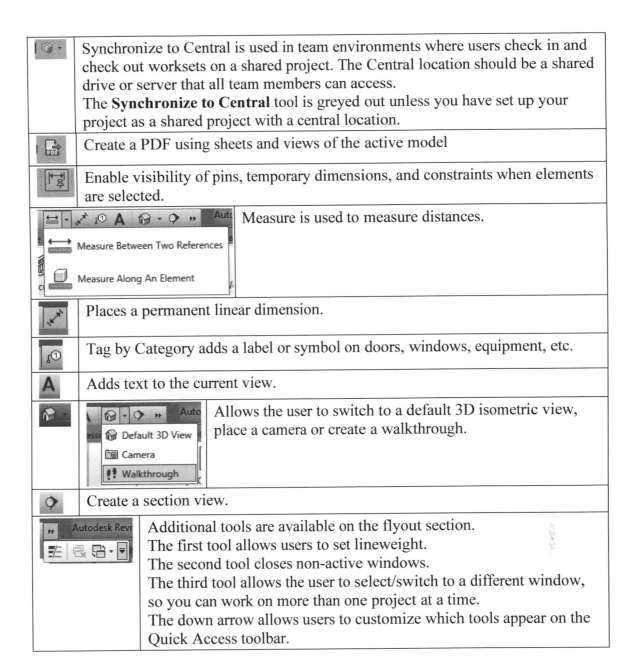	Measure is used to measure distances.
	Places a permanent linear dimension.
	Tag by Category adds a label or symbol on doors, windows, equipment, etc.
	Adds text to the current view.
	Allows the user to switch to a default 3D isometric view, place a camera or create a walkthrough.
	Create a section view.
	Additional tools are available on the flyout section. The first tool allows users to set lineweight. The second tool closes non-active windows. The third tool allows the user to select/switch to a different window, so you can work on more than one project at a time. The down arrow allows users to customize which tools appear on the Quick Access toolbar.

Printing

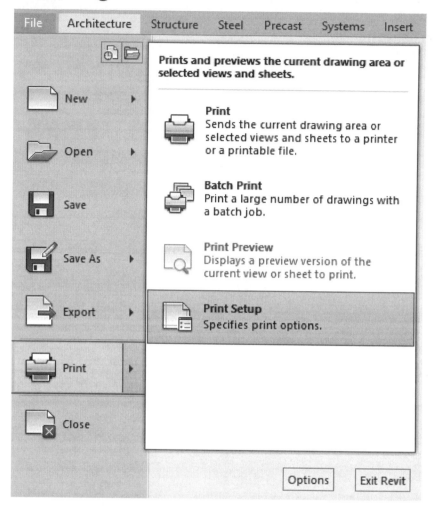

Print is located in the Application Menu as well as the Quick Access Toolbar.

The 2021 release has added the ability to batch print.

The Print dialog is fairly straightforward.
Select the desired printer from the drop-down list, which shows installed printers.

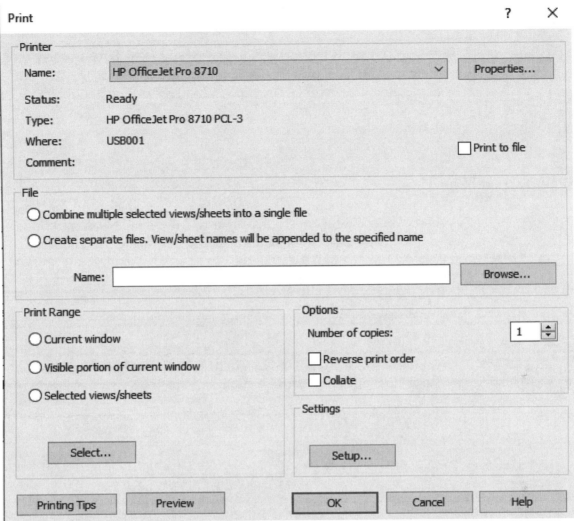

You can set to 'Print to File' by enabling the check box next to Print to File.

The Print Range area of the dialog allows you to print the current window, a zoomed in portion of the window, and selected views/sheets.

Undo

The Undo tool allows the user to select multiple actions to undo. To do this, use the drop down arrow next to the Undo button; you can select which recent action you want to undo. You cannot skip over actions (for example, you cannot undo 'Note' without undoing the two walls on top of it).

Ctrl-Z also acts as UNDO.

Redo

The Redo button also gives you a list of actions, which have recently been undone. Redo is only available immediately after an UNDO. For example, if you perform UNDO, then WALL, REDO will not be active.

Ctrl-Y is the shortcut for REDO.

Viewing Tools

A scroll wheel mouse can replace the use of the steering wheel. Click down on the scroll wheel to pan. Rotate the scroll wheel to zoom in and out.

The Rewind button on the steering wheel takes the user back to the previous view.

Different steering wheels are available depending on whether or not you are working in a Plan or 3D view.

The second tool has a flyout menu that allows the user to zoom to a selected window/region or zoom to fit (extents).

> Orient to a view allows the user to render an elevation straight on without perspective. Orient to a plane allows the user to create sweeps along non-orthogonal paths.

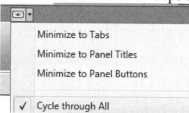

> If you right click on the Revit Ribbon, you can minimize the ribbon to gain a larger display window.

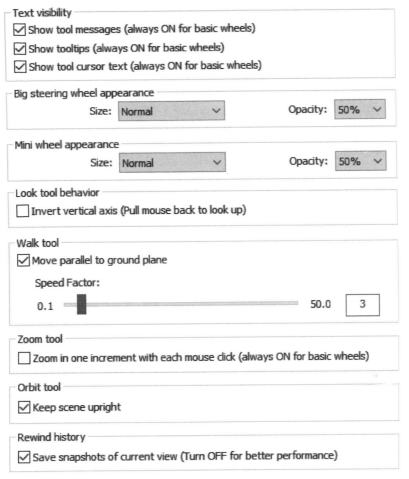

You can control the appearance of the steering wheels by right clicking on the steering wheel and selecting Options.

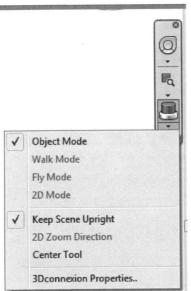

Some users have a 3D Connexion device – this is a mouse that is used by the left hand to zoom/pan/orbit while the right hand selects and edits.

Revit 2023 will detect if a 3D Connexion device is installed and add an interface to support the device.

It takes some practice to get used to using a 3D mouse, but it does boost your speed.

For more information on the
3D mouse, go to
3dconnexion.com.

Exercise 1-1:

Using the Steering Wheel & ViewCube

Drawing Name: *basic_project.rvt*
Estimated Time: 30 minutes

Before learning how to start a project from scratch, we will be using practice files to help you understand Revit's interface and get comfortable with the tools available inside of Revit. You will learn how to start a project from scratch starting in Lesson 3.

This exercise reinforces the following skills:

- ❏ ViewCube
- ❏ 2D Steering Wheel
- ❏ 3D Steering Wheel
- ❏ Project Browser
- ❏ Shortcut Menus
- ❏ Mouse

1. R ▣   Select Home from the QAT or Click **Ctl+D**.

2. The Home screen launches.

Sample Architecture Project

3. Select the Sample Architectural Project file. The file name is
    basic_sample_project.rvt.

    File name: basic_project   *If you do not see the Basic Sample Project listed, you can use the file basic_project included with the Class Files available for download from the publisher's website. To download the Class Files, type the following in the address bar of your web browser:* **SDCpublications.com/downloads/**

4. 3D Views
    - Approach
    - From Yard
    - Kitchen
    - Living Room
    - Section Perspective
    - Solar Analysis
    - {3D}

    In the Project Browser:
    Expand the 3D Views.
    Double left click on the {3D} view.
    This activates the view.
    The active view is displayed in BOLD in the Project Browser.

5. 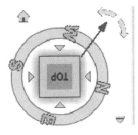 If you have a mouse with a scroll wheel,
    experiment with the following motions:
    If you roll the wheel up and down, you can zoom in and out.
    Hold down the SHIFT key and Click down the scroll wheel at the same time. This will rotate or orbit the model.
    Release the SHIFT key. Click down the scroll wheel. This will pan the model.

6. When you are in a 3D view, a tool called the ViewCube is visible in the upper right corner of the screen.

    The Viewcube orientation mirrors the model orientation.

7. Click on the top of the cube as shown.

    The display changes to a top or plan view.

8. Use the rotate arrows to rotate the view.

9.  Click on the little house (home/default view tool) to return to the default view.

The little house disappears and reappears when your mouse approaches the cube.

10.  Select the Steering Wheel tool located on the View Control toolbar.

11. A steering wheel pops up.

 Notice that as you mouse over sections of the steering wheel they highlight.
Mouse over the Zoom section and hold down the left mouse button. The display should zoom in and out.
Mouse over the Orbit section and hold down the left mouse button. The display should orbit.
Mouse over the Pan section and hold down the left mouse button. The display should pan.

12.

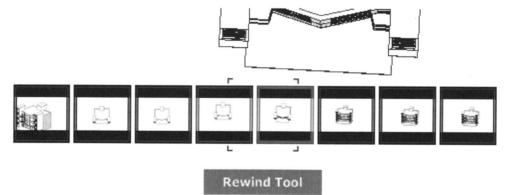

Rewind Tool

Select the Rewind tool and hold down the left mouse button.
A selection of previous views is displayed.
You no longer have to back through previous views. You can skip to the previous view you want.
Select a previous view to activate.
Close the steering wheel by selecting the X in the top right corner.

13.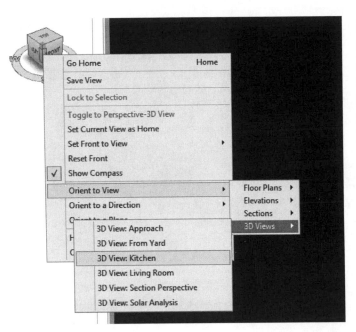

Place the mouse cursor over the View Cube. Right click and a shortcut menu appears. Select **Orient to View→ 3d Views→3D View: Kitchen**.

14.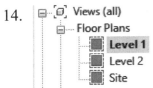

Double click on **Level 1** in the Project Browser. This will open the Level 1 floor plan view.

Be careful not to select the ceiling plan instead of the floor plan.

15. 3 ◀ A1-05

This is an elevation marker. It defines an elevation view.

16.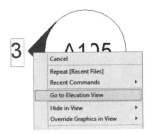

If you right click on the triangle portion of the elevation view, you can select **Go to Elevation View** to see the view associated with the elevation marker.

17.

Locate the section line on Level 1 labeled A104.

18. The question mark symbols in the drawing link to help html pages. Click on one of the question marks.

19. Look in the Properties palette.
In the Learning Links field, a link to a webpage is shown. Click in that field to launch the webpage.

20. A browser will open to the page pertaining to the question mark symbol.

21. Double left click on the arrow portion of the section line.
22. This view has a callout.

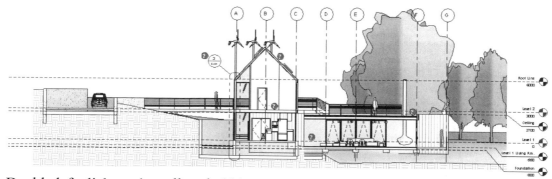

Double left click on the callout bubble.

The bubble is located on the left.

23.

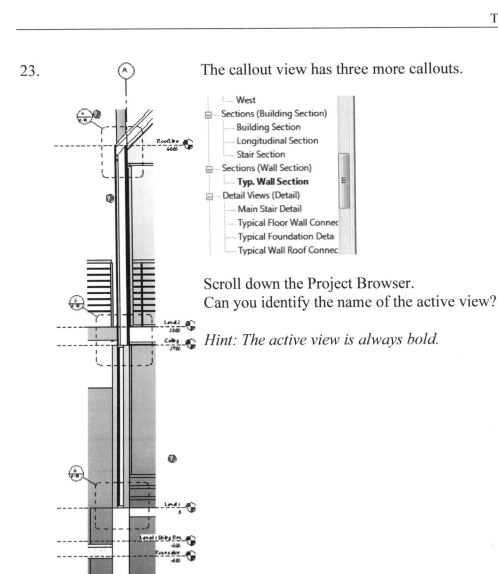

The callout view has three more callouts.

West
Sections (Building Section)
    Building Section
    Longitudinal Section
    Stair Section
Sections (Wall Section)
    Typ. Wall Section
Detail Views (Detail)
    Main Stair Detail
    Typical Floor Wall Connec
    Typical Foundation Deta
    Typical Wall Roof Connec

Scroll down the Project Browser.
Can you identify the name of the active view?

Hint: The active view is always bold.

24.

To the right of the view, there are levels.
Some of the levels are blue and some of
them are black.
The blue levels (like Level 1) are story
levels; they have views associated with
them.
The black levels (like Foundation) are
reference levels; they do not have views
associated with them.
Double left click on the Level 1 bubble.

Level 1
0

Level 1 Living Rm.
-550
Foundation
-800

25. The Level 1 floor plan is opened. *We are back where we started!*

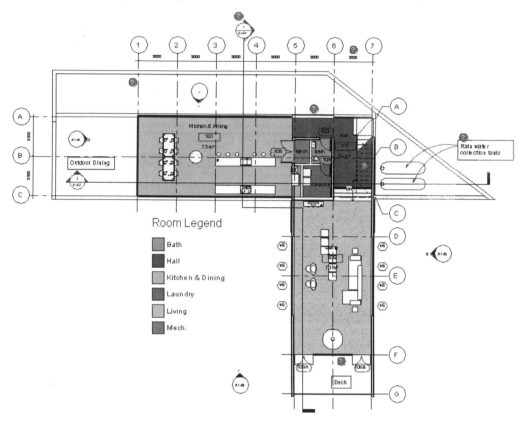

Level 1 is also in BOLD in the project browser.

26.

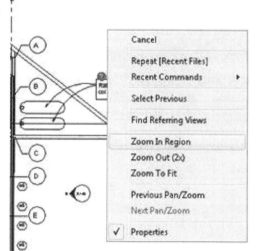

Right click in the window.
On the shortcut menu:
Select **Zoom In Region**.

This is the same as the Zoom Window tool in AutoCAD.

27. 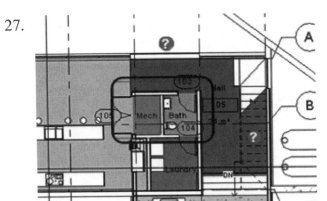 Place a rectangle with your mouse to zoom into the bathroom area.

28. 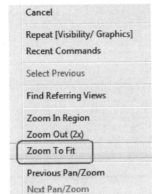 Right click in the window.
On the shortcut menu:
Select **Zoom To Fit**.
This is the same as the Zoom Extents tool in AutoCAD.
You can also double click on the scroll wheel to zoom to fit.

29. Close the file without saving.
Close by Clicking **Ctl+W** or using **File→Close**.

Exercise 1-2:

Changing the View Background

Drawing Name: *basic_project.rvt*
Estimated Time: 5 minutes

This exercise reinforces the following skills:
- ❑ Graphics mode
- ❑ Options

Many of my students come from an AutoCAD background and they want to change their background display to black.

1. 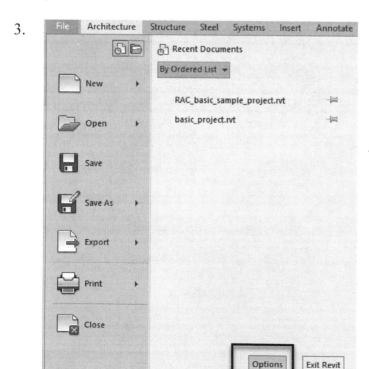   Go to **File→Open→Project**.

2.    Locate the file called *basic_project.rvt*.
This is included with the Class Files you downloaded from the publisher's website.

3. Select the drop-down on the Application Menu.
Select **Options**.

Options manages all the system options for your projects and files.

4.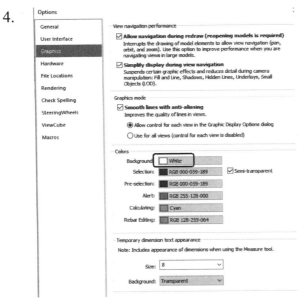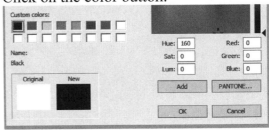

Select **Graphics**.
Locate the color tab next to Background.
Click on the color button.

Select the Black color.
Click **OK** twice.

5. Close the file without saving.

Revit's Project Browser

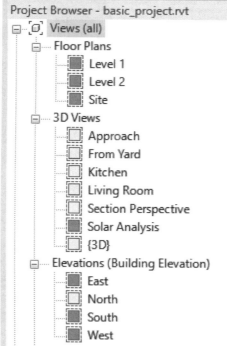

The Project Browser displays a hierarchy for all views, schedules, sheets, families and elements used in the current project. It has an interface similar to the Windows file explorer, which allows you to expand and collapse each branch as well as organize the different categories.

You can change the location of the Project Browser by dragging the title bar to the desired location. Changes to the size and the location are saved and restored when Revit is launched.

You can search for entries in the Project Browser by right-clicking inside the browser and selecting Search from the right-click menu to open a dialog.

The Project Browser is used to navigate around your project by opening different views. You can also drag and drop views onto sheets or families into views.

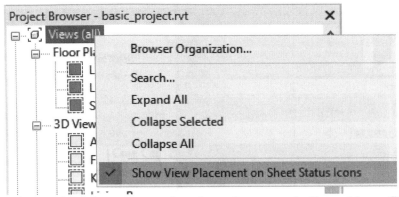

Views which have been placed on sheets are indicated by a filled in rectangle next to the view name. You can turn off the display of sheet status icons by right clicking on the Views.

You can organize the Project Browser using different parameters to organize sheets, schedules, and views.

Revit's Properties Palette

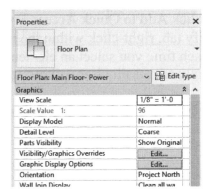

The Properties Palette is a contextually based dialog. If nothing is selected, it will display the properties of the active view. If something is selected, the properties of the selected item(s) will be displayed.

By default, the Properties Palette is docked on the left side of the drawing area. You can modify the location of this palette by dragging the title bar to the desired location.

Datum elements help to define the project context. Examples are grids, levels, and reference planes.

View-specific elements only display in the views in which they are placed. Most annotations fall into this category. Examples are dimensions, text, or tags.

When you place an element into a building project, such as a strut rack, it has two types of properties: Type Properties and Instance Properties. Type Properties do not change regardless of where the rack is placed. Type Properties are values like material, size, and electrical ratings. The Instance Properties are unique to that element, such as the location of the strut rack and the rack ID.

Exercise 1-3:

Closing and Opening the Project Browser and Properties Palette

Drawing Name: *basic_project.rvt*
Estimated Time: 5 minutes

This exercise reinforces the following skills:
- ❏ User Interface
- ❏ Ribbon
- ❏ Project Browser
- ❏ Properties panel

Many of my students will accidentally close the project browser and/or the properties palette and want to bring them back. Other students prefer to keep these items closed most of the time so they have a larger working area.

1. Go to **File→Open→Project**.

2.  Locate the file called *basic_project.rvt*.
    This is included with the Class Files you downloaded from the publisher's website.

3. Close the Properties palette by clicking on the x in the corner.

4. Close the Project Browser by clicking on the x in the corner.

5. Activate the View ribbon.
    Go to the User Interface dropdown at the far right.
    Place a check next to the Project Browser and Properties to make them visible.

6. Close without saving.

Revit's System Browser

| Systems ⌄ | All Disciplines ⌄ | ⊞ 📇 |

Systems	Space Name	Space Number
⊞ [?] Unassigned (2 items)		
☐ Mechanical (0 systems)		
☐ Piping (0 systems)		
⊟ ☐ Electrical (1 systems)		
⊟ (II) Power		
⊟ 🔲 <unnamed>		
🔲 LP1		

The System Browser opens a separate window that displays a hierarchical list of all the components in each discipline in a project, either by the system type or by ones.

The System Browser contains a list of all electrical components in a project and the systems to which they are assigned. Any electrical elements that are not assigned to a system appear in the Unassigned category. Using the System Browser allows you to quickly locate any unassigned electrical components and assign them to the correct system.

Customizing the View of the System Browser

The options in the View bar allow you to sort and customize the display of systems in the System Browser.

- **Systems**: displays components by major and minor systems created for each discipline.

- **Zones**: displays zones and spaces. Expand each zone to display the spaces assigned to the zone.

- **All Disciplines**: displays components in separate folders for each discipline (mechanical, piping, and electrical). Piping includes plumbing and fire protection.

- **Mechanical**: displays only components for the Mechanical discipline.

- **Piping**: displays only components for the Piping disciplines (Piping, Plumbing, and Fire Protection).

- **Electrical**: displays only components for the Electrical discipline.

- **AutoFit All Columns**: adjusts the width of all columns to fit the text in the headings.

Note: You can also double-click a column heading to automatically adjust the width of a column.

- **Column Settings**: opens the Column Settings dialog where you specify the columnar information displayed for each discipline. Expand individual categories (General, Mechanical, Piping, Electrical) as desired, and select the properties that you want to appear as column headings. You can also select columns and click Hide or Show to select column headings that display in the table.

Exercise 1-4:

Using the System Browser

Drawing Name: *system_browser.rvt*
Estimated Time: 10 minutes

This exercise reinforces the following skills:
- ❑ User Interface
- ❑ Ribbon
- ❑ System Browser

This browser allows the user to locate and identify different elements used in lighting, mechanical, and plumbing inside the project.

1.   Go to **File**.

Select **Open→Project**.

2.   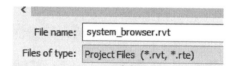 Locate the file called *system_browser.rvt*. *This is included with the Class Files you downloaded from the publisher's website.*

3.   Activate the **View** ribbon.

4.   Go to the User Interface drop-down list.
Place a check next to **System Browser**.

5.   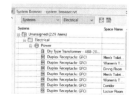 At the top of the Browser, use the Filters to show only the Electrical elements.
Enable **Systems**.
Enable **Electrical.**
Expand the **Unassigned** category.
Expand the **Electrical** category.
Expand the **Power** category.

6. Locate the **Duplex Receptacle** in the Locker Room and highlight.

7. Look in the Properties panel. This receptacle has not been assigned to an electrical panel and it doesn't have a circuit number.

8. Right click on the **Duplex Receptacle** in the Locker Room. Select **Show**.

9. Click **OK**.

10. Select the **Show** button.

11. Click the **Show** button until you see this view.

Click **Close**.

12. Verify in the Project Browser that you are in the Site floor plan.

This is bold in the Project Browser.

13. Zoom out to see the Locker Room.

To close the System Browser, select the x button in the upper right corner of the dialog.

14. Close the file without saving.

Revit's Ribbon

The ribbon displays when you create or open a project file. It provides all the tools necessary to create a project or family.

An arrow next to a panel title indicates that you can expand the panel to display related tools and controls.

By default, an expanded panel closes automatically when you click outside the panel. To keep a panel expanded while its ribbon tab is displayed, click the push pin icon in the bottom-left corner of the expanded panel.

Exercise 1-5:

Changing the Ribbon Display

Drawing Name: *basic_project.rvt*
Estimated Time: 15 minutes

This exercise reinforces the following skills:
- ❑ User Interface
- ❑ Ribbon

Many of my students will accidentally collapse the ribbon and want to bring it back. Other students prefer to keep the ribbon collapsed most of the time so they have a larger working area.

1.

basic_project

Select **basic_project** from the Recent Files window.

2.

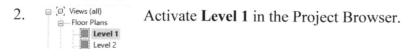

Activate **Level 1** in the Project Browser.

3. Modify ⬒▾   On the ribbon: Locate the two small up and down arrows.

4.   Left click on the white button.

5.      The ribbon collapses to tabs.

6. 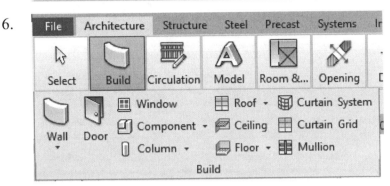 Left click on the word **Architecture**.
Hover the mouse over the word **Build** and the Build tools panel will appear. *The build tools will be grayed out in a 3D view.*

7.   Click on the white button until the ribbon is restored.

8.  The ribbon is full-size again.

9.   Click on the black arrow.

10.   Click on **Minimize to Panel Buttons**.

    Minimize to Tabs
    ✓   Minimize to Panel Titles
    Minimize to Panel Buttons

11.   The ribbon changes to panel buttons.

12.   Left click on a panel button and the tools for the button will display.
Some buttons are grayed out depending on what view is active.

    Room   Area
    Room Separator   Area Boundary
    Tag Room   Tag Area
    Color Schemes
    Area and Volume Computations
    Room & Area

13.   Click on the white button to display the full ribbon.

    Show Full Ribbon

14. Close without saving.

The Modify Ribbon

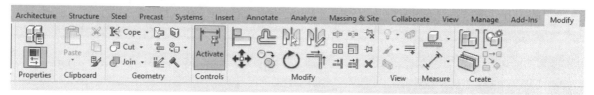

When you select an entity, you automatically switch to Modify mode. A Modify ribbon will appear with different options.

Select a wall in the view. Note how the ribbon changes.

Revit uses three types of dimensions: Listening, Temporary, and Permanent. Dimensions are placed using the relative positions of elements to each other. Permanent dimensions are driven by the value of the temporary dimension. Listening dimensions are displayed as you draw or place an element.

Exercise 1-6:

Temporary, Permanent and Listening Dimensions

Drawing Name: dimensions.rvt
Estimated Time: 30 minutes

This exercise reinforces the following skills:
- ❑ Project Browser
- ❑ Scale
- ❑ Graphical Scale
- ❑ Numerical Scale
- ❑ Temporary Dimensions
- ❑ Permanent Dimensions
- ❑ Listening Dimensions
- ❑ Type Selector

1. Browse to *dimensions.rvt* in the Class Files you downloaded from the publisher's website. Save the file to a folder. Open the file.

The file has four walls.
The horizontal walls are 80′ in length.
The vertical walls are 52′ in length.

We want to change the vertical walls so that they are 60′ in length.

2. 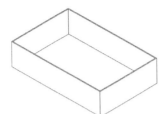 Select the House icon on the Quick Access toolbar.

3. *This switches the display to a default 3D view.*

You see that the displayed elements are walls.

4. 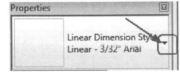 Double left click on **Level 1** under Floor Plans in the Project Browser.

The view display changes to the Level 1 floor plan.

5. 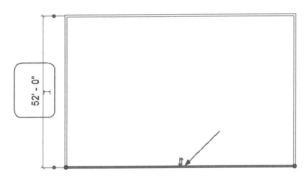 Select the bottom horizontal wall.

A temporary dimension appears showing the vertical distance between the selected wall and the wall above it.

6. A small dimension icon is next to the temporary dimension.

Left click on this icon to place a permanent dimension.

Left click anywhere in the window to release the selection.

7. Select the permanent dimension extension line, not the text.

A lock appears. If you left click on the lock, it would prevent you from modifying the dimension.

If you select the permanent dimension, you cannot edit the dimension value, only the dimension style.

8. The Properties palette shows the dimension style for the dimension selected.

Select the small down arrow.
This is the Type Selector.

Linear Dimension Style
Linear - 3/32″ Arial

The Type Selector shows the dimension styles available in the project.
If you do not see the drop-down arrow on the right, expand the properties palette and it should become visible.

9.

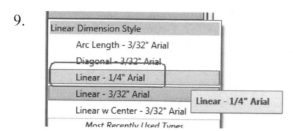

Left click on the **Linear - 1/4″ Arial** dimension style to select this dimension style.

Left click anywhere in the drawing area to release the selection and change the dimension style.

10.

The dimension updates to the new dimension style.

11.

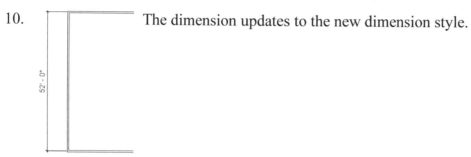

Select the right vertical wall.

12.

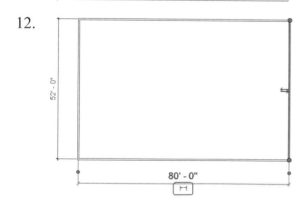

A temporary dimension appears showing the distance between the selected wall and the closest vertical wall.

Left click on the permanent dimension icon to place a permanent dimension.

13. A permanent dimension is placed.

Select the horizontal permanent dimension extension line, not the text.

14. Use the Type Selector to change the dimension style of the horizontal dimension to the **Linear - 1/4″ Arial** dimension style.

Left click anywhere in the drawing area to release the selection and change the dimension style.

Linear Dimension Style

Arc Length - 3/32" Arial

Diagonal - 3/32" Arial

Linear - 1/4" Arial

Linear - 3/32" Arial   Linear - 1/4" Arial

Linear w Center - 3/32" Arial

Most Recently Used Types

15. You have placed two *permanent* dimensions and assigned them a new dimension type.

52' - 0"

80' - 0"

16. Hold down the Control key and select both **horizontal** (top and bottom) walls so that they highlight.

   NOTE: *We select the horizontal walls to change the vertical wall length, and we select the vertical walls to change the horizontal wall length.*

17. Select the **Scale** tool located on the Modify Walls ribbon.

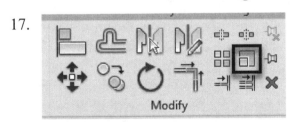

Modify

In the Options bar, you may select Graphical or Numerical methods for resizing the selected objects.

18. Modify | Walls   ⊙ Graphical ○ Numerical Scale: 2   Enable **Graphical**.

The Graphical option requires three inputs.
   Input 1: Origin or Base Point
   Input 2: Original or Reference Length
   Input 3: Desired Length

19. 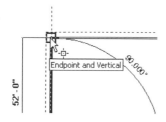 Select the left lower endpoint for the first input – the origin.

20. 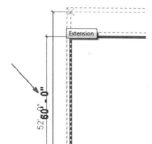 Select the left upper endpoint for the second input – the reference length.

21. Extend your cursor until you see a dimension of 60′.

The dimension you see as you move the cursor is a *listening dimension.*

You can also type 60' and Click ENTER.

22. 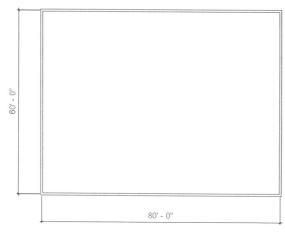 Left click for the third input – the desired length.
Left click anywhere in the window to release the selection and exit the scale command.
Note that the permanent dimension updates.

23. 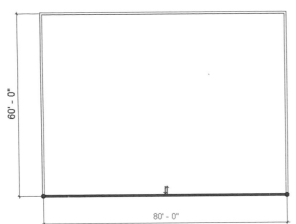 Select the bottom horizontal wall to display the temporary dimension.

24. 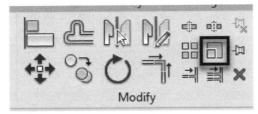 Left click on the vertical temporary dimension.
An edit box will open. Type **50 0**.
Revit does not require you to enter units.
Left click to enter and release the selection.

25.  The permanent dimension updates.

Note if you select the permanent dimension, you cannot edit the dimension value, only the dimension style.

26. Now, we will change the horizontal dimension using the *Numerical* option of the Resize tool.
Hold down the Control key and select the left and right vertical walls.

27.  Select the **Scale** tool.

28. Modify | Walls ○ Graphical ● Numerical Scale: 0.5

Enable **Numerical**
Set the Scale to **0.5**.
This will change the wall length from **80'** to **40'**.
The Numerical option requires only one input for the Origin or Base Point.

29.  Select the left lower endpoint for the first input – the origin.

30. The walls immediately adjust. Left click in the drawing area to release the selection.
Note that the permanent dimension automatically updates.

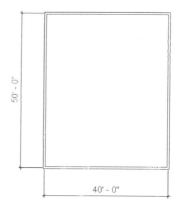

31. Close without saving.

 Model Group ▾

The Group tool works in a similar way as the AutoCAD GROUP. Basically, you are creating a selection set of similar or dissimilar objects, which can then be selected as a single unit. Once selected, you can move, copy, rotate, mirror, or delete them. Once you create a Group, you can add or remove members of that group. Existing groups are listed in your browser panel and can be dragged and dropped into views as required. Typical details, office layouts, bathroom layouts, etc. can be grouped, and then saved out of the project for use on other projects.

Tips & Tricks If you hold down the Control key as you drag a selected object or group, Revit will automatically create a copy of the selected object or group.

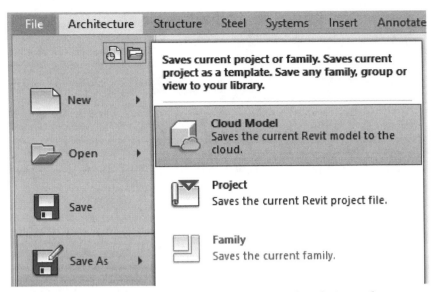

Revit has an option to save projects to the Cloud. In order to save to the cloud, the user needs to have a subscription for their software.

This option allows you to store your projects to a secure Autodesk server, so you can access your files regardless of your location. Users can share their project with each other and even view and comment on projects without using Revit. Instead, you use a browser interface.

Autodesk hopes that if you like this interface you will opt for their more expensive option – BIM360, which is used by many AEC firms to collaborate on large projects.

The Collaborate Ribbon

Collaborate tools are used if more than one user is working on the same project. The tools can also be used to check data sets coming in from other sources.

➤ If you plan to export to .dwg or import existing .dwg details, it is important to set your import/export setting. This allows you to control the layers that Revit will export to and the appearance of imported .dwg files.

➤ Clicking the ESC Key twice will always take you to the Modify command.

➤ As you get more proficient with Revit, you may want to create your own templates based on your favorite settings. A custom template based on your office standards (including line styles, rendering materials, common details, door, window and room schedules, commonly used wall types, door types and window types and typical sheets) will increase project production.

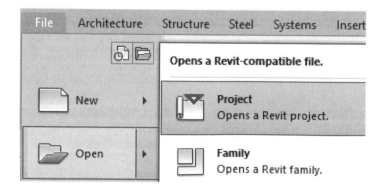

When you want to start a new project/building, you go to File→New→Project or use the hot key by Clicking Control and 'N' at the same time.

When you start a new project, you use a default template (default.rte). This template creates two Levels (default floor heights) and sets the View Window to the Floor Plan view of Level 1.

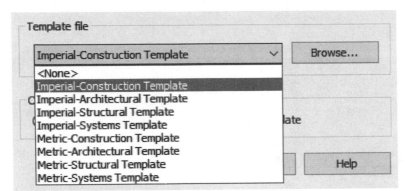

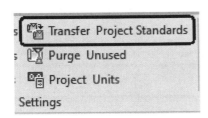

You can transfer the Project settings of an old project to a new one by opening both projects in one session of Revit, then with your new project active, select **Manage→Settings→Transfer Project Standards**.
Check the items you wish to transfer, and then click 'OK'.

Transfer Project Settings is useful if you have created a custom system family, such as a wall, floor, or ceiling, and want to use it in a different project.

The View Ribbon

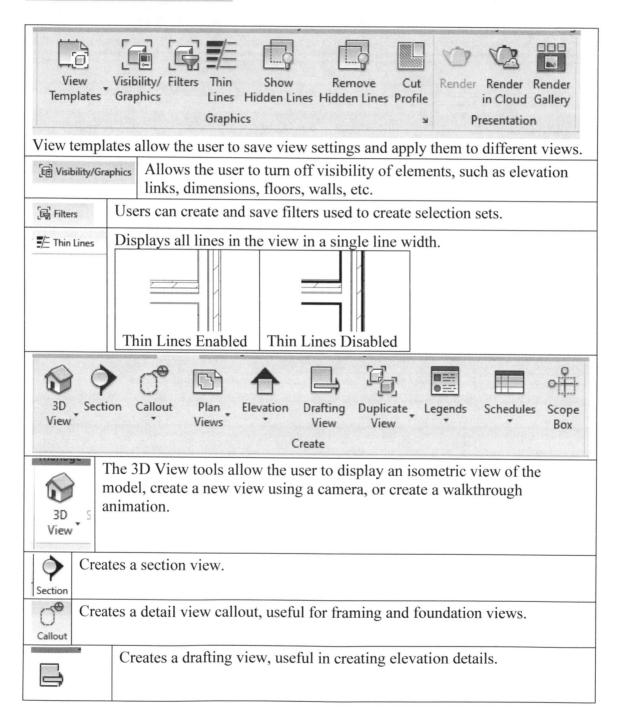

View templates allow the user to save view settings and apply them to different views.

Visibility/Graphics	Allows the user to turn off visibility of elements, such as elevation links, dimensions, floors, walls, etc.
Filters	Users can create and save filters used to create selection sets.
Thin Lines	Displays all lines in the view in a single line width.

Thin Lines Enabled   Thin Lines Disabled

3D View	The 3D View tools allow the user to display an isometric view of the model, create a new view using a camera, or create a walkthrough animation.
Section	Creates a section view.
Callout	Creates a detail view callout, useful for framing and foundation views.
	Creates a drafting view, useful in creating elevation details.

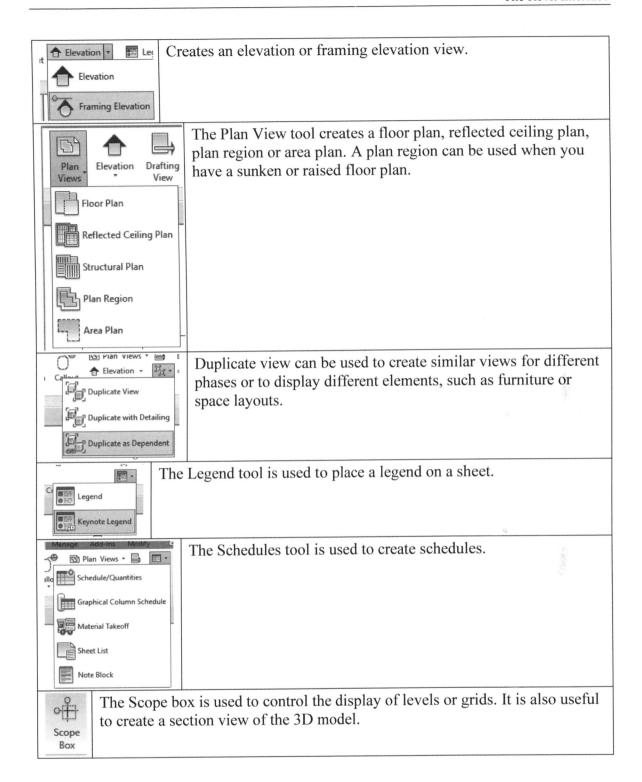

Elevation / Framing Elevation	Creates an elevation or framing elevation view.
Plan Views / Elevation / Drafting View — Floor Plan, Reflected Ceiling Plan, Structural Plan, Plan Region, Area Plan	The Plan View tool creates a floor plan, reflected ceiling plan, plan region or area plan. A plan region can be used when you have a sunken or raised floor plan.
Duplicate View / Duplicate with Detailing / Duplicate as Dependent	Duplicate view can be used to create similar views for different phases or to display different elements, such as furniture or space layouts.
Legend / Keynote Legend	The Legend tool is used to place a legend on a sheet.
Schedule/Quantities, Graphical Column Schedule, Material Takeoff, Sheet List, Note Block	The Schedules tool is used to create schedules.
Scope Box	The Scope box is used to control the display of levels or grids. It is also useful to create a section view of the 3D model.

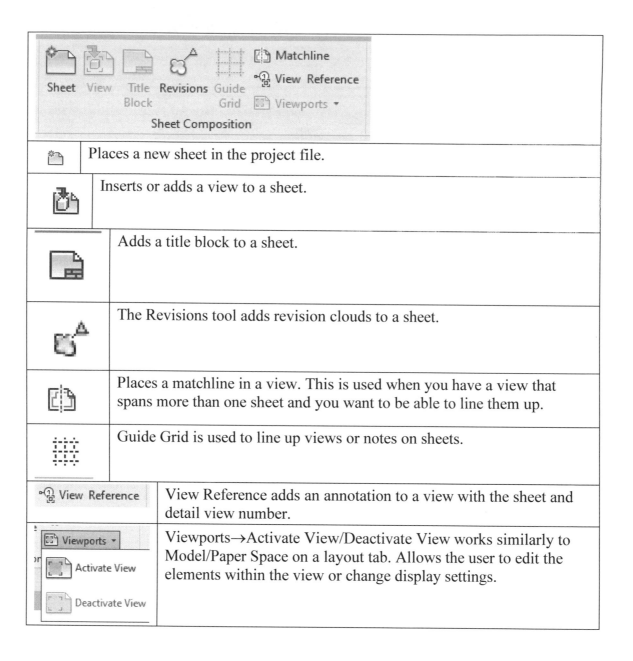

	Places a new sheet in the project file.
	Inserts or adds a view to a sheet.
	Adds a title block to a sheet.
	The Revisions tool adds revision clouds to a sheet.
	Places a matchline in a view. This is used when you have a view that spans more than one sheet and you want to be able to line them up.
	Guide Grid is used to line up views or notes on sheets.
View Reference	View Reference adds an annotation to a view with the sheet and detail view number.
Viewports ▾ / Activate View / Deactivate View	Viewports→Activate View/Deactivate View works similarly to Model/Paper Space on a layout tab. Allows the user to edit the elements within the view or change display settings.

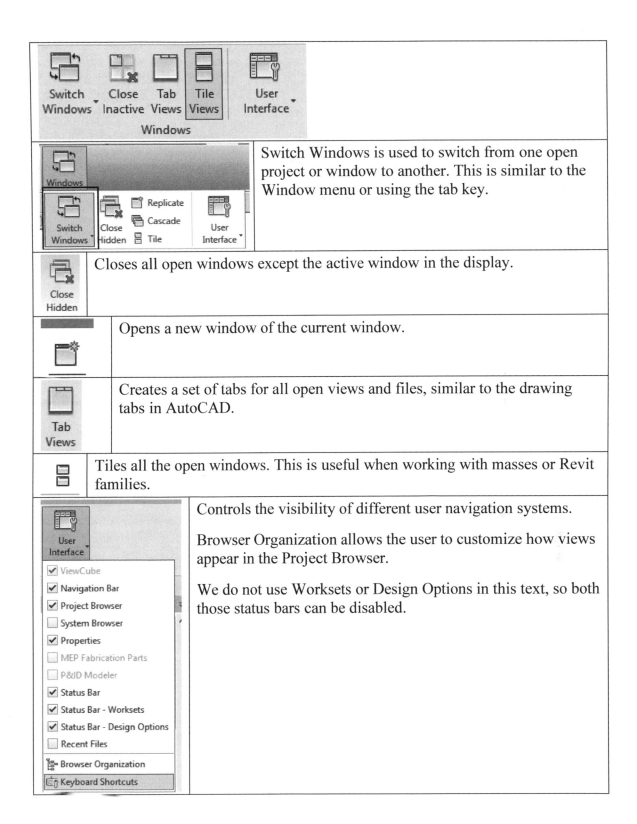

	Switch Windows is used to switch from one open project or window to another. This is similar to the Window menu or using the tab key.
	Closes all open windows except the active window in the display.
	Opens a new window of the current window.
	Creates a set of tabs for all open views and files, similar to the drawing tabs in AutoCAD.
	Tiles all the open windows. This is useful when working with masses or Revit families.
	Controls the visibility of different user navigation systems. Browser Organization allows the user to customize how views appear in the Project Browser. We do not use Worksets or Design Options in this text, so both those status bars can be disabled.

Creating standard view templates (exterior elevations, enlarged floor plans, etc.) in your project template or at the beginning of a project will save lots of time down the road.

The Manage Ribbon

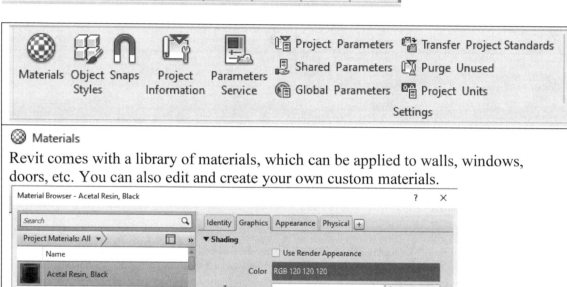

⬡ Materials

Revit comes with a library of materials, which can be applied to walls, windows, doors, etc. You can also edit and create your own custom materials.

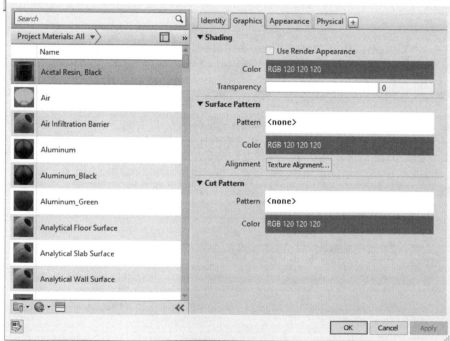

Any image file can be used to create a material.

My favorite sources for materials are fabric and paint websites.

Object Styles is used to control the line color, line weight, and line style applied to different Revit elements. It works similarly to layers in AutoCAD, except it ensures that all the similar elements, like doors, look the same.	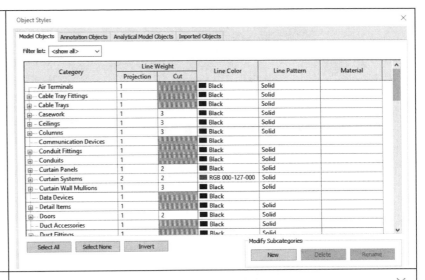
Snaps in Revit are similar to object snaps in AutoCAD. You can set how your length and dimensions snap as well as which object types your cursor will snap to when making a selection. Like AutoCAD, it is best not to make your snaps all-encompassing so that you snap to everything as that defeats the purpose and usefulness of snaps.	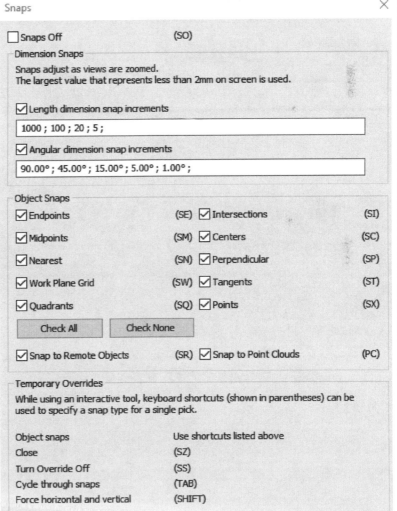

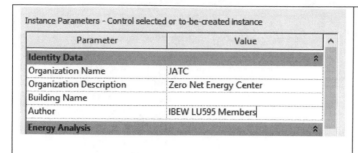

Project Information

Project Information will automatically fill in on the title blocks for each sheet.

Use the Parameters Service to manage shared parameters and import them into a Revit session. This is similar to creating a shared parameters file and placing the file on a server where all the team members can access it.

Parameters Service

Parameters Service is a cloud based paid service.

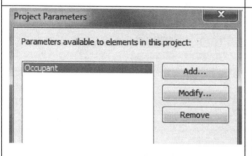

Project Parameters

Project parameters may be used to create custom parameters to be added to a title block or to organize your project browser, families or views.

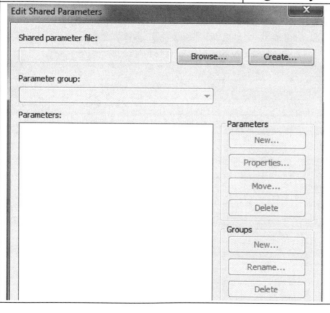

Shared Parameters

Shared Parameters are used to create labels for Revit families and for use in schedules.

This dialog allows you to set up parameters that can be shared across family categories for use in schedules.

For example, a window assembly and a window both have width, height, glass type, etc.

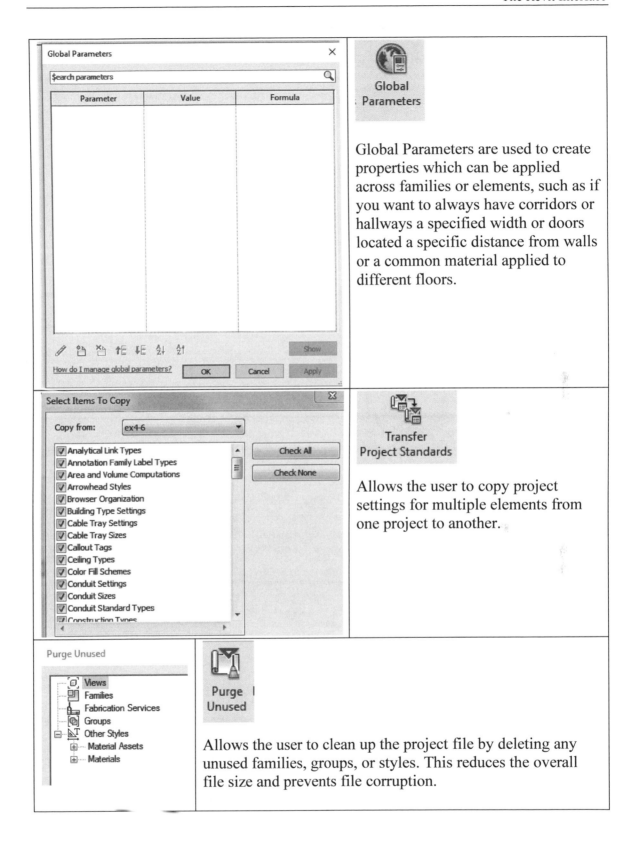

Global Parameters are used to create properties which can be applied across families or elements, such as if you want to always have corridors or hallways a specified width or doors located a specific distance from walls or a common material applied to different floors.

Transfer Project Standards

Allows the user to copy project settings for multiple elements from one project to another.

Purge Unused

Allows the user to clean up the project file by deleting any unused families, groups, or styles. This reduces the overall file size and prevents file corruption.

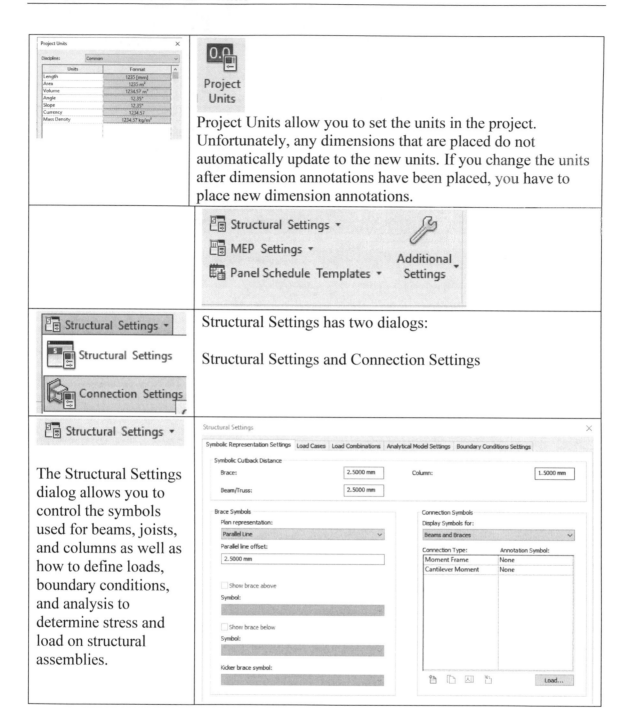	**Project Units** Project Units allow you to set the units in the project. Unfortunately, any dimensions that are placed do not automatically update to the new units. If you change the units after dimension annotations have been placed, you have to place new dimension annotations.
	Structural Settings ▾ MEP Settings ▾ Panel Schedule Templates ▾ Additional Settings ▾
Structural Settings ▾ Structural Settings Connection Settings	Structural Settings has two dialogs: Structural Settings and Connection Settings
Structural Settings ▾ The Structural Settings dialog allows you to control the symbols used for beams, joists, and columns as well as how to define loads, boundary conditions, and analysis to determine stress and load on structural assemblies.	

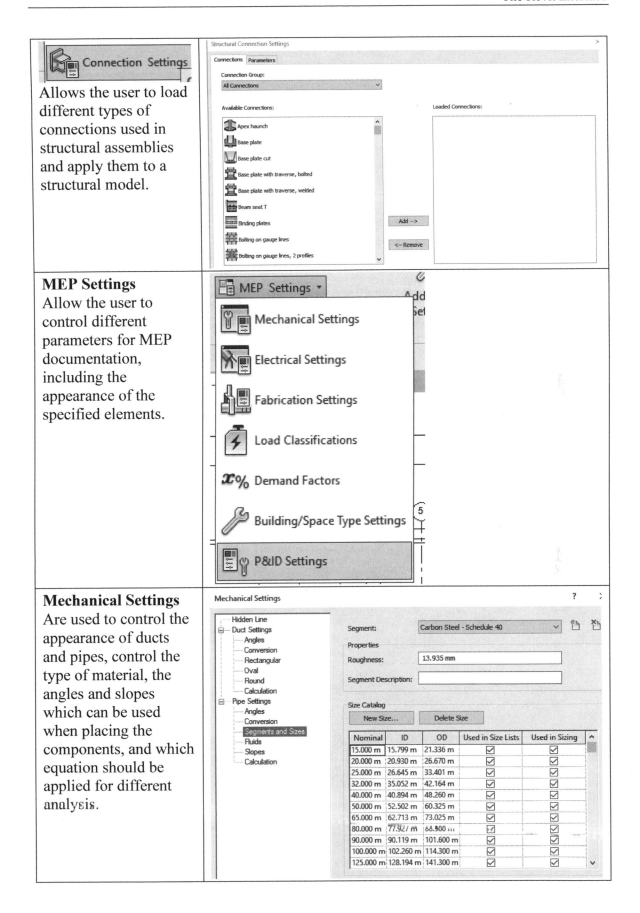 **Connection Settings** Allows the user to load different types of connections used in structural assemblies and apply them to a structural model.	
MEP Settings Allow the user to control different parameters for MEP documentation, including the appearance of the specified elements.	
Mechanical Settings Are used to control the appearance of ducts and pipes, control the type of material, the angles and slopes which can be used when placing the components, and which equation should be applied for different analysis.	

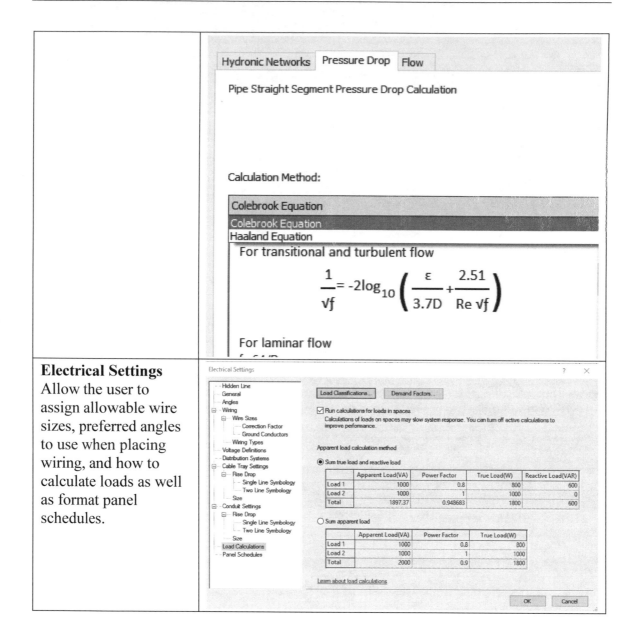

Electrical Settings
Allow the user to assign allowable wire sizes, preferred angles to use when placing wiring, and how to calculate loads as well as format panel schedules.

Fabrication Settings
Determines which libraries are loaded and available for ductwork, electrical, and piping.

Load Classifications
Allows the user to select or create load classification types to be used in load calculations.

Demand Factors
Allows the user to select or create demand factor types to be used in load calculations.

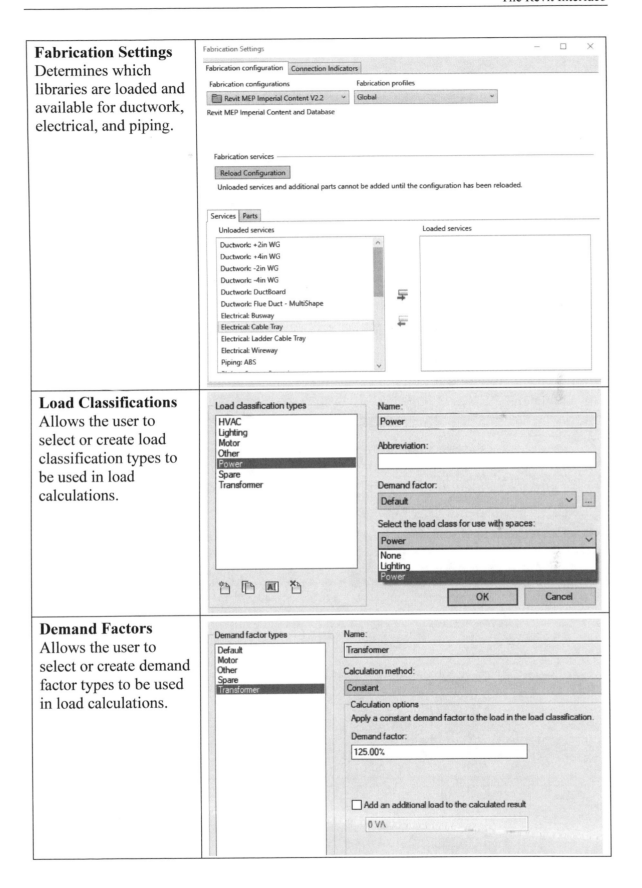

Building/Space Type Settings

Are used for energy analysis. Users can assign different values to different building or space types or create their own custom types.

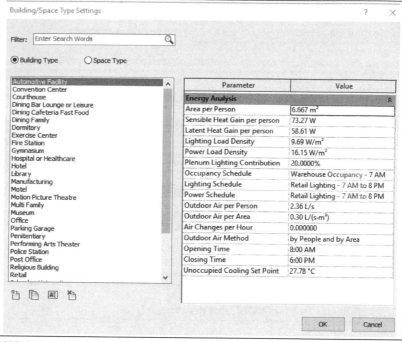

To facilitate modeling an AutoCAD P&ID (piping and instrumentation diagrams) drawing in Revit, you'll need to map P&ID components and schematic lines to real-world 3D model components, such as a P&ID valve symbol with a Revit family, such as a ball valve.

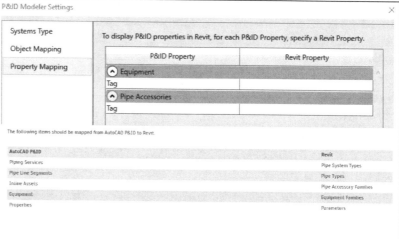

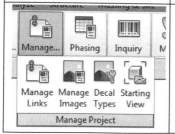

Manage Project is similar to the External Reference Manager in AutoCAD. This is used when you have inserted external CAD files, such as topo files or architectural floor plans to be used in the model.

Starting View sets the view you want to open automatically whenever you open a file.

The Additional Settings menu allows the user to customize the interface.

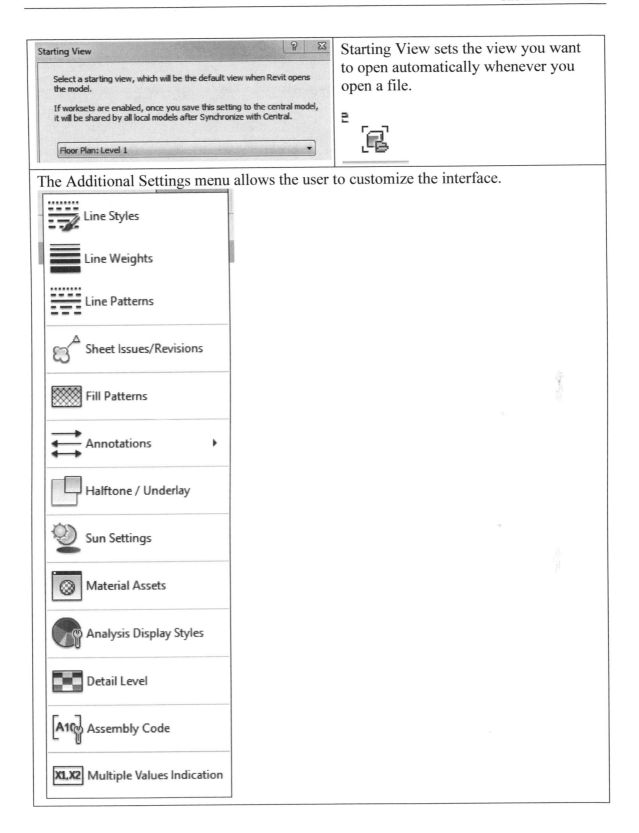

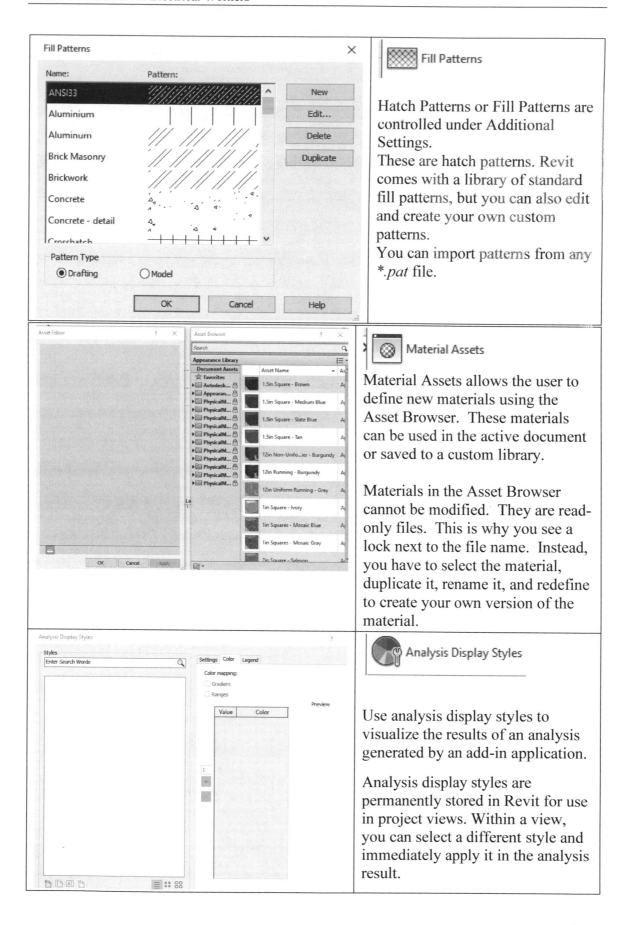

Fill Patterns

Hatch Patterns or Fill Patterns are controlled under Additional Settings.

These are hatch patterns. Revit comes with a library of standard fill patterns, but you can also edit and create your own custom patterns.

You can import patterns from any *.pat* file.

Material Assets

Material Assets allows the user to define new materials using the Asset Browser. These materials can be used in the active document or saved to a custom library.

Materials in the Asset Browser cannot be modified. They are read-only files. This is why you see a lock next to the file name. Instead, you have to select the material, duplicate it, rename it, and redefine to create your own version of the material.

Analysis Display Styles

Use analysis display styles to visualize the results of an analysis generated by an add-in application.

Analysis display styles are permanently stored in Revit for use in project views. Within a view, you can select a different style and immediately apply it in the analysis result.

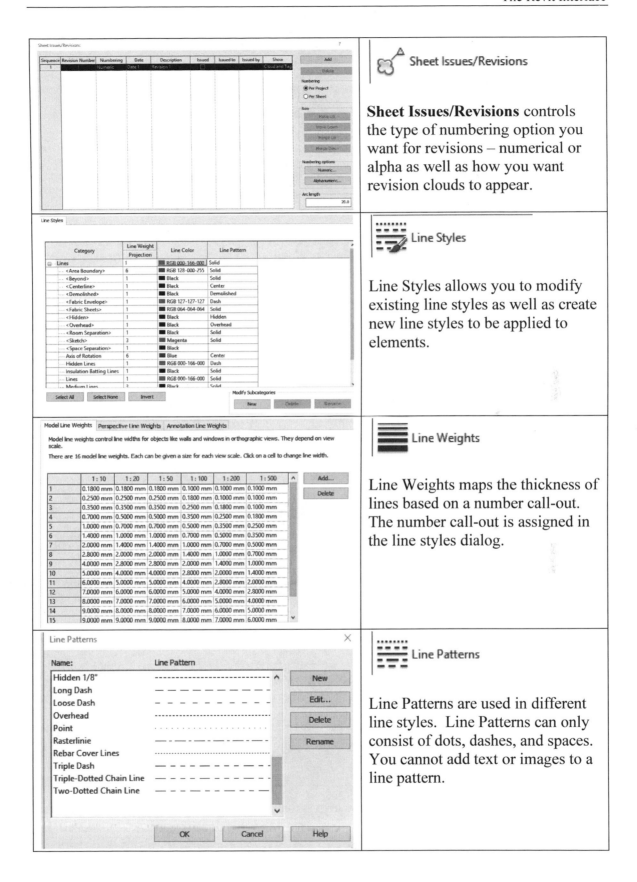

Sheet Issues/Revisions

Sheet Issues/Revisions controls the type of numbering option you want for revisions – numerical or alpha as well as how you want revision clouds to appear.

Line Styles

Line Styles allows you to modify existing line styles as well as create new line styles to be applied to elements.

Line Weights

Line Weights maps the thickness of lines based on a number call-out. The number call-out is assigned in the line styles dialog.

Line Patterns

Line Patterns are used in different line styles. Line Patterns can only consist of dots, dashes, and spaces. You cannot add text or images to a line pattern.

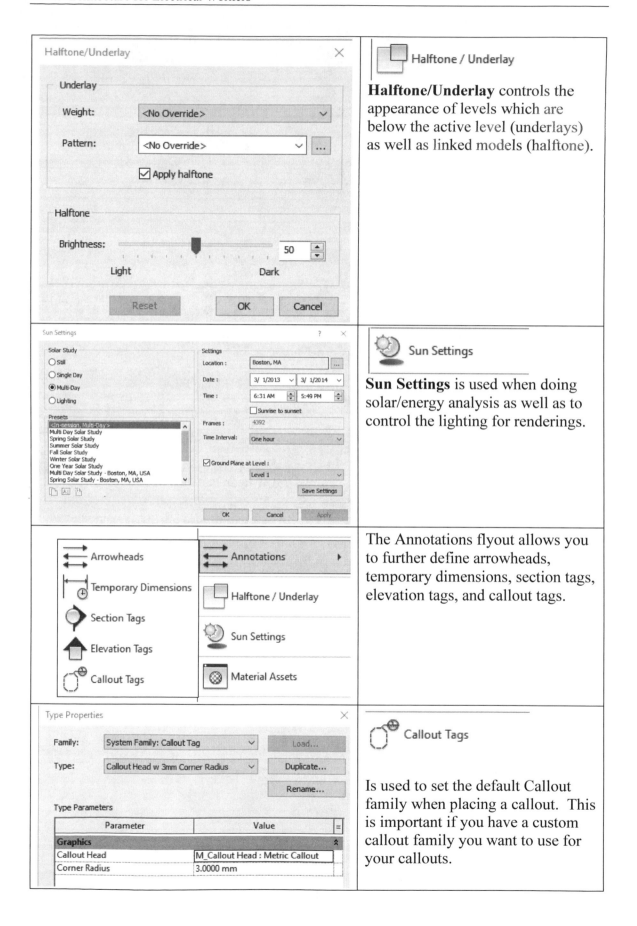

Halftone / Underlay

Halftone/Underlay controls the appearance of levels which are below the active level (underlays) as well as linked models (halftone).

Sun Settings

Sun Settings is used when doing solar/energy analysis as well as to control the lighting for renderings.

The Annotations flyout allows you to further define arrowheads, temporary dimensions, section tags, elevation tags, and callout tags.

Callout Tags

Is used to set the default Callout family when placing a callout. This is important if you have a custom callout family you want to use for your callouts.

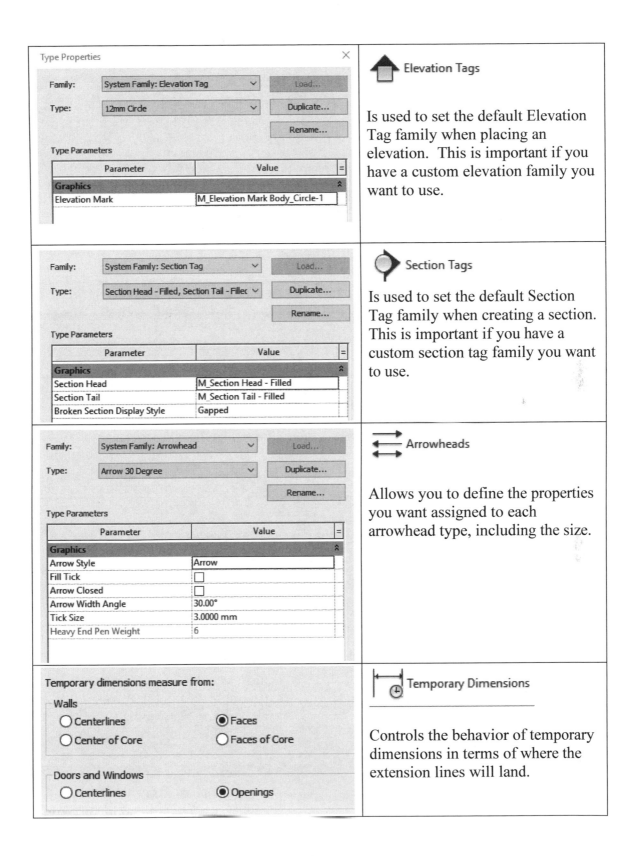

Elevation Tags

Is used to set the default Elevation Tag family when placing an elevation. This is important if you have a custom elevation family you want to use.

Section Tags

Is used to set the default Section Tag family when creating a section. This is important if you have a custom section tag family you want to use.

Arrowheads

Allows you to define the properties you want assigned to each arrowhead type, including the size.

Temporary Dimensions

Controls the behavior of temporary dimensions in terms of where the extension lines will land.

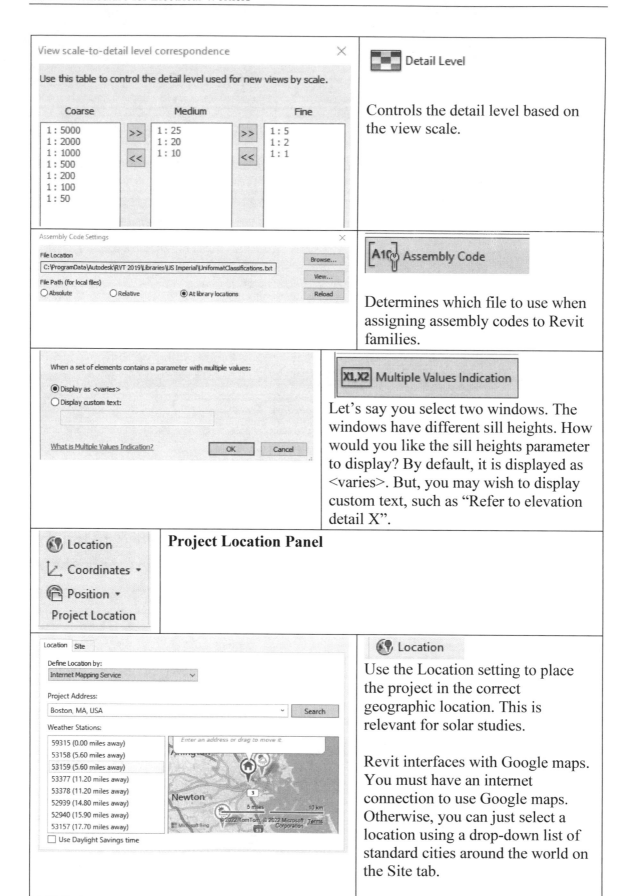

Detail Level

Controls the detail level based on the view scale.

Assembly Code

Determines which file to use when assigning assembly codes to Revit families.

Multiple Values Indication

Let's say you select two windows. The windows have different sill heights. How would you like the sill heights parameter to display? By default, it is displayed as <varies>. But, you may wish to display custom text, such as "Refer to elevation detail X".

Project Location Panel

Location

Use the Location setting to place the project in the correct geographic location. This is relevant for solar studies.

Revit interfaces with Google maps. You must have an internet connection to use Google maps. Otherwise, you can just select a location using a drop-down list of standard cities around the world on the Site tab.

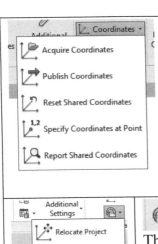

Coordinates ▾

The Coordinates tools are used when you have placed a linked CAD file into a project and allows the user to acquire coordinates from the linked file or transfer (publish) the Revit coordinates.
Shared coordinates are used to recognize the position of multiple linked files.

Position ▾

The position tools are used to position the model to reflect the correct geographical location and position relative to the sun.

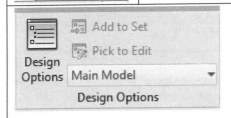

Design Options are used to design different options for a building model. This allows the user to create different versions of the same building to present to a client.
Elements may be added to different option sets and then displayed.
This is similar to a Configuration Manager.

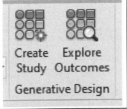

Generative Design is used for creating conceptual models of projects. You input the design constraints, such as spatial requirements and materials. Revit will then generate two or more possible solutions using masses. The user then modifies the most appealing solution and develops it into an architectural model.

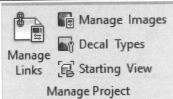

The Manage Project panel is used to manage different file types as well as set the starting/default view when you open a project file.

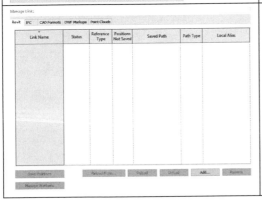

Manage Links

The Manage Links dialog is similar to the External References dialog in AutoCAD. You can manage any external files, unload, reload, and remove them.

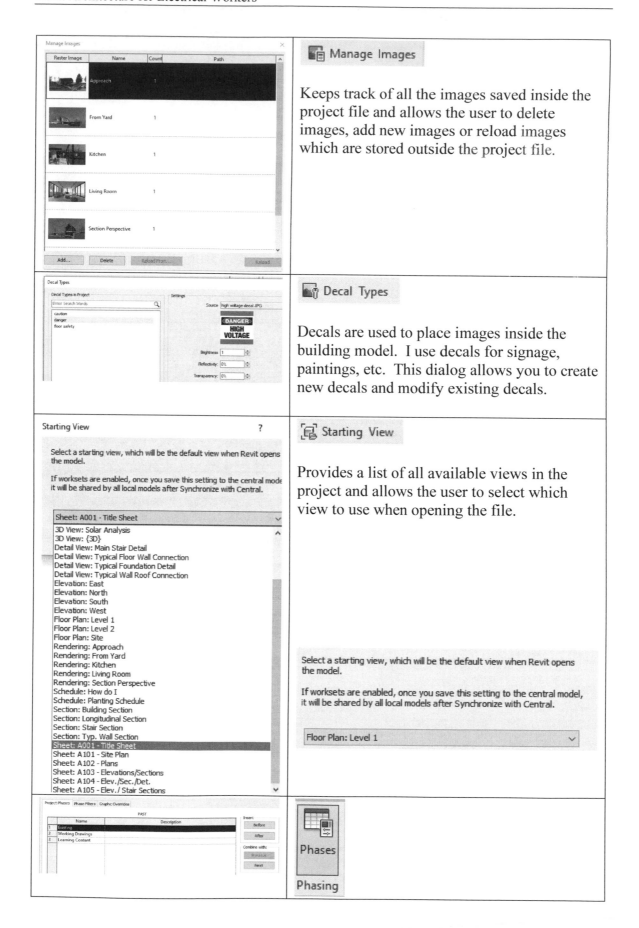

Manage Images

Keeps track of all the images saved inside the project file and allows the user to delete images, add new images or reload images which are stored outside the project file.

Decal Types

Decals are used to place images inside the building model. I use decals for signage, paintings, etc. This dialog allows you to create new decals and modify existing decals.

Starting View

Provides a list of all available views in the project and allows the user to select which view to use when opening the file.

Select a starting view, which will be the default view when Revit opens the model.

If worksets are enabled, once you save this setting to the central model, it will be shared by all local models after Synchronize with Central.

Floor Plan: Level 1

Phases

Phasing

<table>
<tr><td></td><td>Phases are used to manage the different phases of a building project. For example, you can assign elements, such as walls, to As-Built, Demo, or New Construction. The Phases can then be used to control what is visible in a view or in a schedule.</td></tr>
</table>

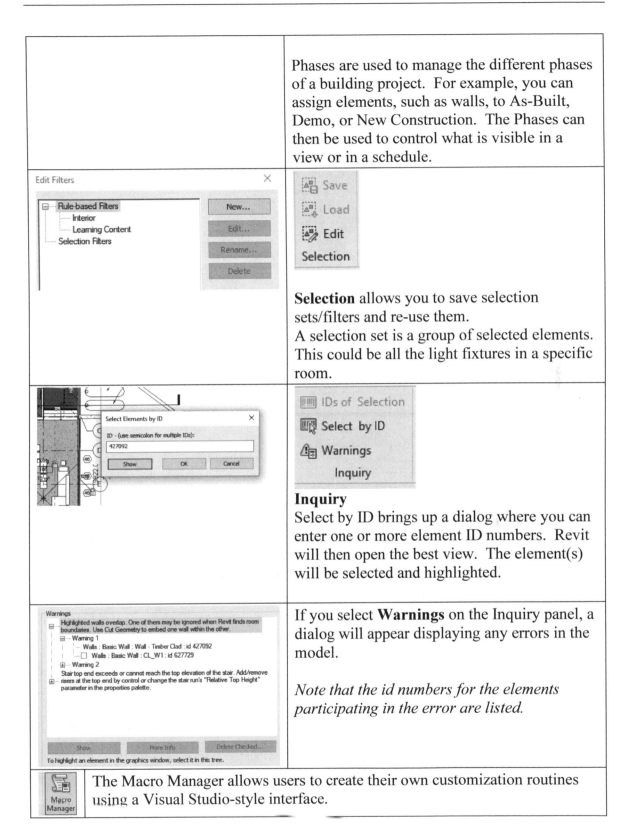

Selection allows you to save selection sets/filters and re-use them.
A selection set is a group of selected elements. This could be all the light fixtures in a specific room.

Inquiry
Select by ID brings up a dialog where you can enter one or more element ID numbers. Revit will then open the best view. The element(s) will be selected and highlighted.

If you select **Warnings** on the Inquiry panel, a dialog will appear displaying any errors in the model.

Note that the id numbers for the elements participating in the error are listed.

The Macro Manager allows users to create their own customization routines using a Visual Studio-style interface.

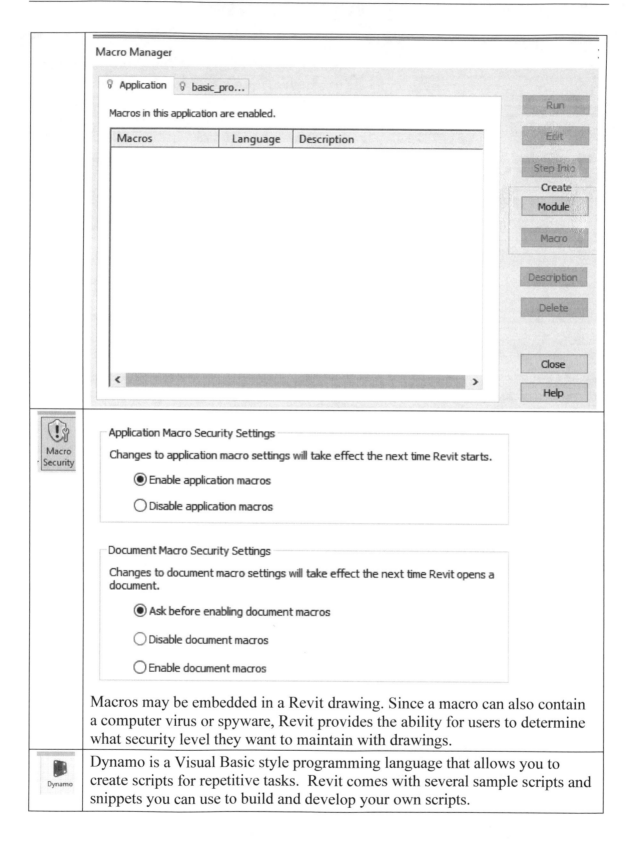

Macros may be embedded in a Revit drawing. Since a macro can also contain a computer virus or spyware, Revit provides the ability for users to determine what security level they want to maintain with drawings.

Dynamo is a Visual Basic style programming language that allows you to create scripts for repetitive tasks. Revit comes with several sample scripts and snippets you can use to build and develop your own scripts.

➤ By default, the Browser lists all sheets and all views.
➤ If you have multiple users of Revit in your office, place your family libraries and rendering libraries on a server and designate the path under the File Locations tab under Options; this allows multiple users to access the same libraries and materials.

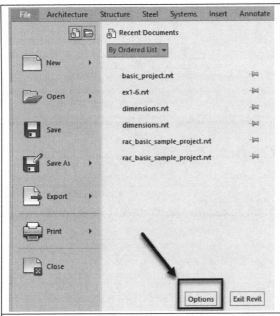

To access the System Options:

Go to the Revit menu and select the Options button located at the bottom of the dialog.

Save Reminder interval:

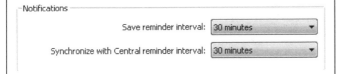

Notifications

You can select any option from the drop-down. This is not an auto-save feature. Your file will not be saved unless you select the 'Save' button. All this does is bring up a reminder dialog at the selected interval.

Username	This indicates the username that checks in and checks out documents. You can only change the username when there are no open projects. The username can also be used as a property for sheets.
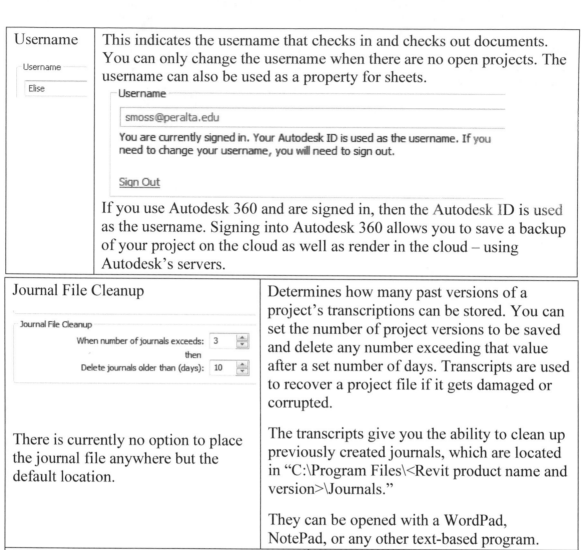 Username Elise	Username smoss@peralta.edu You are currently signed in. Your Autodesk ID is used as the username. If you need to change your username, you will need to sign out. Sign Out If you use Autodesk 360 and are signed in, then the Autodesk ID is used as the username. Signing into Autodesk 360 allows you to save a backup of your project on the cloud as well as render in the cloud – using Autodesk's servers.
Journal File Cleanup Journal File Cleanup When number of journals exceeds: 3 then Delete journals older than (days): 10 There is currently no option to place the journal file anywhere but the default location.	Determines how many past versions of a project's transcriptions can be stored. You can set the number of project versions to be saved and delete any number exceeding that value after a set number of days. Transcripts are used to recover a project file if it gets damaged or corrupted. The transcripts give you the ability to clean up previously created journals, which are located in "C:\Program Files\<Revit product name and version>\Journals." They can be opened with a WordPad, NotePad, or any other text-based program.
Worksharing Update Frequency Less Frequent More Frequent Every 5 seconds	If you are in a worksharing environment, where you and other team members are working on the same project, you can set how often the common project is updated.
 View Options Default view discipline: Architectural Architectural Structural Mechanical Electrical Plumbing Coordination OK	The View options set the display style for views depending on discipline.

The User Interface Options

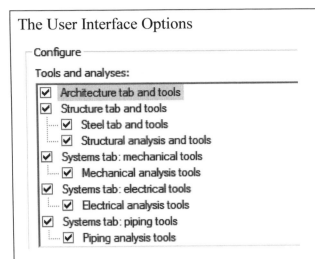

Enabling the Tools and analyses controls the visibility of the ribbons. You can hide the ribbons for those tools which you don't need or use by unchecking those ribbons.

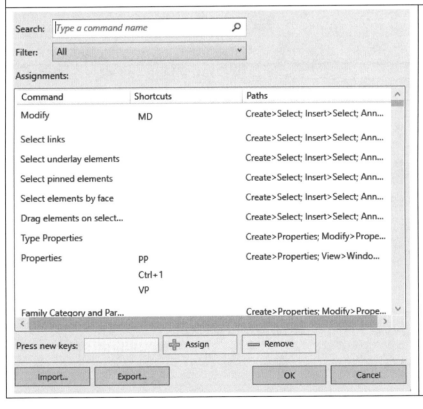

Keyboard Shortcuts (also located under User Interface on the View ribbon) allow you to assign keyboard shortcuts to your favorite commands.

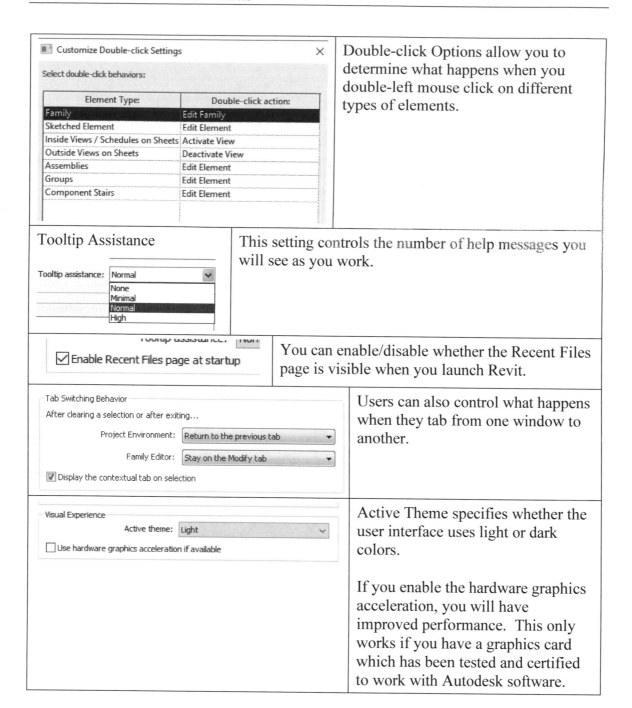

Double-click Options allow you to determine what happens when you double-left mouse click on different types of elements.

Tooltip Assistance

This setting controls the number of help messages you will see as you work.

You can enable/disable whether the Recent Files page is visible when you launch Revit.

Users can also control what happens when they tab from one window to another.

Active Theme specifies whether the user interface uses light or dark colors.

If you enable the hardware graphics acceleration, you will have improved performance. This only works if you have a graphics card which has been tested and certified to work with Autodesk software.

The Graphics tab

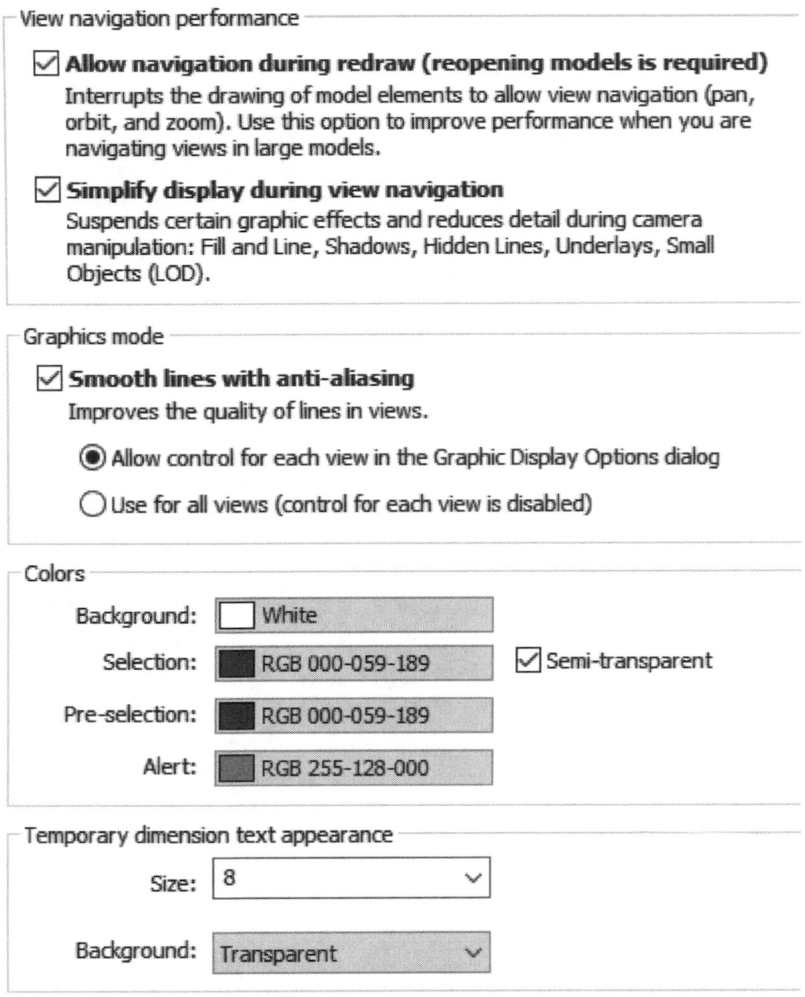

View navigation performance

☑ **Allow navigation during redraw (reopening models is required)**

Interrupts the drawing of model elements to allow view navigation (pan, orbit, and zoom). Use this option to improve performance when you are navigating views in large models.

☑ **Simplify display during view navigation**

Suspends certain graphic effects and reduces detail during camera manipulation: Fill and Line, Shadows, Hidden Lines, Underlays, Small Objects (LOD).

Graphics mode

☑ **Smooth lines with anti-aliasing**

Improves the quality of lines in views.

◉ Allow control for each view in the Graphic Display Options dialog

○ Use for all views (control for each view is disabled)

Colors

Background:	☐ White	
Selection:	■ RGB 000-059-189	☑ Semi-transparent
Pre-selection:	■ RGB 000-059-189	
Alert:	■ RGB 255-128-000	

Temporary dimension text appearance

Size: 8 ⌄

Background: Transparent ⌄

Background	Allows you to set the background color of the display window.
Selection Color	Set the color to be used when an object(s) is selected. You can also enable to be semi-transparent.
Pre-selection Color	Set the color to be used when the cursor hovers over an object.
Alert Color	Sets the color for elements that are selected when an error occurs.
Temporary Dimension Text	Users can set the font size of the temporary dimension text and change the background of the text box.

Hardware Setup 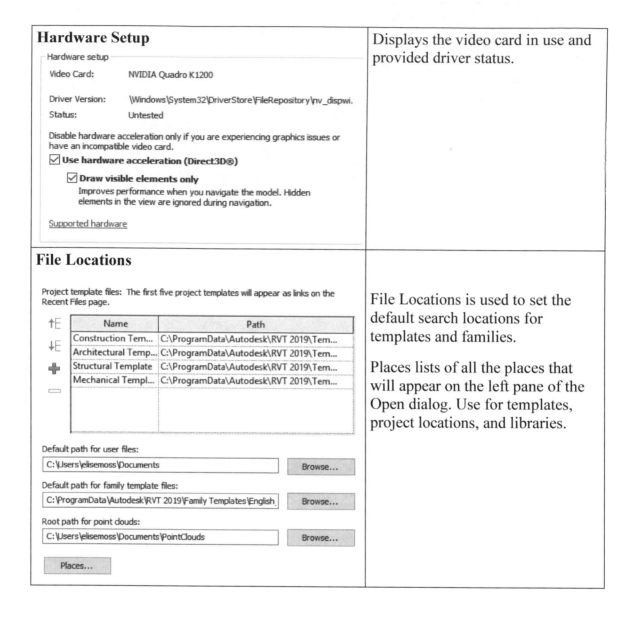	Displays the video card in use and provided driver status.
File Locations	File Locations is used to set the default search locations for templates and families. Places lists of all the places that will appear on the left pane of the Open dialog. Use for templates, project locations, and libraries.

Rendering 	Rendering controls where you are storing your AccuRender files and directories where you are storing your materials. This allows you to set your paths so Revit can locate materials and files easily. If you Click the Get More RPC button, your browser will launch to the Archvision website. You must download and install a plug-in to manage your RPC downloads. There is a free exchange service for content, but you have to create a login account.

Check Spelling 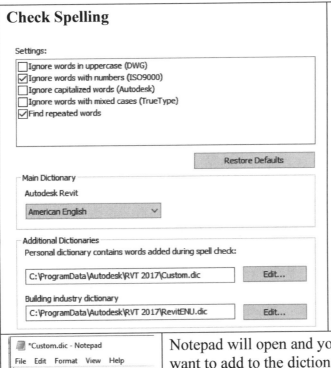	Check Spelling allows you to use your Microsoft Office dictionary as well as any custom dictionaries you may have set up. These dictionaries are helpful if you have a lot of notes and specifications on your drawings. To add a word to the Custom Dictionary, Click the **Edit** button.
*Custom.dic - Notepad File Edit Format View Help JBOX HRC HOMERUN CAN	Notepad will open and you can just type in any words you want to add to the dictionary. Save the file.

Steering Wheels

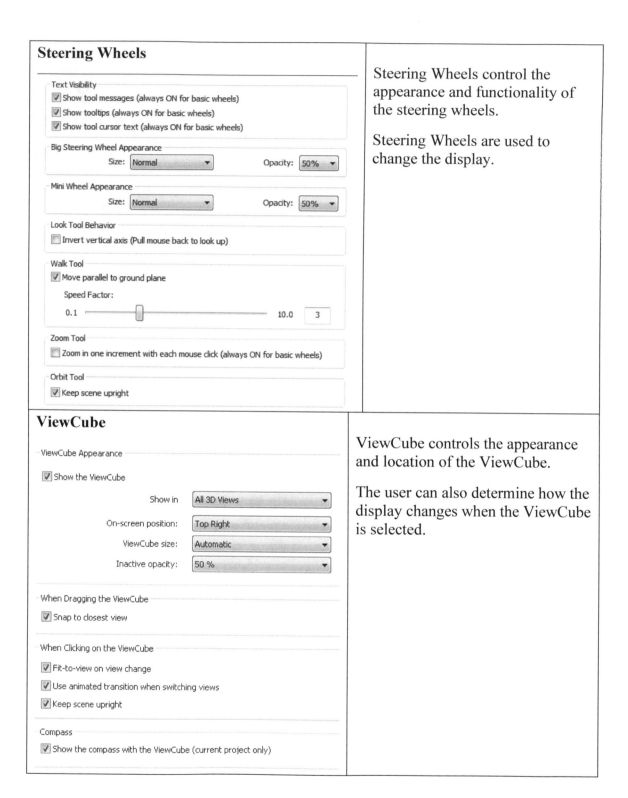

Steering Wheels control the appearance and functionality of the steering wheels.

Steering Wheels are used to change the display.

ViewCube

ViewCube controls the appearance and location of the ViewCube.

The user can also determine how the display changes when the ViewCube is selected.

Macros

Application Macro Security Settings

Changes to application macro settings will take effect the next time Revit starts.

- ◉ Enable application macros
- ○ Disable application macros

Document Macro Security Settings

Changes to document macro settings will take effect the next time Revit opens a document.

- ◉ Ask before enabling document macros
- ○ Disable document macros
- ○ Enable document macros

Macros is where the user sets the security level for Macros.

To boost productivity, store all your custom templates on the network for easy access and set the pointer to the correct folder.

The Help Menu

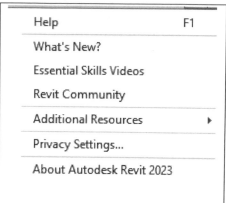

Revit Help (also reached by function key F1) brings up the Help dialog.

What's New allows veteran users to quickly come up to speed on the latest release.

Essential Skills Videos is a small set of videos to help get you started. They are worth watching, especially if your computer skills are rusty.

Revit Community is an internet-based website with customer forums where you can search for solutions to your questions or post your questions. Most of the forums are monitored by Autodesk employees who are friendly and knowledgeable.

Additional Resources launches your browser and opens to a link on Autodesk's site. Autodesk Building Solutions take you to a YouTube channel where you can watch video tutorials.

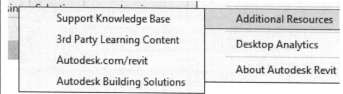

Privacy Settings are used by Autodesk to track your usage of their software remotely using your internet connection. You can opt out if you do not wish to participate.

About Autodesk Revit 2023 reaches a splash screen with information about the version and build of the copy of Revit you are working with. If you are unsure about which Service Pack you have installed, this is where you will find that information.

Exercise 1-7:

Setting File Locations

Drawing Name: Close all open files
Estimated Time: 5 minutes

This exercise reinforces the following skills:
- □ Options
- □ File Locations

1. Close all open files or projects.

2.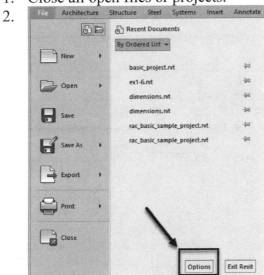

   Go to the **File Menu**.

   Select the **Options** button at the bottom of the window.

3. ┌─ File Locations   Select the **File Locations** tab.

4. Default path for user files:

   C:\Users\Elise\Documents   Browse...

   In the **Default path for users** section, pick the **Browse** button.

5. Default path for user files:

   C:\IBEW LU 595 Class\

   Navigate to the local or network folder where you will save your files. When the correct folder is highlighted, pick **Open**. Your instructor or CAD manager can provide you with this file information.

I recommend to my students to bring a flash drive to class and back up each day's work onto the flash drive. That way you will never lose your valuable work. Some students forget their flash drive. For those students, make a habit of uploading your file to Google Drive, Dropbox, Autodesk 360, or email your file to yourself.

Exercise 1-8:

Adding a Template to the Template list

Drawing Name: Close all open files
Estimated Time: 5 minutes

This exercise reinforces the following skills:
- Options
- File Locations
- Project Templates
- Recent Files

1.

   Select **Options** on the File Menu.

2. Activate File Locations.

   Click the **Plus** icon to add a project template.

   Project templates: The templates disp

	Name	
	Imperial-Constru...	C:\Pro
	Imperial-Architec...	C:\Pro
	Imperial-Structur...	C:\Pro
	Imperial-Systems...	C:\Pro
	Metric-Construct...	C:\Pro
	Metric-Architect...	C:\Pro
	Metric-Structural...	C:\Pro

3.

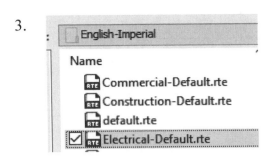

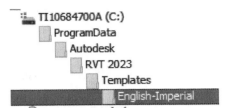

Locate the *Electrical-Default.rte* template in the English_Imperial directory.

Click **Open**.

4.

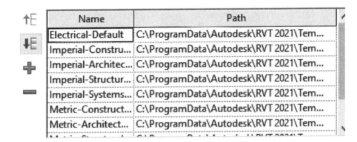

Note the file is now listed.
Select the Up arrow to move it to be ahead of the Construction template.

5.

Project templates: The templates display in a list when you create a new project.

	Name	Path
	Electrical-Default	C:\ProgramData\Autodesk\RVT 2021\Tem...
	Imperial-Constru...	C:\ProgramData\Autodesk\RVT 2021\Tem...
	Imperial-Architec...	C:\ProgramData\Autodesk\RVT 2021\Tem...
	Imperial-Structur...	C:\ProgramData\Autodesk\RVT 2021\Tem...
	Imperial-Systems...	C:\ProgramData\Autodesk\RVT 2021\Tem...
	Metric-Construct...	C:\ProgramData\Autodesk\RVT 2021\Tem...
	Metric-Architect...	C:\ProgramData\Autodesk\RVT 2021\Tem...

It is now listed above the Construction template. Organize the remaining templates in your preferred order.
Click **OK**.

Exercise 1-9:

Turning Off the Visibility of Ribbons

Drawing Name: Close all open files
Estimated Time: 5 minutes

This exercise reinforces the following skills:
- Options
- User Interface
- Ribbon Tools

1.    Select **Options** on the Application Menu.

2.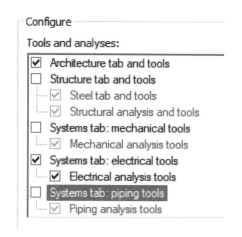

   Highlight **User Interface**.

   Uncheck/disable Structure and tab tools, System tab: mechanical tools and System tab: piping tools.

   Click **OK**.

3. Notice that the available ribbons update to only display the tools you are interested in.

Revit Families

Revit projects use Revit families.

There are three types of families:

- System
- In-Place
- Loadable

System families are specific to a project. You can copy system families from one project to another, but they are not stand-alone files, like loadable/model families. Examples of system families are walls, conduits, wires, and ceilings.

In-Place families are elements which are created "on the fly" using massing tools. Users often create an in-place family for a feature that is unique to a project. A generator or electrical equipment that is specialized may be created using massing tools, so that users can see the amount of space it takes up in a project.

Loadable families are the most common type of family. Examples include cable trays, power devices, and electrical equipment. These are external files which are inserted/loaded into a project and placed in the desired location. These families can be counted, and their properties can be organized in schedules. These elements can be created from scratch using the Family Editor using family templates. They can be created and loaded into a project, as well as deleted or saved from a project.

Revit families are defined using parameters. There are two types of parameters: Type and Instance.

Revit elements are defined by a hierarchy.

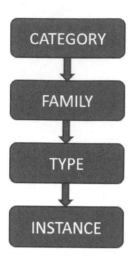

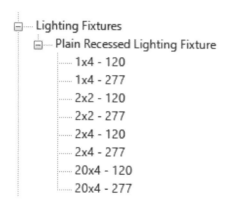

In the Revit Project Browser under the Families folder, you see the families organized into categories. Lighting Fixtures is a Category. The Plain Recessed Lighting Fixture is a Family. A family is an element that represents a specific component used in a project. Each Family can have several different types. This lighting fixture has different types which are defined by size and voltage. The family type doesn't change regardless of where it is placed in the project. If you place a 1x4 lighting fixture in the living room or the bedroom, it is still a 1x4 lighting fixture.

Every time you place or define a family in a project, you are creating an "instance" of that family. Location is an instance parameter. Hardware or finish can be unique to each family places, so these can also be instance properties. Type properties are properties that are common to all elements of that type. Instance properties are properties that are unique to each individual element.

Throughout the rest of the text, we will be creating new types of families. Here are the basic steps to creating a new family.

1. Select the element you want to define (switch gear, receptacle, panel, etc.).
2. Select **Edit Type** from the Properties pane.
3. Select **Duplicate**.
4. Rename: Enter a new name for your family type.
5. Redefine: Edit the structure, assign new materials, and change the dimensions.
6. Reload or Reassign: Assign the new type to the element.

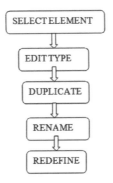

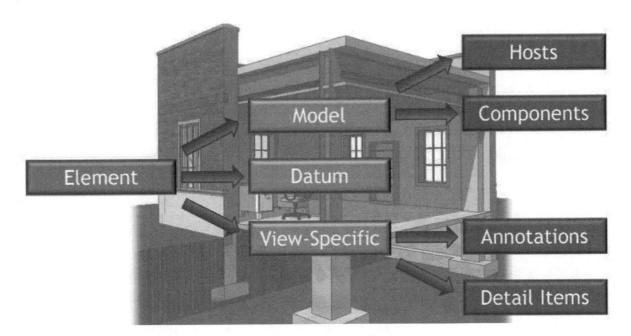

Elements are the building blocks of any Revit Project. Everything used in a Revit project is considered an element.

There are three classes of Revit families:

- Model
- Datum
- View-Specific

Model families are families which you can physically touch if you were walking through a building, such as walls or electrical panels. A host model family is an element which can be used to hold or place other components. For example, a wall can host a door, window, or electrical panel. An element which is placed on a host is considered a component. Datums are levels, grids, and survey points. They are used to constrain the project. View-specific families are annotations, like dimensions or text, and detail items, like filled regions

Non-hosted families can be placed anywhere in the view. They are typically placed aligned to the elevation of the view. If you need to offset them from the elevation, place the element in the view, then select it, then change the offset from elevation value in the Properties palette to the desired location.

Hosted families must be placed on a surface or work plane and the surface must be visible in the view. If a fixture needs to be placed on a ceiling, the view needs to be a ceiling plan. Check on the ribbon to specify the type of face to be used for placement. If the placement face is deleted, any elements hosted by the face will also be deleted. If the placement face is moved, then the elements will also move.

Electrical Devices

The workflow to add electrical devices.

1. Select a category of family to add to the model on the ribbon.
2. Use the Type selector to choose the exact type in the category.
3. Place it as required on a vertical, horizontal, or work plane face.
4. Adjust the instance properties of the family in the Properties palette.
5. Tag if needed.

Some guidelines when working with Revit families to help you work efficiently:

- Familiarize yourself with the content libraries which come with the software as well as the libraries used within your company. Then, when you are looking for a specific family, you might be able to use or modify an existing family.
- When you modify a Revit family or element, save the family under a new name and save it to a custom library location, preferably on your company's server. This will make the family type available across projects as well as to other users in your company.
- Avoid accidentally selecting elements in a view so they don't get modified.
- If you hover your cursor over an element, a small dialog will appear informing you of the family and type.

Lighting Fixtures

Lighting fixtures follow the same workflow as other electrical devices and are powered the same way. They also have the ability to calculate an average estimated illumination level for spaces.

When loading light fixture families, make sure to pull them from the MEP folder, not the Architectural folder.

Exercise 2-1:

Working with Revit Families and Elements

Drawing Name: *elements.rvt*
Estimated Time: 20 minutes

This exercise reinforces the following skills:

- ❑ Identifying elements and their families in a project
- ❑ Place a Component

1. 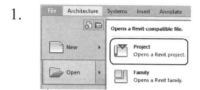 Open *elements.rvt.*

    This file is downloaded from the publisher's website.

2. 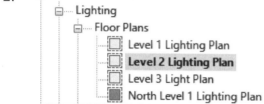 Open the **Level 2 Lighting Plan**.

3. 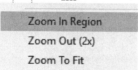 Right click and select **Zoom In Region**.

4. 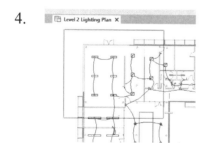 Draw a rectangle/region in the upper left corner of the building.

5. Hover the cursor over one of the lighting fixtures labeled 'C'.
    Note that the family and type are displayed.

6. Pan down.
    Hover the cursor over one of the lighting fixtures labeled 'B'.
    Note that the family and type are displayed.
    The 'B' and 'C' lighting fixtures are the same family, but different types.

7. Select the 'C' lighting fixture that is located on the upper left.

8. 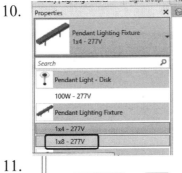 *Note that the Family and Type are displayed in the Properties panel.*

9. Select the small down arrow located at the top of the Properties panel.

This is called the Type Selector.

10. 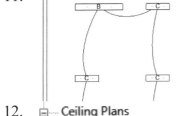 Select the **1x8-277v** from the Type Selector list.

Click **ESC** to release the selection.

11. *Notice that the element is updated to the new type which you selected.*

12. Ceiling Plans
    Level 1
    Level 2
    Level 3

Open the **Level 2** view under Ceiling Plans.

Look in the Coordination category.

Note that the lighting fixture which was changed in the previous view updated in this view.

Revit has bi-directional associativity. This means that if you make a change to a model element in one view, the change is propagated throughout the model. All the views update.

This only applies to model elements – NOT to annotation elements. Annotation elements, like dimensions and notes, are view-specific.

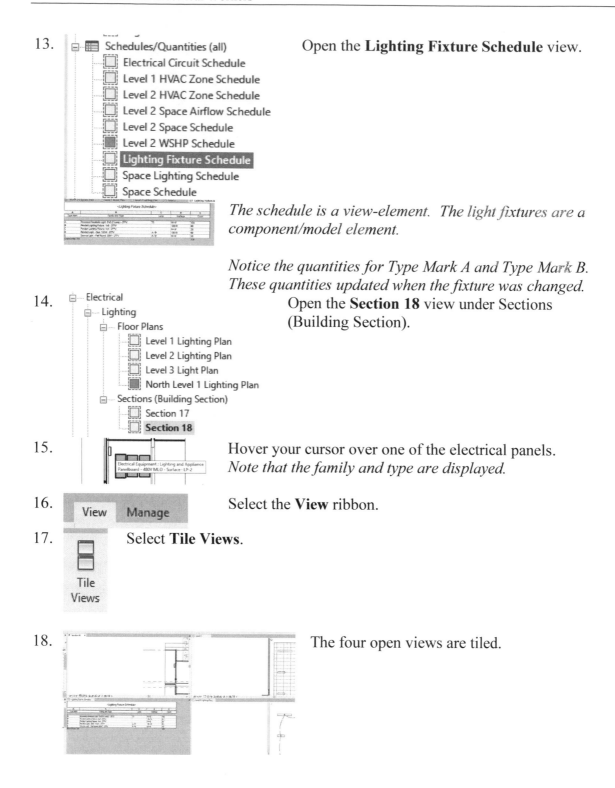

13. Open the **Lighting Fixture Schedule** view.

The schedule is a view-element. The light fixtures are a component/model element.

Notice the quantities for Type Mark A and Type Mark B. These quantities updated when the fixture was changed.

14. Open the **Section 18** view under Sections (Building Section).

15. Hover your cursor over one of the electrical panels. *Note that the family and type are displayed.*

16. Select the **View** ribbon.

17. Select **Tile Views**.

18. The four open views are tiled.

19. 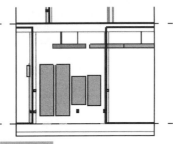 Click in the upper left window to activate it.
Double click on the mouse wheel to Zoom All.
Repeat for the other windows.

20. In the window with the schedule:

Put your mouse in the field for Family and Type for the 'C' type mark fixture.

Notice how the lighting fixture is highlighted in the other windows.

21. Click **Tab Views** on the View ribbon.

22. The windows close to a single active window. There are tabs that allow you to switch between the open windows.

23. Close without saving.

Many electricians need to determine the location of the stud framing in a wall as they are routing their wiring between the studs. Most Revit projects created by electrical workers are defined by using a host project which links to the files provided by the architect and the other sub-contractors. It is helpful to be able to select elements that reside in the linked file, so you can identify them and determine how they are defined.

Exercise 2-2:

Identifying a Wall in a Linked File

Drawing Name: *simple-building.rvt*
Estimated Time: 10 minutes

This exercise reinforces the following skills:

- ❏ System Families
- ❏ Revit Links
- ❏ Walls

1. 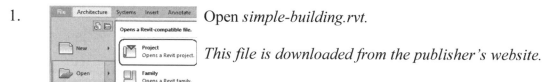 Open *simple-building.rvt*.

   This file is downloaded from the publisher's website.

2. *This project uses a Linked Revit project.*
   Open the **Insert** ribbon.

   Select **Manage Links**.

3. The linked file is named *Simple-Building-Arch.rvt*.

Link Name	Status	
Simple-Building-Arch.rvt	Not Found	C

   Because the files are now in a new location, the link has to be re-established.

   Highlight the file name.

   Select **Reload From.**

4. File name: Simple-Building-Arch.rvt
   Files of type: RVT Files (*.rvt)

   Locate the file in the downloaded files from the publisher's website.

   Select **Open**.
   The file now shows as loaded.

   Click **OK** to close the dialog.

5. | Link Name | Status | |
   |---|---|---|
   | Simple-Building-Arch.rvt | Loaded | C |

6.

In the lower right hand corner of the display are selection tools.

The link tool allows you to select elements which are linked to the host file.

The select underlay tool allows you to select elements which are part of the linked file.

Click on these icons to enable.

7.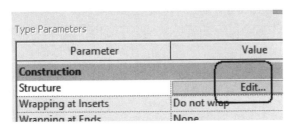

See if you can select the right vertical wall using the TAB key.

8.

On the Properties palette, select **Edit Type**.

9.

Parameter	Value
Construction	
Structure	Edit...
Wrapping at Inserts	Do not wrap
Wrapping at Ends	None

Select **Edit** next to Structure.

10.

Family:	Basic Wall
Type:	Generic - 8"
Total thickness:	0' 8"
Resistance (R):	0.0000 (h·ft²·°F)/BTU
Thermal Mass:	0.0000 BTU/°F

Sample Height: 20' 0"

Layers

EXTERIOR SIDE

	Function	Material	Thickness	Wraps	Structural Material
1	Core Boundary	Layers Above	0' 0"		
2	Structure [1]	<By Category	0' 8"		
3	Core Boundary	Layers Below	0' 0"		

INTERIOR SIDE

You see how the wall has been defined by the architect.

Notice that it has an exterior side and an interior side.

The Core boundary is the boundary around the stud or framing. Anything outside of the core boundary is considered a wrapped layer and is usually a finish, like gypsum board or siding.

11. Click **OK** twice to close the dialogs.
12. Left click in the window to release the selection.
13. Close the file without saving.

Exercise 2-3:

Place a Lighting Fixture and a Switch

Drawing Name: *simple-building.rvt*
Estimated Time: 10 minutes

This exercise reinforces the following skills:

- ❑ System Families
- ❑ Revit Links
- ❑ Walls

1.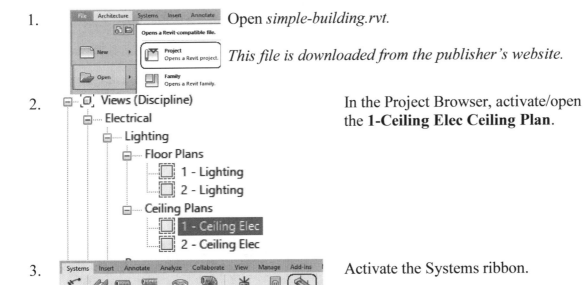

   Open *simple-building.rvt*.

   This file is downloaded from the publisher's website.

2. In the Project Browser, activate/open the **1-Ceiling Elec Ceiling Plan**.

3. Activate the Systems ribbon.

   Select **Lighting Fixture**.

4. If the Work Plane dialog opens:

   Enable **Name**.

   Select **Level 1**.

   Click **OK**.

5.

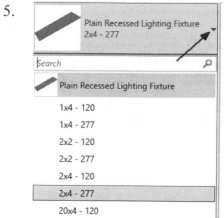

Use the Type Selector to select **Plain Recessed Lighting Fixture:2x4 – 277**.

The Type Selector is the small down arrow located to the right of the family name. The Type Selector is used when a family has more than one version available.

6.

The ribbon has changed to a contextual style.
Enable **Place on Face.**
This means the fixture will be hosted by a work plane.

7.

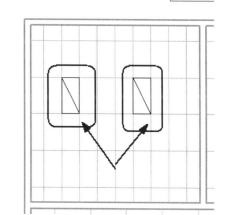

Use the SPACEBAR to rotate the light fixture.

Place two lighting fixtures in the upper left room.

8.

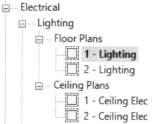

In the Project Browser, activate/open the **1-Lighting** floor plan.

Notice you see the lighting fixtures you placed in the ceiling plan. This is because Revit is a BIM software. If you make a change to the model, any relevant views will update.

9.

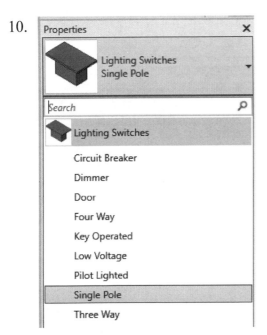

Activate the Systems ribbon.

Select **Lighting** under the **Device** drop-down list.

10.

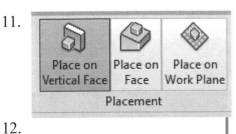

Use the Type Selector to select **Lighting Switches – Single Pole**.

11.

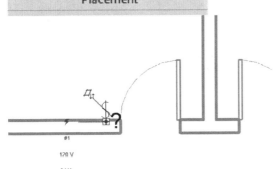

Enable **Place on Vertical Face** on the ribbon.

12.

Place the switch to the left of the door in the upper left room where the lighting fixtures were placed.

Right click and select Cancel to exit the command.

13.

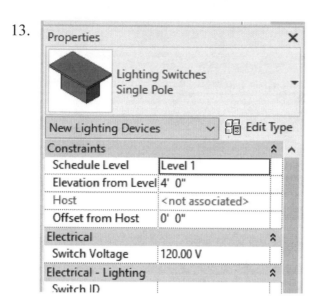

Select the switch that was just placed.

Note that the Elevation from the Level is set to 4'-0".

Save the project as *ex2-3.rvt*.

Exercise 2-4:

Select and Modify a Component

Drawing Name: *modify.rvt*
Estimated Time: 5 minutes

This exercise reinforces the following skills:

❏ Filter
❏ Type Selector

1. 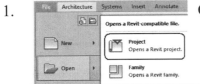 Open or continue working in *ex2-3.rvt*.

2. 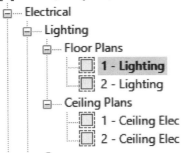 If you open the modify file, remember you have to reload the linked file.

3. 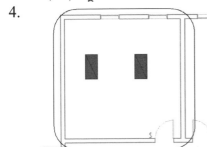 In the Project Browser, activate/open the **1-Lighting** floor plan.

4. Left click at the upper left corner of the room with the lighting fixtures. Hold down the left mouse button to create a window.

   Left click at the lower right corner of the room with the lighting fixtures.

   This creates a selection group using a window.

5. 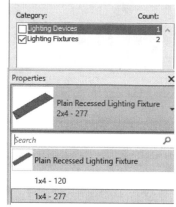 Select the **Filter** tool from the ribbon.

6. Uncheck any items in the list except for **Lighting Fixtures.**

Click **OK.**

7. 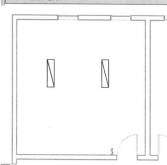 Use the Type Selector to change the lighting fixtures to **1x4 – 277**.

Click **ESC** to release the selection.

8. The lighting fixtures update.

Save the project as *ex2-4.rvt.*

Exercise 2-5:

Copy a Component

Drawing Name: *copy.rvt*
Estimated Time: 5 minutes

This exercise reinforces the following skills:

- ❑ Copy
- ❑ Type Selector

1. Open *copy.rvt*.

    You may have to use Manage Links to reload the linked file.

2. In the Project Browser, activate/open the **1-Ceiling Elec** Ceiling plan.

3. 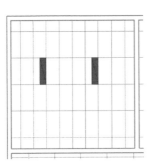 Window around the first room to select the two lighting fixtures.

4. 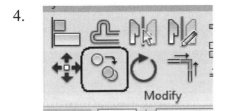 Select the **Copy** tool on the ribbon.

5.

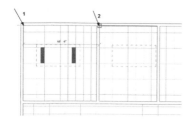

Select the top left corner of the first room as the base point.

Select the top left corner of the second room as the target point.

Right click and select Cancel to exit the command.

6.

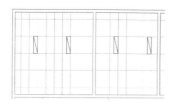

Save the file as *ex2-5.rvt*.

Exercise 2-6:

Mirror a Component

Drawing Name: *mirror.rvt*
Estimated Time: 5 minutes

This exercise reinforces the following skills:

❑   Mirror→Draw Axis

1.    Open *mirror.rvt.*

You may have to use Manage Links to reload the linked file.

2.    In the Project Browser, activate/open the **1-Ceiling Elec** Ceiling plan.

3.   Window around the lighting fixtures to select them.

4.   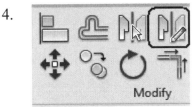 Select the **Mirror→Draw Axis** tool.

5.   Select the top of the center line of the wall.

Select the bottom of the center line of the wall.

6.   The light fixtures are mirrored to the other rooms.

Save as *ex2-6.rvt.*

Exercise 2-7:

Align a Component

Drawing Name: *align.rvt*
Estimated Time: 5 minutes

This exercise reinforces the following skills:

❑   Align

1.  Open *align.rvt*.

    You may have to use Manage Links to reload the linked file.

2. 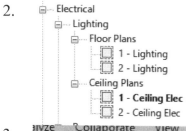 In the Project Browser, activate/open the **1-Ceiling Elec** Ceiling plan.

3. 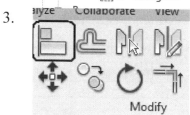 Select the **ALIGN** tool on the Modify ribbon.

4. Select the ceiling grid line to the left of the lighting fixture.

    Select the left edge of the lighting fixture.

5.

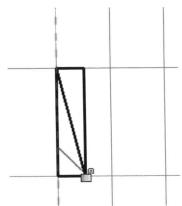

The lighting fixture's position adjusts.

6. Use the ALIGN tool to adjust the position of the lighting fixtures so they are aligned to the ceiling grid.

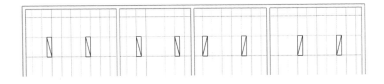

Save as *ex2-7.rvt*.

Exercise 2-8:

Draw, Modify, and Offset Cable Trays

Drawing Name: *cable_trays.rvt*
Estimated Time: 30 minutes

This exercise reinforces the following skills:

- ❑ Cable Trays
- ❑ Offset
- ❑ Trim
- ❑ Split
- ❑ Options Bar
- ❑ Properties

1.   Open *cable_trays.rvt*.

You may have to use Manage Links to reload the linked file.

2.   In the Project Browser, activate/open the **1-Power** floor plan.

3.   Activate the Systems ribbon.

Select the **Cable Tray** tool.

4. 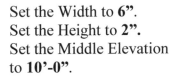 Select the **Trough Cable Tray** from the Type Selector.

5. On the Options bar:

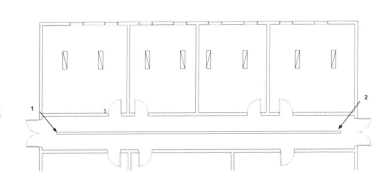

Set the Width to **6"**.
Set the Height to **2"**.
Set the Middle Elevation to **10'-0"**.

6. Left click to place the start of the cable tray at the left side of the corrider (indicated by 1).

Left click to place the end of the cable tray at the right side of the corrider (indicated by 2).

7. Click **ESC** to complete placing the tray but remain in the Cable Tray command.

8. 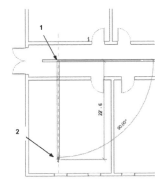 Place a second cable tray perpendicular to the first cable tray and into the Mech/Elec room (lower left).

The new cable tray should be 22' 6" long.

Right click and select **Cancel** to exit the command.

9. This is what the layout should look like.

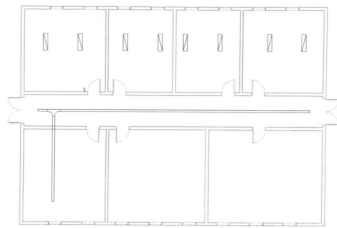

10. Select the ALIGNED DIMENSION tool from the QAT.

11. Place a dimension between the end of the horizontal cable tray and the right side of the perpendicular cable tray.

4' - 2"

12. Select the vertical cable tray.

    The temporary dimension appears.

4' - 2"

10' 0'

13. Change the horizontal dimension to **4' 0"**.

4' - 0"

10' 0'

14. 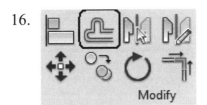 Change the vertical dimension to **20' 6"**.

15. Select the vertical cable tray so it is highlighted.

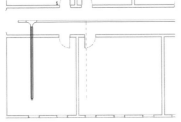

16. Select the **Offset** tool.

Modify

17. On the Options bar:

○ Graphical ● Numerical   Offset: 15' 0"   ☑ Copy

Enable **Numerical**.
Set the Offset to **15'-0"**.
Enable **Copy**.

18. 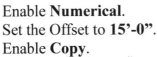 Left pick the cable tray.

Use the Preview to determine the cable tray is placed to the right.

19. Repeat the Offset to place a total of five vertical cable trays.

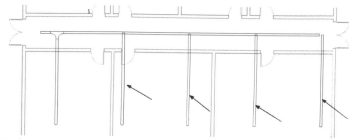

20.  Select the **Trim to Corner** tool on the Modify ribbon.

21. Select the horizontal cable tray and the far right cable tray.

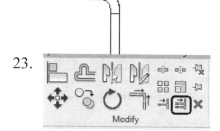

22. The two cable trays are joined.

23. 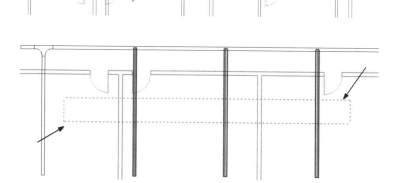 On the Modify ribbon:

Select the **Trim/Extend Multiple** tool.

24. Select the lower line of the horizontal cable tray to define the boundary of the trim.

25. Left pick to the right of the fourth vertical cable tray to start the selection/crossing window.

Left pick to the left of the second vertical cable tray to complete the selection/crossing window.

26. This is the cable tray layout so far.

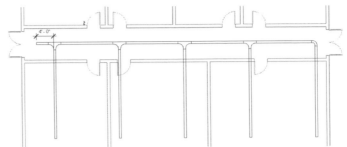

27.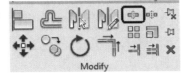

Select the **Split** tool on the Modify ribbon.

28.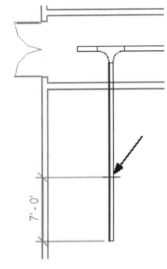

Use the temporary dimension to split the cable tray 7'-0" from the end.

Right click and select Cancel twice to exit the command.

29.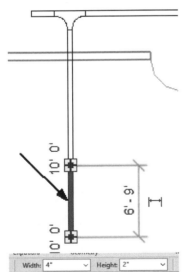

Select the lower part of the cable tray.

30. On the Options bar:

Set the Width to **4"**.

Left click in the display window to release the selection.

31. Save as *ex2-8.rvt*.

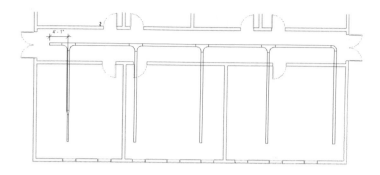

Exercise 2-9:

Place Light Fixtures and Switches (reprised)

Drawing Name: *i_elec_circuits.rvt*
Estimated Time: 20 minutes

This exercise reinforces the following skills:

- ❑ Place Components
- ❑ Copy

1.  Open *i_elec_circuits.rvt.*

2. 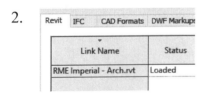 *This file uses RME Imperial-Arch as the linked file. Use the Manage Links tool on the Insert tab to reload the file, if necessary.*

3. 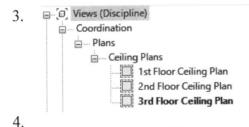 Verify that you are in the **3ʳᵈ Floor Ceiling Plan** view.

4. 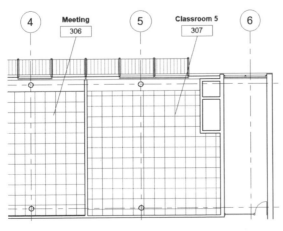 Use **Zoom In Region** to change the display to focus on Classroom 5 Room 307 between Grids 5 and 6.

5.

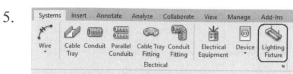

Activate the Systems tab of the ribbon.

Select **Lighting Fixture**.

6.

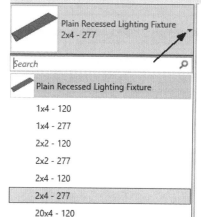

Use the Type Selector to select **Plain Recessed Lighting Fixture:2x4 – 277**.

The Type Selector is the small down arrow located to the right of the family name. The Type Selector is used when a family has more than one version available.

7.

The ribbon has changed to a contextual style.

Enable **Place on Face.**

This means the fixture will be hosted by a work plane.

8.

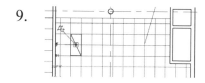

Use the SPACE BAR to rotate the component prior to placing.

Left click to place in the upper left area of the room.

Right click and select Cancel to exit the command.

9.

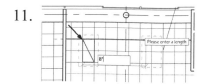

Select the light fixture you just placed.

10.

Select the **Copy** tool from the ribbon.

11.

Left click to select the upper left corner of the light fixture as the base point.
Move the cursor to the right and type **8'** for the distance.
Click ENTER or left click to complete the copy.

12.   Repeat to place a third light fixture 8' to the right of the copied/second light fixture.

Exit the COPY command.

13.   Use the CTL key to select the three light fixtures.
Use the COPY tool to place a copy of the three fixtures 8' below them.

14.   Repeat to place a third row of light fixtures 8' below the second row.

Exit the COPY command.

15.   *The room should show a total of nine light fixtures.*

16.   Open the **3ʳᵈ Floor Lighting** floor plan.

- Electrical
  - Lighting
    - Floor Plans
      - 1st Floor Lighting
      - 2nd Floor Lighting
      - 3rd Floor Lighting Plan

17.    Switch to the Architecture ribbon.

Select **Component→Place a Component**.

18.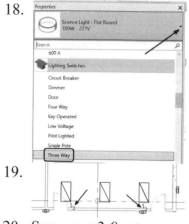

Use the Type Selector to locate the **Lighting Switches→Three Way**.

Notice that the available components are sorted alphabetically.

19. Place a light switch next to each door located at the south wall of the room.

20. Save as *ex2-9.rvt*.

Exercise 2-10:

Adding and Modifying Equipment, Devices and Fixtures

Drawing Name: *adding_elements.rvt*
Estimated Time: 20 minutes

This exercise reinforces the following skills:

- ❏ Place Components
- ❏ Electrical Equipment
- ❏ Load Families
- ❏ Type Selector
- ❏ Naming Equipment

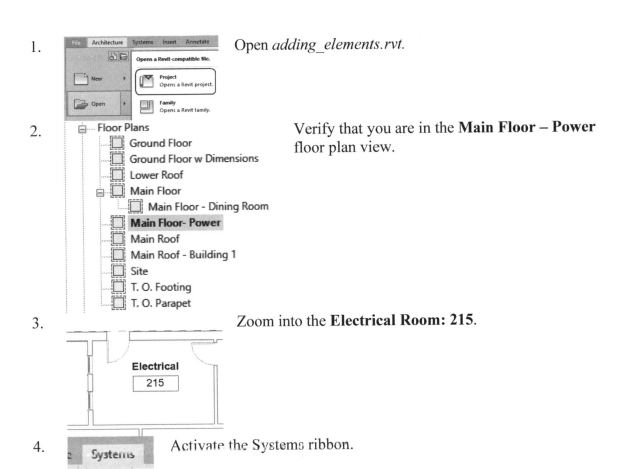

1.   Open *adding_elements.rvt*.

2.   Verify that you are in the **Main Floor – Power** floor plan view.

3.   Zoom into the **Electrical Room: 215.**

4.   Activate the Systems ribbon.

5. 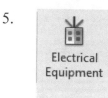   Select the **Electrical Equipment** tool.

6.   No Electrical Equipment family is loaded in the project. Would you like to load one now?

   Yes   No

   Click **Yes**.

7.      Browse to the *Distribution* folder under *US Imperial\Electrical\MEP\Electric Power*.

8.      Hold down the CTL key to make a multiple selection.

   Select *Lighting and Appliance Panelboard – 208V MLO* and *Lighting and Appliance Panelboard – 480V MLO*.

   Click **Open**.

9.   On the Properties palette:

   Verify that the *Lighting and Appliance Panelboard – 208V MLO-225A* is active.

10.   *This is a face-based component, meaning it needs a vertical or horizontal face to be placed. Since it is a hosted component, if the wall where it is placed moves, it will also move.*

   Enable **Place on Vertical Face** on the ribbon.

11.   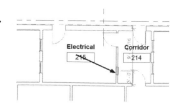   Place it on the east wall of Room 215.

   Type ESC out of the command.

12.

Select the panel that was just placed.

On the Properties palette, note the elevation from the level.

Left click anywhere in the display window to release the selection.

13.

Return to the Systems ribbon.

Select the **Electrical Equipment** tool.

Using the Type Selector:
Select *Lighting and Appliance Panelboard – 480V MLO 250A.*

Note that by default it is located at the same elevation as the other panelboard.

14.

Electrical

215

Place the *Lighting and Appliance Panelboard – 480V MLO 250A* above the other panelboard close to the door with a space between the two elements.

15.

Return to the Systems ribbon.

Select the **Electrical Equipment** tool.

16.

Select **Load Family**.

17.

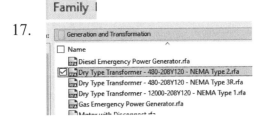

Browse to the *Generation and Transformation* folder.

Select the *Dry Type Transformer – 480-208Y120-NEMA Type 2.*

Click **Open**.

18.

Use the Type Selector to select the **45 kVA** type.

19.

Electrical

215

Electrical Equipment : Lighting and Appliance
Panelboard - 208V MLO : 225 A

Place the transformer next to the panelboards in Room 215.

Use the SPACEBAR to rotate the element prior to placing.

Escape out of the command.
Remember that if you hover the cursor over an element a tooltip will appear showing you the element information.

20.

General	
Mounting	Recessed
Enclosure	Type 1
Panel Name	LP1
Location	Electrical 215

Select the 208V MLO panelboard.

In the Properties palette:
Scroll down to the General area.
Type **LP1** in the Panel Name.

21.

General	
Mounting	Recessed
Enclosure	Type 1
Panel Name	HP1

Select the 480V MLO panelboard.

In the Properties palette:
Scroll down to the General area.
Type **HP1** in the Panel Name.

22.

Properties

Dry Type Transformer -
480-208Y120 - NEMA Type 2
45 kVA

Electrical Equipment (1)		Edit Type
Image		
Comments		
Mark	4	
Phasing		
Phase Created	New Constru...	
Phase Demolished	None	
General		
Enclosure	Type 2	
Mounting		
Panel Name	T1	
Location	Electrical 215	
Electrical - Circuiting		

Select the Transformer.

In the Properties palette:
Scroll down to the General area.
Type **T1** in the Panel Name.

23. Save as *ex2-10.rvt*.

Room Aware

In most instances, families placed within a room are associated with the room in a schedule. There are a few circumstances in which a family instance does not report its room correctly. When families such as furniture, doors, windows, casework, specialty equipment and generic models are placed in a project, sometimes parts of their geometry are located outside a room, space, or within another family, which results in no calculable values being reported.

Room Calculation Points are used to make loadable families "room aware." Once you enable the room calculation point, the family will display the correct room in the project schedule.

To enable and modify the Room Calculation Point to reorient room aware families:

1. Select the family instance in the drawing area.
2. Click Modify | <Element> tab ➤ Mode panel ➤ ⬛Edit Family.
3. In the Family Editor, open a floor plan view of the family.
4. On the Properties palette, select the Room Calculation Point parameter in the Other section. The point is now visible in the drawing area as a green dot.
5. Select the Room Calculation Point and move it to a location that will not be obscured by geometry when placed in the project.
6. Click Modify tab ➤ Family Editor panel ➤ ⬛Load into Project.

Exercise 2-11:

Making a Component Room Aware

Drawing Name: *room aware.rvt*
Estimated Time: 15 minutes

This exercise reinforces the following skills:

❑   Schedules
❑   Room Calculation Point
❑   Edit Family

1. 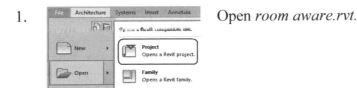   Open *room aware.rvt*.

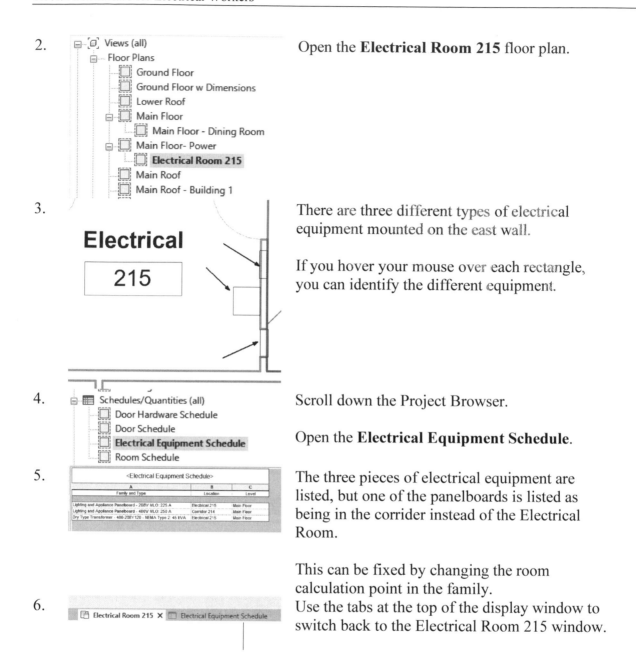

2. Open the **Electrical Room 215** floor plan.

3. There are three different types of electrical equipment mounted on the east wall.

   If you hover your mouse over each rectangle, you can identify the different equipment.

4. Scroll down the Project Browser.

   Open the **Electrical Equipment Schedule**.

5. The three pieces of electrical equipment are listed, but one of the panelboards is listed as being in the corrider instead of the Electrical Room.

   This can be fixed by changing the room calculation point in the family.

6. Use the tabs at the top of the display window to switch back to the Electrical Room 215 window.

7.

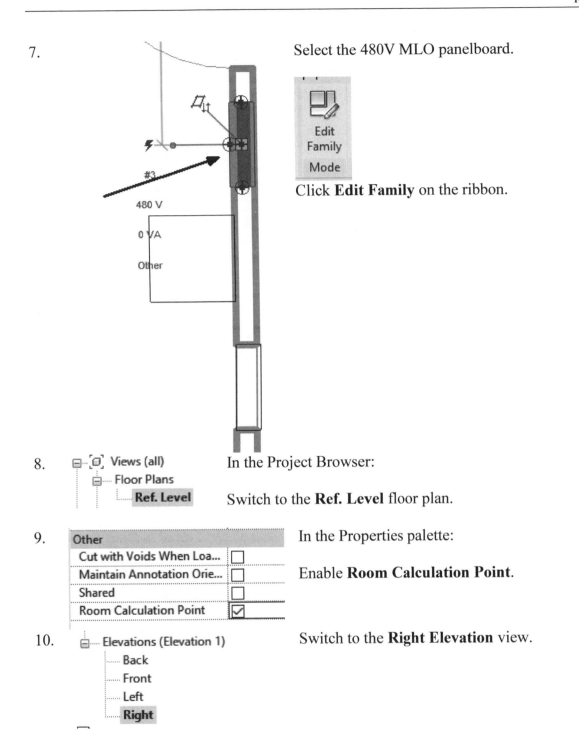

Select the 480V MLO panelboard.

Click **Edit Family** on the ribbon.

8. ⊟ ⌐◻⌐ Views (all)   In the Project Browser:
    ⊟ ⸱⸱⸱ Floor Plans
             Ref. Level   Switch to the **Ref. Level** floor plan.

9.

Other	
Cut with Voids When Loa…	☐
Maintain Annotation Orie…	☐
Shared	☐
Room Calculation Point	☑

In the Properties palette:

Enable **Room Calculation Point**.

10. ⊟ ⸱⸱⸱ Elevations (Elevation 1)   Switch to the **Right Elevation** view.
        ⸱⸱⸱⸱ Back
        ⸱⸱⸱⸱ Front
        ⸱⸱⸱⸱ Left
        ⸱⸱⸱⸱ **Right**

11.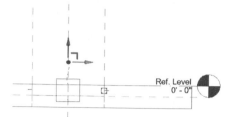

Select the Room Calculation Point in the view.

Use the MOVE tool to adjust the position of the Room Calculation Point to below the Ref Level.

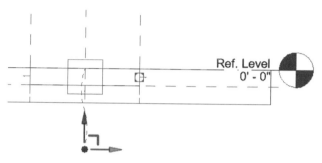

12.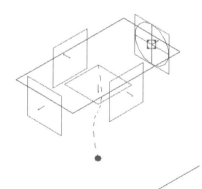

Switch to a 3D view.

Save the family to your work folder.

13.

Click **Load into Project and Close** on the ribbon.

o   Load into
Project and Close

14.

You are trying to load the family Lighting and Appliance Panelboard - 480V MLO, which already exists in this project. What do you want to do?

→ Overwrite the existing version

→ Overwrite the existing version and its parameter values

Click **Overwrite the existing version and its parameter values**.

15.

X Electrical Equipment Schedule X Electrical Room 215

<Electrical Equipment Schedule>

A	B	C
Family and Type	Location	Level
Lighting and Appliance Panelboard - 208V MLO: 225 A	Electrical 215	Main Floor
Lighting and Appliance Panelboard - 480V MLO: 250 A	Electrical 215	Main Floor
Dry Type Transformer - 480-208Y120 - NEMA Type 2: 45 kVA	Electrical 215	Main Floor

Click the tab for the Electrical Equipment Schedule.

The panelboard now is listed in the correct location.

16.   Save as *ex2-11.rvt*.

Exercise 2-12:

Adding Receptacles

Drawing Name: *receptacles.rvt*
Estimated Time: 15 minutes

This exercise reinforces the following skills:

- ❏   Place Components
- ❏   Electrical Fixture
- ❏   Load Families
- ❏   Type Selector

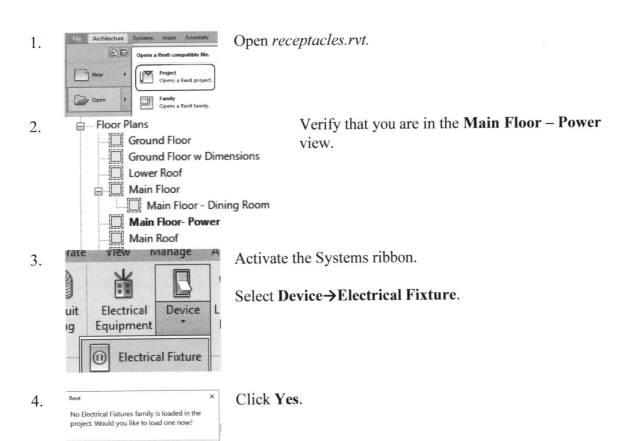

1.   Open *receptacles.rvt*.

2.   Verify that you are in the **Main Floor – Power** view.

3.   Activate the Systems ribbon.

        Select **Device→Electrical Fixture**.

4.   Click **Yes**.

5.

Libraries
English-Imperial
Electrical
MEP
Electric Power
Terminals

Browse to the *Terminals* folder under *Electric Power*.

6.

File name: Duplex Receptacle.rfa
Files of type: All Supported Files (*.rfa, *.adsk)

Select the *Duplex Receptacle*.
Click **Open**.

7.

Properties
Duplex Receptacle
GFCI
New Electrical Fixtures   Edit Type

Select the **GFCI** type from the Properties palette.

8.

GFCI

Men's Toilet Room
212

Place a GFCI receptacle next to each of the sinks in the lavatories.

Note that the receptacles are hosted by walls. You do not see the models unless you hover over a wall.

Cancel out of the command.

9.

Women's Toilet Room
213

Select one of the receptacles that was placed.

Note the elevation of the receptacle in the Properties palette.

10.

uit   Electrical   Device   L
ig   Equipment

Electrical Fixture

Select **Device→Electrical Fixture**.

11.

Properties

Duplex Receptacle
Standard

Select the **Standard** type from the Properties palette.

Note the elevation of the receptacle in the Properties palette.

12.

Sleeping Quarters Sleeping Quarters Sleeping Quarters Sleeping Quarters
201   202   203   204

Place four receptacles in each of the sleeping quarters as shown.

13. Save as *ex2-12.rvt*.

Exercise 2-13:

Create a New Family Type

Drawing Name: *receptacle_family.rvt*
Estimated Time: 20 minutes

This exercise reinforces the following skills:

- ❏ Place Components
- ❏ Electrical Fixtures
- ❏ Load Families
- ❏ Type Selector
- ❏ Type Properties
- ❏ New Type

1.   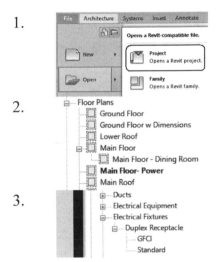 Open *receptacle_family.rvt*.

2.   Open the **Main Floor – Power** floor plan.

3.   In the Project Browser:

Scroll down to the *Families* category.
Locate the *Electrical Fixtures* folder.
Locate the **Duplex Receptacle** family.

There are currently two types:

- GFCI
- Standard

4.   Highlight the **Duplex Receptacle** family.
Right click and select **New Type**.

5.   Name the new type **TV**.

6.   Right click on the **TV** type.

Select **Type Properties**.

7.   Set the Default Elevation to **7' 6"**.
Set the Load to **600.00 VA**.

8.   Scroll down to the Other category.
Locate the **Label** field.
Set the Value to **TV**.
Click **OK**.

9.   Highlight the **Duplex Receptacle** family.
Right click and select **New Type**.

10.   Name the new type **EWC**.

11. Right click on the **EWC** type.

Select **Type Properties**.

12. Set the Default Elevation to **4' 0"**.
Set the Load to **600.00 VA**.

13. Scroll down to the Other category.
Locate the **Label** field.
Set the Value to **EWC**.
Click **OK**.

14. Using the left mouse button, drag and drop the EWC receptacle family to place it on the wall outside the **Room 212 - Men's Lavatory**.

15. Using the left mouse button, drag and drop the EWC receptacle family to place it on the west wall in **Room 205 – Ready Room**.

16. Using the left mouse button, drag and drop the TV receptacle family to place it on the south wall in **Room 205 – Ready Room**.

17. Save as *ex2-13.rvt*.

Revit doesn't have a lot of MEP families which work well with creating legends. In this exercise, we create a detail component family which can be used when creating an electrical symbol legend.

Exercise 2-14:

Create a Detail Component Family

Drawing Name: none
Estimated Time: 10 minutes

This exercise reinforces the following skills:

- ❑ Families
- ❑ Detail Components
- ❑ Detail Lines
- ❑ Dimensions

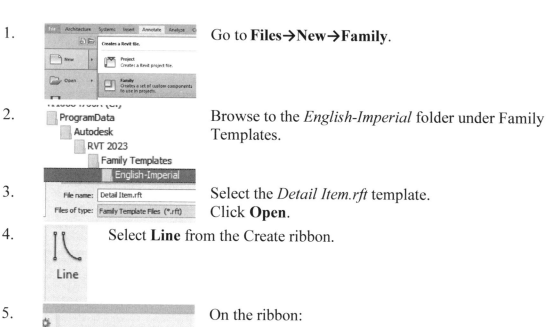

1.   Go to **Files→New→Family**.

2.   Browse to the *English-Imperial* folder under Family Templates.

3.   Select the *Detail Item.rft* template.
   Click **Open**.

4.   Select **Line** from the Create ribbon.

5.   On the ribbon:

   Set the line style to **Heavy Lines**.

6.

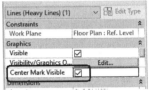

Draw a 1' radius circle above the horizontal reference plane.

7.

Select the Circle.
On the Properties palette:
Enable **Center Mark Visible**.

8.

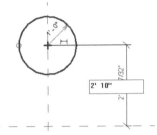

Use the **ALIGNED** dimension tool on the Create ribbon to position the circle 2' 10" above the horizontal reference plane.

Place the dimension.
Select the circle.
Change the temporary dimension.
The aligned dimension will update.

9.

Select **Line** from the Create ribbon.

10.

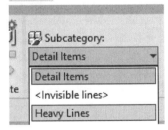

On the ribbon:

Set the line style to **Heavy Lines**.

11.

Draw a vertical line from the horizontal reference plane to intersect with the circle.

Use the temporary dimension to position the line 6" left of the circle's center.

12.

Select the vertical line.
Select Mirror →Pick Axis.
Select the vertical reference plane/dashed green line.

13.

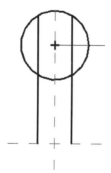

A second vertical line is placed.

14.   Save as *receptacle_symbol.rfa*.

Exercise 2-15:

Create a Detail Item Family

Drawing Name: new family
Estimated Time: 20 minutes

This exercise reinforces the following skills:
- ❑ Detail Component families

1. 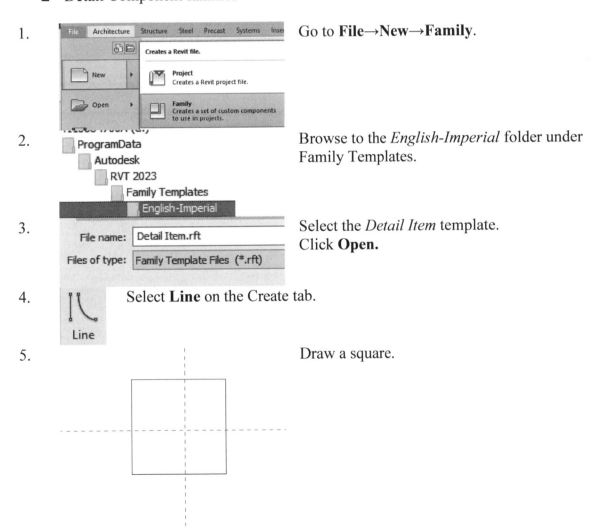 Go to **File→New→Family**.

2. Browse to the *English-Imperial* folder under Family Templates.

3. Select the *Detail Item* template.
    Click **Open.**

4. Select **Line** on the Create tab.

5. Draw a square.

6.

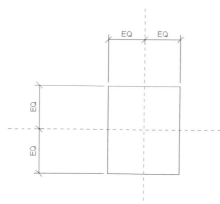

Place a continuous dimension in the vertical and horizontal direction to center the square on the insertion point.

To place the dimension, select one line, then the center reference planc, then the next line, then left click to place.

Left click on the EQ icon to set the dimension equal.

7.

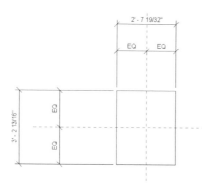

Place an overall horizontal dimension and an overall vertical dimension.

To place the dimensions, select the outside lines, then left click to add the dimension.

8.

Select the horizontal dimension.

2' - 7 19/32"

EQ   EQ

9.

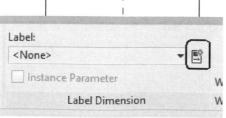

On the ribbon:

Select **New** to add a label parameter to the dimension.

10.

Type **Width** in the Name field.
Enable **Instance**.
Click **OK**.

11.

You should see Width in front of the dimension value now.

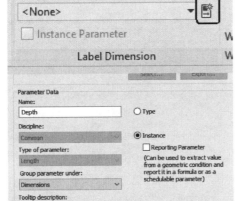

12.

Select the vertical dimension.

13.

On the ribbon:
Select **New** to add a label parameter to the dimension.

14.

Type **Depth** in the Name field.
Enable **Instance**.
Click **OK**.

15.

You should see Depth in front of the dimension value now.

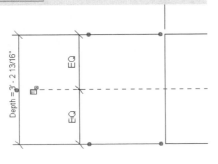

16. Select **Family Types** from the ribbon.

17. Set the Default values for the Depth and Width to **3' 0"**.
    Click **Apply.**

18. The values in the display window should update.
    If the values do not update, you did not place the dimensions correctly. You will need to delete the dimensions and try again.

    Close the Family Types dialog.

19. Select **Line** on the Create tab.

20. Set the line style to **Medium Lines**.

21. Disable **Chain** on the Options bar.

22. Draw two lines to indicate a center mark.

23. Save as *switchboard-section.rfa*.

    This family will be used in an exercise in Lesson 9.

Lab Exercises

Create the following legend symbols:

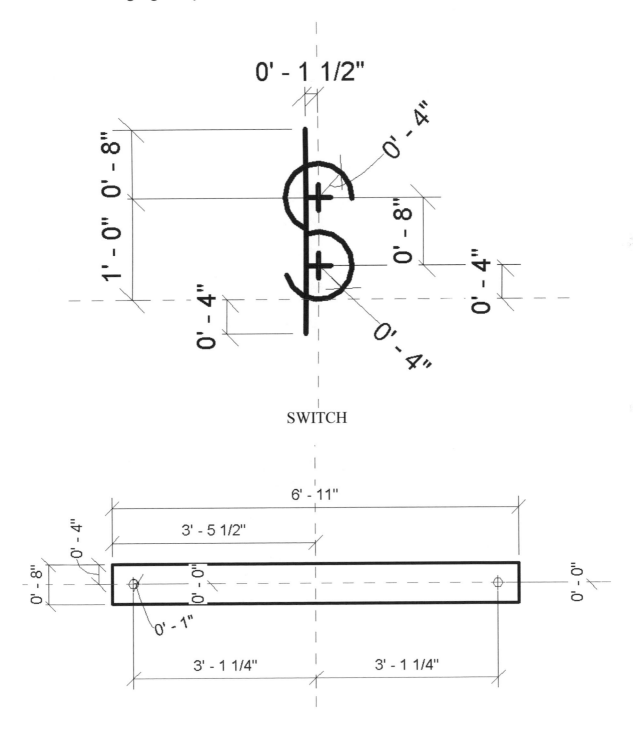

SWITCH

PENDANT LIGHT FIXTURE

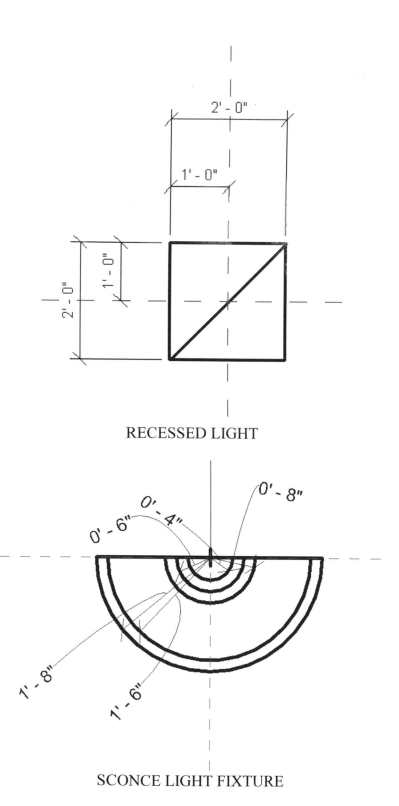

RECESSED LIGHT

SCONCE LIGHT FIXTURE

Lesson
03

Revit Systems

Electrical Settings

The basis of electrical systems in Revit is defined in the Electrical Settings dialog. This is most easily accessed by typing the shortcut (ES). There is an icon in the Settings panel on the Manage tab of the ribbon or it can be accessed by clicking the cleverly hidden southeast pointing arrow found in the bottom right-hand corner of the Electrical panel of the Systems tab.

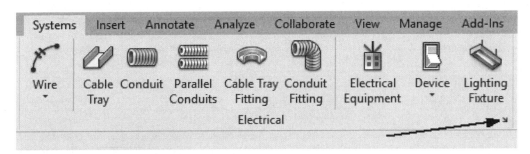

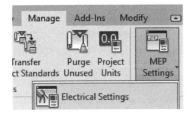

The left-hand side of the Electrical Settings dialog lets the user navigate differing settings for Wiring, Voltage Definitions, Distribution Systems, and more. Simply select the area of interest on the left and its setting will appear on the right. The out-of-the-box settings in the Electrical-Default template should suffice for beginners to the software. That being said, users should verify that the required voltages are defined and the required distribution systems are present before proceeding.

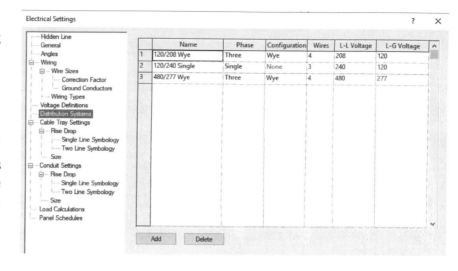

Exercise 3-1:

Space Lighting Calculations

Drawing Name: *space_lighting.rvt*
Estimated Time: 5 minutes

This exercise reinforces the following skills:

- ☐ Spaces
- ☐ Properties

1.  Open *space_lighting.rvt*.

2. *This file uses a linked Revit file.*

   Go to the Insert ribbon.

   Click **Manage Links**.

   Use **Reload From** to select the new file location.

3. In the Project Browser, activate/open the **3rd Floor Lighting** Plan.

4. Use **Zoom In Region** to change the display to focus on **Classroom 5 – Room 307** between Grids 5 and 6.

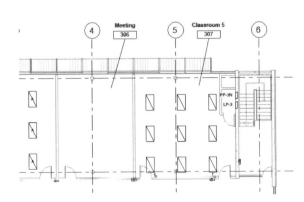

5. 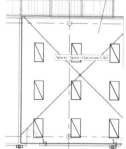 Hover your mouse in the center of Classroom 5 – Rm 307.

You should see a large blue X and the Space label.

Left click to select the space.

6.  Look in the Properties palette.

Locate the value for the Average Estimated Illumination based on the number of light fixtures in the room.

7. Close the file without saving.

When performing lighting calculations, verify that the spaces are defined properly. If the spaces have not been set to be "room bounding", they will ignore the ceiling heights. This occurs for accurate heating and cooling loads. To ensure that lighting calculations are accurate, spaces should have an upper limit to match the ceiling heights.

Exercise 3-2:

Managing Spaces

Drawing Name: *Architecture_Model.rvt*
Estimated Time: 30 minutes

This exercise reinforces the following skills:

- Spaces
- Properties
- Linking Files

1. 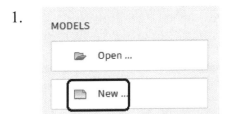 Start a **New** project.

2. 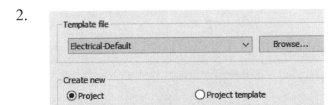 Select the *Electrical – Default* template.

   Click **OK**.

3.  Activate the Insert ribbon.

   Select **Manage Links**.

4. 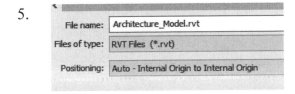 Select the **Add** button at the bottom right of the dialog.

5. Select *Architecture_Model.rvt*.

   Set the Positioning to **Auto – Internal Origin to Internal Origin**.

   Click **Open**.

   Close the dialog.

6.

Properties	×
Linked Revit Model Architecture_Model.rvt	
RVT Links (1)	Edit Type
Identity Data	
Name	1
Other	
Shared Site	<Not Shared>

Left click on the inserted file.

Select **Edit Type** on the Properties palette.

7.

Type Parameters

Parameter	
Constraints	
Room Bounding	☑

Enable **Room Bounding**.

This ensures any lighting calculations performed on the model are accurate.

Click **OK** to close the dialog.

Left click anywhere in the display window to release the selection.

8.

odel Room Room Tag Area Area Tag
oup Separator Room Boundary Area

Color Schemes
Area and Volume Computations

Activate the Architecture ribbon.

Expand the ribbon on the Room & Area panel.

Select **Area and Volume Computations**.

9.

Computations | Area Schemes

Volume Computations

Volumes are computed at finish faces.

○ Areas only (faster)

◉ Areas and Volumes

Enable **Areas and Volumes**.

Click **OK**.

10.

Architecture Systems Insert Annotate Analyze Collal

Space Space Space Space Zone   Panel   Schedule/
       Separator Tag Naming   Schedules Quantities

Spaces & Zones ▼   Reports & Schedules »

Activate the Analyze ribbon.

Select **Space**.

11.

No Space Tags family is loaded in the project. Would you like to load one now?

| Yes | No |

If this dialog appears:

Click **No**.

12.

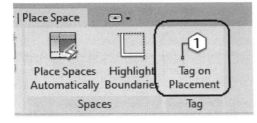

Toggle off **Tag on Placement** on the ribbon.

If this is enabled, spaces will automatically be tagged/labeled.

13.

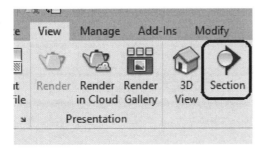

Set the Offset to **8'-0"** on the Options bar.

Left click inside the large curved room on the right side of the building.

Escape out of the command.

14.

Activate the View ribbon.

Select the **Section** tool.

You can also select the Section tool from the Standard toolbar at the top of the window.

15.

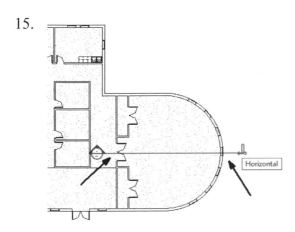

Left click to the left of the curved room with the space to start the section line.

Left click to the right and outside of the curved room to end the section line.

16.

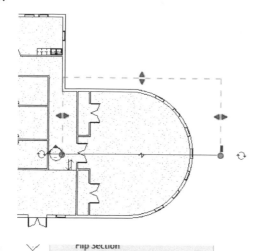

Use the grip arrows to extend the section above the room.

Left click anywhere in the display window to exit the command.

17.

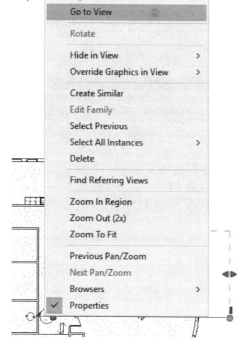

Double left click on the section bubble to open the section view.

OR

Right click on the section line and select **Go to View**.

18.

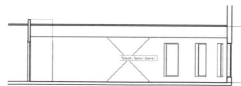

Hover your mouse over the room to "wake up" the space.

You see that the space has a height of 8' -0", but the ceiling is higher.

19.	**Spaces (1)** ⌄ Edit Type Floor Reflectance 20.0000% **Electrical - Loads** Design HVAC Load per area 0.00 W/ft² Design Other Load per area 0.00 W/ft² **Dimensions** Area 1102.54 SF Perimeter 133' 3 143/256" Unbounded Height 8' 0" Volume **8820.44 CF** Computation Height 0' 0" **Mechanical - Flow** Specified Supply Airflow 0.00 CFM Calculated Supply Airflow Not Computed Actual Supply Airflow 0.00 CFM Return Airflow Specified	Left click on the space to select it. Locate the **Volume** value in the Properties palette under Dimensions.

20.	**Spaces (1)** ⌄ **Constraints** Level Level 1 Upper Limit Level 1 Limit Offset 10' 0" Base Offset 0' 0" **Electrical - Lighting**	Change the Limit Offset to **10'-0"**. *Note the space now goes up to the ceiling.*

21.	Design Other Load per area 0.00 W/ft² **Dimensions** Area 1102.54 SF Perimeter 133' 3 143/256" Unbounded Height 10' 0" Volume 11025.55 CF Computation Height 0' 0"	Scroll down to the Volume value and notice it has updated.

22.	**Spaces (1)** ⌄ **Constraints** Level Level 1 Upper Limit Level 1 Limit Offset 12' 0" Base Offset 0' 0" **Electrical - Lighting**	Change the Limit Offset to **12'-0"**. Note the space still goes up to the ceiling but shows a green dashed line to indicate the new offset.

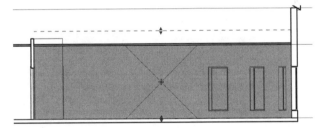

23.	Design Other Load per area 0.00 W/ft² **Dimensions** Area 1102.54 SF Perimeter 133' 3 143/256" Unbounded Height 10' 0" Volume 11025.55 CF Computation Height 0' 0"	Scroll down to the Volume value and notice it has remained the same value. *This confirms that the ceiling is acting as a room boundary.*

24.	⊟ Lighting
	   ⊟ Floor Plans
	     ⬚ **1 - Lighting**
	     ⬚ 2 - Lighting
		Return to the **1-Lighting** floor plan.

25.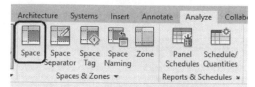

Activate the Analyze ribbon.

Select **Space**.

26.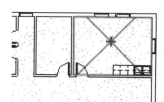

Set the Offset to 1**5'-0"** on the Options bar.

Left click inside the upper corner room on the right side of the building.

Cancel out of the command.

27.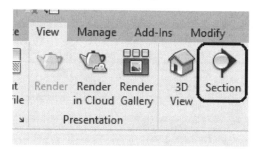

Activate the View ribbon.

Select the **Section** tool.

28.

Left click to the left of the room with the space to start the section line.

Left click to the right and outside of the room to end the section line.

29.

| Flip Section |
| Go to View |
| Rotate |
| Hide in View > |
| Override Graphics in View > |
| Create Similar |
| Edit Family |
| Select Previous |
| Select All Instances > |
| Delete |
| Find Referring Views |
| Zoom In Region |
| Zoom Out (2x) |
| Zoom To Fit |
| Previous Pan/Zoom |
| Next Pan/Zoom |
| Browsers > |
| ✓ Properties |

Double left click on the section bubble to open the section view.

OR

Right click on the section line and select **Go to View**.

30.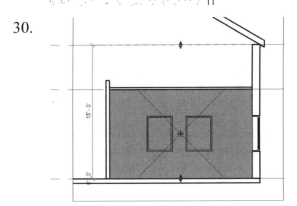

Hover your mouse over the room to "wake up" the space.

You see that the ceiling is still acting as a boundary for the space.

31. Save as *ex3-2.rvt*.

Exercise 3-3:

Creating a Distribution System

Drawing Name: *systems.rvt*
Estimated Time: 15 minutes

This exercise reinforces the following skills:

- ❑ Electrical Settings
- ❑ Voltage Definitions
- ❑ Distribution Systems

1. 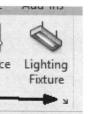 Open *systems.rvt.*

   This file is downloaded from the publisher's website.

2. Activate the Systems ribbon.

   Launch the **Electrical Settings** dialog.

3.

	Name	Value	Minimum	Maximum
1	120	120.00 V	110.00 V	130.00 V
2	208	208.00 V	200.00 V	228.00 V
3	240	240.00 V	220.00 V	250.00 V
4	277	277.00 V	260.00 V	280.00 V
5	480	480.00 V	460.00 V	490.00 V

Highlight **Voltage Definitions**.

Use the **Add** button to add five voltage definitions for 120V, 208V, 240V, 277V, and 480V.

The values you set in the Voltage Definitions are the values available to the Distribution Systems.

You do not need to type V – just the values.

2	120V	120.00 V	110.00 V	130.00 V
3	208V	208.00 V	200.00 V	220.00 V
4	240V	240.00 V	230.00 V	250.00 V
5	277V	277.00 V	260.00 V	280.00 V
6	480V	480.00 V	460.00 V	490.00 V

4. 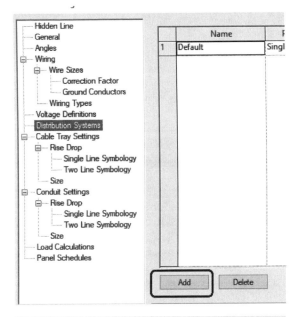 Highlight **Distribution Systems**.

Click **Add**.

5.

	Name	Phase	Configuration	Wires	L-L Voltage	L-G Voltage
1	120/240 Delta	Three	Delta	4	240V	120V
2	Default	Single	None	2	None	None

In the Name field: Type **120/240 Delta**.

In Phase: Select **Three** from the drop-down list.

In Configuration: Select **Delta** from the drop-down list.

Set the Wires to **4**.

Set the L-L value to **240**.

Set the L-G value to **120**.

Hint: You can use the drop-down lists to select the values.

The configuration names are case sensitive, so use sentence case or the drop-down list.

6.

	Name	Phase	Configuration	Wires	L-L Voltage	L-G Voltage	
1	120/240 Delta	Three	Delta	4	240V	120V	
2	Default	Single	None	2	None	None	
3	120/208 Wye	Three	Wye	4	208V	120V	

Click **Add** to add a second distribution system.

In the Name field: Type **120/208 Wye**.

In Phase: Select **Three** from the drop-down list.

In Configuration: Select **Wye** from the drop-down list.

Set the Wires to **4**.

Set the L-L value to **208**.

Set the L-G value to **120**.

7.

	Name	Phase	Configuration	Wires	L-L Voltage	L-G Voltage	
1	120/240 Delta	Three	Delta	4	240V	120V	
2	Default	Single	None	2	None	None	
3	120/208 Wye	Three	Wye	4	208V	120V	
4	480/277 Wye	Three	Wye	4	480V	277V	

Click **Add** to add a third distribution system.

In the Name field: Type **480/277 Wye**.

In Phase: Select **Three** from the drop-down list.

In Configuration: Select **Wye** from the drop-down list.

Set the Wires to **4**.

Set the L-L value to **480**.

Set the L-G value to **277**.

8.   Click **OK** to close the Electrical Settings dialog.

9.

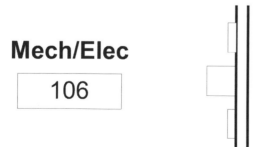

Activate the **Main Floor- Power** floor plan.

Zoom into **Room 106 – Mech/Elec**.

10.

Mech/Elec

106

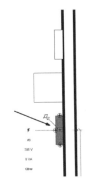

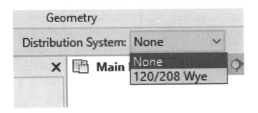

Select the lower panelboard.

On the Options bar: Assign it to the **120/208 Wye Distribution System**.

Left click in the display window to release the selection.

11.

Mech/Elec

106

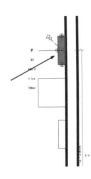

Select the high voltage panelboard.

On the Options bar: Assign it to the **480/277 Wye Distribution System**.

Left click in the display window to release the selection.

Notice that only the applicable distribution systems appear in the drop-down list.

12.

Mech/Elec

106

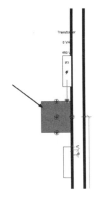

Select the transformer.

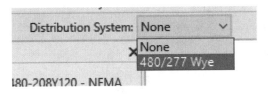

On the Options bar: Assign it to the **480/277 Wye Distribution System**.

Left click in the display window to release the selection.

13. Save as *ex3-3.rvt*.

Exercise 3-4:

Define a Power System

Drawing Name: *power_system.rvt*
Estimated Time: 30 minutes

This exercise reinforces the following skills:

- Distribution Systems
- Panels
- Power Connections
- Circuits

1. 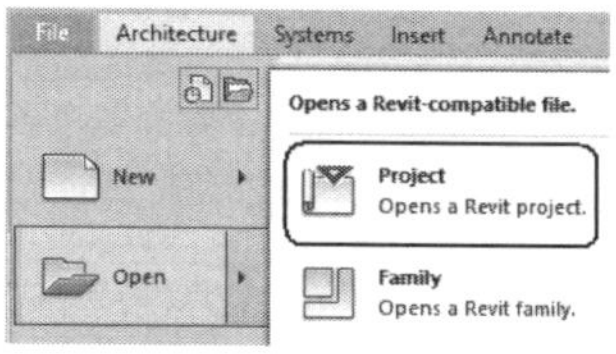 Open *power_system.rvt*

2. 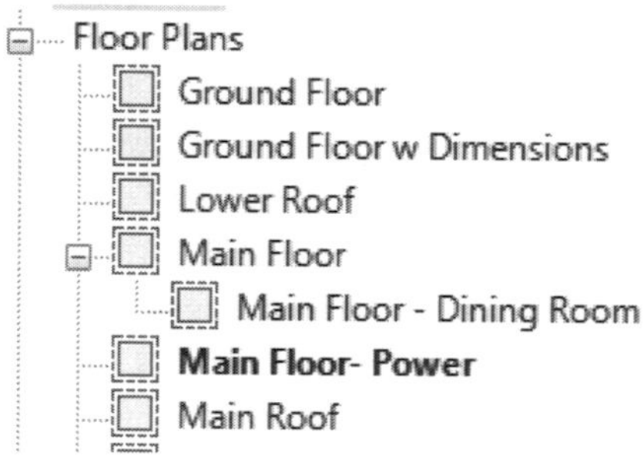 Open the **Main Floor- Power** floor plan.

3. 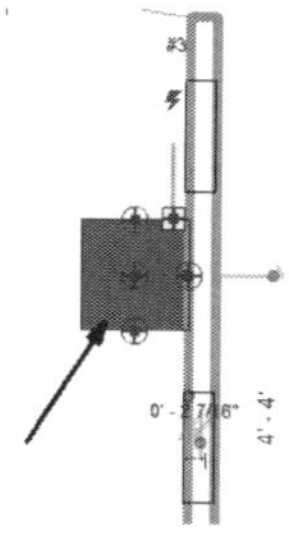 Zoom into **Room 215- Electrical**.

   Select the transformer.

4. 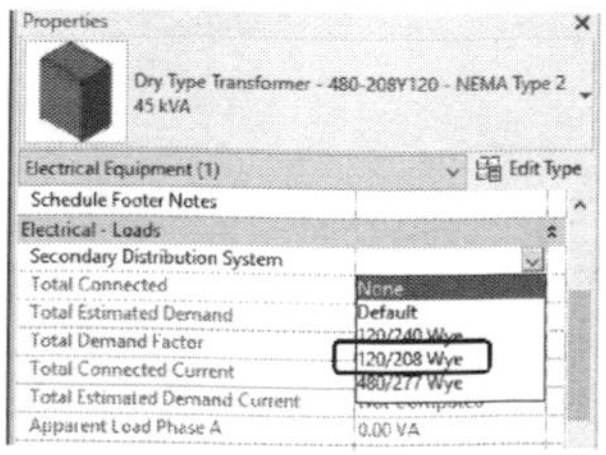 On the Properties palette:

   Under Electrical Loads:

   Set the Secondary Distribution System to **120/208 Wye**.

   Release the selection.

5.

Select the low power panel board located below the transformer.

6.

On the Electrical Circuits ribbon:

Select the **Select Panel** tool.

Select the transformer.

7.

On the Electrical Circuits ribbon:

You see that T1 has been assigned to the panel.

8.

An arrow appears next to the low power panelboard representing the connection between the low voltage panel and the transformer.

9.

Select the high power panel board located above the transformer.

10.

Select the **Power** tool on the ribbon.

11.  Select the transformer.

A lightning bolt appears next to the high power panel to represent the connection.

Release the selection.

12.  Select the four receptacles in **Room 201 – Sleeping Quarters**.

Hold down the CTL key to select all the receptacles.

13. Select the **Power** tool.

14. 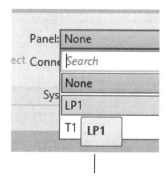 Assign to Panel **LP1** on the ribbon.

15.  A dashed line appears showing the receptacles' connections.

16. 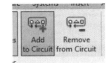 Look in the Properties palette.

Electrical Circuits (1)	
Electrical - Loads	
Circuit Number	1
Connection Type	Breaker
Load Name	Receptacle
Panel	LP1
System Type	Power
Load Classification	Receptacle
Number of Poles	1
Rating	20.00 A
Frame	400.00 A
Voltage	120.00 V
Apparent Load	720.00 VA
Apparent Load Phase A	720.00 VA
Apparent Load Phase B	0.00 VA

True Current Phase A	0.00 A
True Current Phase B	0.00 A
True Current Phase C	0.00 A
Voltage Drop	1.09 V
Power Factor	1.000000
Power Factor State	Lagging

Check the voltage drop based on the connections.

17. Select **Edit Circuit** on the Electrical Circuits ribbon.

18.  Verify that **Add to Circuit** is enabled on the ribbon.

Elements which are not included in the circuit are faded. Once they are selected, they are no longer faded.

19. 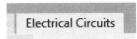 Add the duplex receptacles in Rooms 202 and 203.

You should see this dialog box.

20. Enable **Remove From Circuit** on the ribbon.

Select the receptacles in Room 203 to remove them from the circuit.

21. Select **Finish Editing Circuit** on the ribbon.

22. Select one of the receptacles in Room 201.

Select the Electrical Circuits tab on the ribbon.

In the Properties palette: review the Voltage Drop.

Release the selection.

23. Select the low power panelboard LP1.

On the Properties palette: review the loads.

24. Save as *ex3-4.rvt*.

Exercise 3-5:

Define an Electrical Circuit

Drawing Name: *elec_circuit2.rvt*
Estimated Time: 10 minutes

This exercise reinforces the following skills:

- ❑ Loadable Families
- ❑ Place a Component

1.   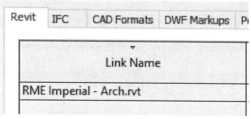   Open *elec_circuit2.rvt*

2.   Use **Insert→Manage Links** to Reload the *RME Imperial – Arch.rvt* file from the downloaded files.

Manage Links

| Revit | IFC | CAD Formats | DWF Markups | P |

Link Name

RME Imperial - Arch.rvt

3.   In the Project Browser, activate/open the **3rd Floor Lighting** Plan.

Electrical
    Lighting
        Floor Plans
            1st Floor Lighting
            2nd Floor Lighting
            3rd Floor Lighting Plan

4.

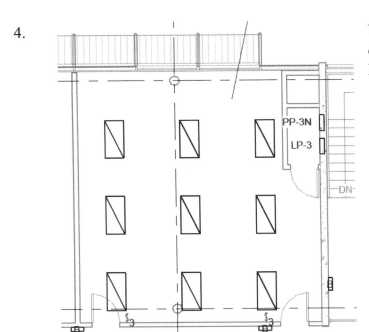

Use **Zoom In Region** to change the display to focus on Classroom 5 Room 307 between Grids 5 and 6.

5.

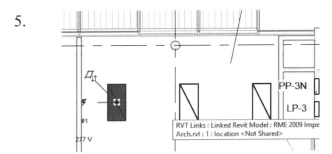

Select the upper left light fixture.

It doesn't really matter which light fixture you select.

6.

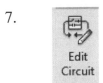

Select the **Power** tool on the ribbon.

7.

Select **Edit Circuit** on the ribbon.

8.

Verify that Add to Circuit is enabled on the ribbon.

9.

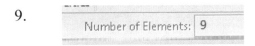

Hold down the Ctrl key and select the remaining light fixtures.

*The Options bar should display that the Number of Elements selected is **9.***

10.

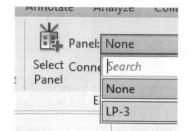

On the ribbon, select the **LP-3** panel from the panel list.

11.

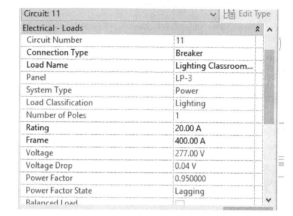

On the Properties palette:

Note that the circuit is now assigned to LP-3 panel.

Note the Circuit Number, the voltages and amperages.

12.

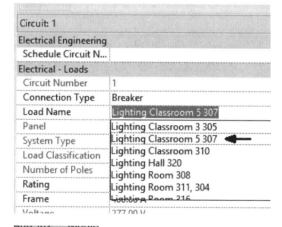

Set the Load Name to **Lighting Classroom 5 307**.

13.

On the ribbon, select **Finish Editing Circuit**.

14.

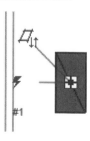

Select a lighting fixture.

15.

Offset from Host	0' 0"
Electrical - Lighting	
Calculate Coefficient of Utilization	☑
Coefficient of Utilization	0.645167
Switch ID	
Electrical - Loads	
Panel	LP-3
Circuit Number	11
Identity Data	
Image	
Comments	
Mark	88

On the Properties palette:

Notice that the lighting fixture is assigned to Panel LP-3.

16.   Save as *ex3-5.rvt*.

Rooms and Spaces

Rooms and spaces are independent components used for different purposes. Rooms are architectural components used to maintain information about occupied areas. Spaces are exclusively used for the MEP disciplines to analyze volume. They contain parameters that maintain information about the areas in which they have been placed. This information is used for performing a heating and cooling loads analysis.

Spaces can be placed (added) and unplaced, and deleted. Unplacing spaces is not the same as deleting spaces. Spaces are immediately assigned to the Default zone when they are initially added to a project. Spaces can be viewed in a section view. Spaces cannot be viewed or placed in elevation or 3D views.

Spaces should be placed throughout the model, including unoccupied areas such as plenums areas. Spaces that are created (manually or automatically) in an area that contains a room arc created as occupied (Occupiable parameter selected).

Zones

Zones define spaces that can be controlled by environmental control systems, such as heating, cooling, and humidity control systems. This lets you perform load balancing and analysis procedures on a building model. Initially, all spaces are assigned to a single default zone until/unless more zones are defined.

Duplicating Views

Revit allows views to be duplicated in three ways:

- Duplicate
- Duplicate with Detailing
- Duplicate as Dependent

Remember there are three types of elements in Revit:

- Model
- Datum
- View-Specific

If you duplicate a view, model elements and datums will be included in that view, but view-specific elements, like dimensions and tags, will not be copied into the new view.

If you duplicate with detailing, all three element types will be copied into the new view.

If you duplicate as dependent, you link the parent and child view, so all three element types are visible in both views. If you add annotations to either the parent or the child view, it will be visible in the linked view.

Exercise 3-6:

Adding Space Tags

Drawing Name: *space_analysis.rvt*
Estimated Time: 15 minutes

This exercise reinforces the following skills:

- ❑ Visibility/Graphics Overrides
- ❑ Adding Space Tags
- ❑ Duplicate View

1.  Open *space_analysis.rvt*

2. 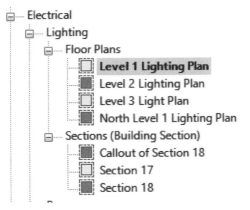 In the Project Browser:

   Locate the **Level 1 Lighting Plan**.

3. 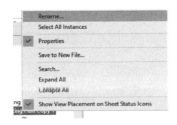 Right click and select **Duplicate View→Duplicate with Detailing.**

   This creates a new view which includes any annotation elements, like tags or dimensions.

4. Rename the copied view: **Level 1 Lighting Analysis**

5.

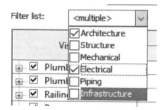

Type **VV** to open the Visibility/Graphics Overrides dialog.

Enable **Architecture** and **Electrical** in the Filter list.

6.

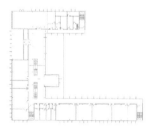

On the Model Categories tab:

Disable **Wires**.

Disable **Lighting Fixtures**.

If you click **Apply** on the dialog, you can see a preview of what the floor plan looks like with these elements hidden.

7.

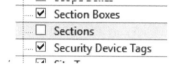

On the Annotation Categories tab:

Disable **Sections**.

Apply allows you to preview the changes you made before you commit. Click OK if you are satisfied with the new setting.

Click **OK** to close the dialog.

8.

Notice how the view has changed.

9.

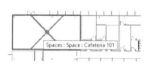

If you hover over the rooms, you see that spaces have already been added to the rooms.

10.

Type **VV** to open the Visibility/Graphics Overrides.

On the Model Categories tab:

Locate Spaces.

Enable **Interior**.

Click **OK**.

11. 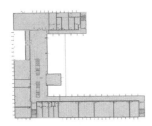 Notice that all the spaces are now indicated by a color fill.

12. Activate the **Analyze** ribbon.

13. Select the **Space Tag** tool from the ribbon.

14. 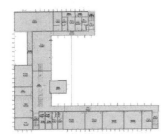 Left click in each space to place a tag.

Save as *ex3-6.rvt*.

Color Schemes

Color Schemes can be added to floor plan views and section views based on a specific value or range of values. You can apply a different color scheme to each view.

Use color schemes to color and apply fill patterns to:

- rooms
- areas
- spaces and zones
- pipes and ducts

Color Fill Legends

For views that use color schemes, color fill legends provide a key to the color representation.

Exercise 3-7:

Creating a Color Scheme For Lighting Loads

Drawing Name: *color_scheme.rvt*
Estimated Time: 15 minutes

This exercise reinforces the following skills:

- ❑ Color Schemes
- ❑ Color Scheme Definitions

1. 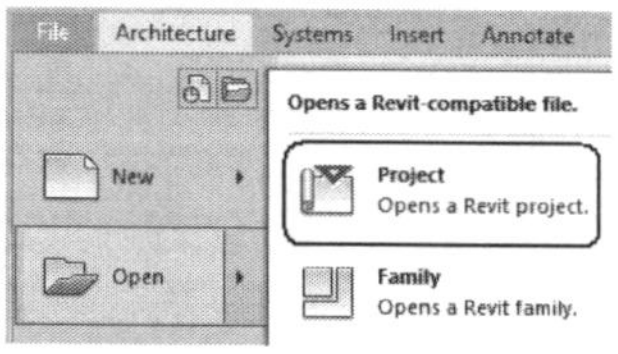   Open *color_scheme.rvt*

2. 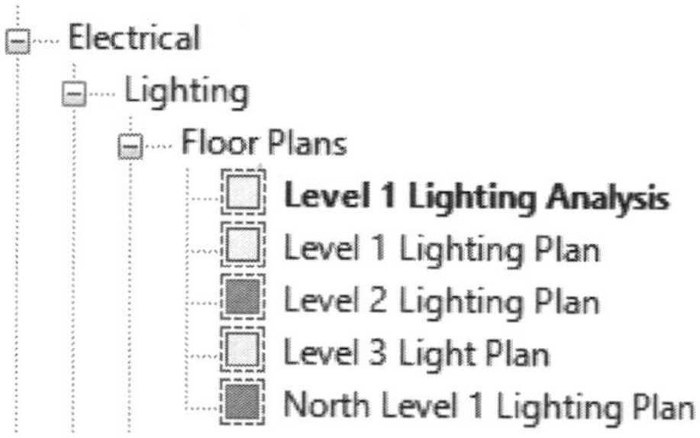   In the Project Browser:

Open the **Level 1 Lighting Analysis** view.

3. 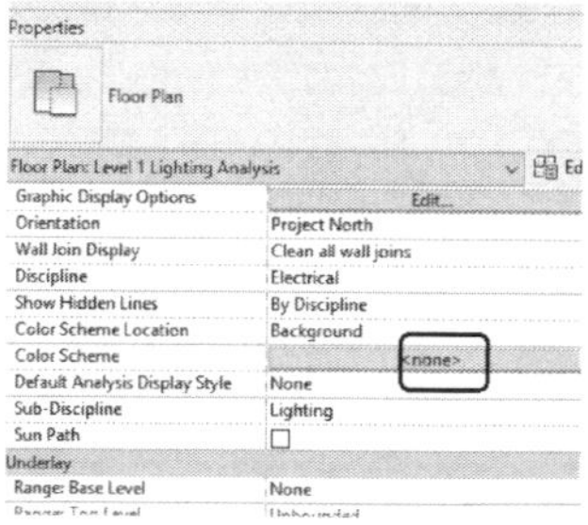   On the Properties palette:

Select the button next to **Color Scheme**.

4.

Under Category:

Select **Spaces**.

Highlight **Schema 1**.

Select **Duplicate** at the bottom left of the dialog.

5.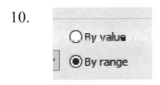

Name the scheme **LPD**.

Click **OK**.

6.

Change the Title to: **LPD (W/sq ft)**.

7.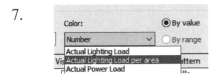

Under Color:

Select **Actual Lighting Load per area**.

8.

Click **OK**.

9.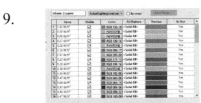

All the different values that currently exist are shown.

We can modify the table to make it a little more meaningful.

10.

Enable **By Range**.

Now, the different colors will be sorted by minimum and maximum values.

11.

Left click in the second row.

Select the + symbol to the left of the row list to add a new value.

12. Edit the Scheme so that there are nine rows as shown.

Click **OK**.

At Least	Less Than	Caption
	0.20 W/ft²	Less than 0.20 W/ft²
0.20 W/ft²	0.30 W/ft²	0.20 W/ft² - 0.30 W/ft²
0.30 W/ft²	0.35 W/ft²	0.30 W/ft² - 0.35 W/ft²
0.35 W/ft²	0.40 W/ft²	0.35 W/ft² - 0.40 W/ft²
0.40 W/ft²	0.45 W/ft²	0.40 W/ft² - 0.45 W/ft²
0.45 W/ft²	0.50 W/ft²	0.45 W/ft² - 0.50 W/ft²
0.50 W/ft²	0.80 W/ft²	0.50 W/ft² - 0.80 W/ft²
0.80 W/ft²	1.00 W/ft²	0.80 W/ft² - 1.00 W/ft²
1.00 W/ft²	1.25 W/ft²	1.00 W/ft² - 1.25 W/ft²
1.25 W/ft²		1.25 W/ft² or more

13.

Select the **Annotate** ribbon.

Select the **Color Fill Legend** tool.

Left pick to place the legend in the view.

14.

Less than 0.20 W/ft²

0.20 W/ft² - 0.30 W/ft²

0.30 W/ft² - 0.35 W/ft²

0.35 W/ft² - 0.40 W/ft²

0.40 W/ft² - 0.45 W/ft²

0.45 W/ft² - 0.50 W/ft²

0.50 W/ft² - 0.80 W/ft²

0.80 W/ft² - 1.00 W/ft²

1.00 W/ft² - 1.25 W/ft²

1.25 W/ft² or more

The legend is used to help you identify what the different colors mean.

15. Save as *ex3-7.rvt*.

Exercise 3-8:

Project Energy Settings

Drawing Name: *project_settings.rvt*
Estimated Time: 10 minutes

This exercise reinforces the following skills:

- ❑ Energy Settings
- ❑ Space Properties

1. 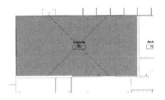 Open *project_settings.rvt*

2.
```
⊟ Electrical
   ⊟ Lighting
      ⊟ Floor Plans
           Level 1 Lighting Analysis
           Level 1 Lighting Plan
           Level 2 Lighting Plan
           Level 3 Light Plan
           North Level 1 Lighting Plan
```

   In the Project Browser:

   Open the **Level 1 Lighting Analysis** view.

3.  Select the Cafeteria space in the upper left corner of the building.

Properties
R
Spaces (1)

   Verify that you selected the space and not the room by checking the Properties palette.

4.

Electrical - Loads	
Design HVAC Load per ar...	0.00 W/ft²
Design Other Load per area	0.00 W/ft²
Actual Lighting Load	1536.00 VA
Actual Other Load	0.00 VA
Actual Power Load	2160.00 VA

   Note that the Actual Lighting Load and Actual Power Load is listed based on the light fixtures in the space.

5.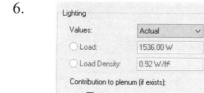

Scroll down to **Electrical Loads**.

Click **Edit**.

6.

Change the Values to **Actual** for both Lighting and Power.

Click **OK**.

7.

Locate the **Space Type** field under the Energy Analysis category.

Left click in the field and select the ... button to the right.

8.

Notice the settings for this space.

9.

Locate the **Lighting Schedule** field.

Left click in the field and select the ... button to the right.

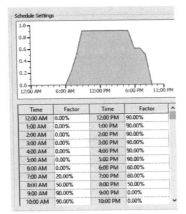

10. Click **OK**.

11. Close all dialogs.

Save as *ex3-8.rvt*.

Lab Exercises

Open *distribution_lab.rvt.*

Electrical Settings

		Name	Value	Minimum	
1		120	120.00 V	110.00 V	130.00 V
2		208	208.00 V	200.00 V	228.00 V
3		240	240.00 V	220.00 V	250.00 V
4		277	277.00 V	260.00 V	280.00 V
5		480	480.00 V	460.00 V	490.00 V

Tree items shown:
- Hidden Line
- General
- Angles
- Wiring
  - Wire Sizes
    - Correction Factor
    - Ground Conductors
  - Wiring Types
- Voltage Definitions
- Distribution Systems
- Cable Tray Settings
  - Rise Drop

Create the Voltage Definitions shown.

Create two distribution systems: **120/208 Wye Distribution System** and **480/277 Wye Distribution System** using the Voltage definitions

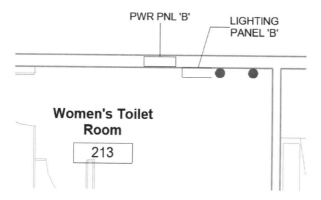

Open Main Floor Annotated under Electrical→Power.

Assign the panels in Room 213 to the correct distribution systems.

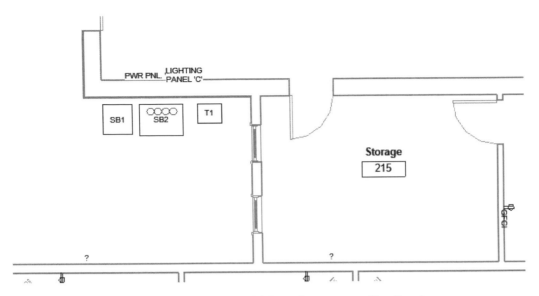

Assign the panels near STORAGE – 215 to the correct distribution systems.

Assign the T1 transformer to the correct distribution system.

Assign SB1 and SB2 to the correct distribution systems.

Open the **Ground Floor** ceiling plan under Electrical→Lighting.

Select the lighting fixtures in Lg Office – 106.

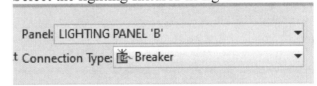

Assign them to Lighting Panel 'B'.

Notes:

Lesson
04

Wiring

In this lesson, you learn the process of defining a wire type and recommended practices for creating and placing wiring.

Wiring depicts the connections between electrical components, such as lighting fixtures, switches, and panels. You can place wiring using an automatic method or a manual method. I tend to prefer the manual method simply because it provides more control.

Wiring can be displayed with or without symbols and tick marks.

A home run is the electrical cable that carries power from the main circuit breaker panel to the first electrical box, outlet, or switch in the circuit. You can designate home runs by manually drawing wires using the Wire tool. If you are using the Wire tool and have not selected an electrical component as the endpoint of a wire, Revit automatically adds a home run symbol to the wire.

Revit includes a library of standard wire sizes. Each wire size is assigned amperage and physical diameter.

Process for Creating New Wire Types

SPECIFY WIRE TYPE SETTINGS

↓

ADD ANY NEW WIRE TYPE

↓

DEFINE PROPERTIES FOR NEW WIRE TYPES

↓

SAVE

Guidelines for Creating and Placing Wiring

- Specify wire settings with available wire sizes.
- Specify wire settings in project template files to save time and maintain consistency across projects.
- Transfer wire settings from one project to another using the Transfer Project Standard option on the Manage ribbon.
- Turn off wire tick marks universally to ensure consistency across all project views by selecting Never for the wiring settings in the Electrical Settings dialog box. You can also turn these settings off in a specific view by controlling the settings in the Visibility/Graphics Overrides dialog.
- Apply tags to circuit numbers.
- Manually wire devices from different circuits. This helps in creating multiple home runs.

Revit considers wires as annotation elements. This means they are view-specific. If you add wires in one view, they are not displayed in related views. If you want to display wiring in all views, then use conduits – which are model elements. Both wires and conduits are system families. This means any family types of wires or conduits you create in a project reside only in that project. To copy them to a new project, you need to use the Transfer Project Standards tool on the Manage ribbon.

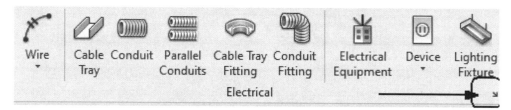

Wires are defined in the Electrical Settings dialog. You can access this dialog by typing ES or clicking in the lower right corner of the Electrical panel on the Systems tab.

Wiring

Setting	Value
Ambient Temperature	86 °F
Gap of Wiring Crossing	0' - 0 1/16"
Hot Wire Tick Mark	
Ground Wire Tick Mark	
Neutral Wire Tick Mark	
Slanted Line across Tick Marks	No
Show Tick Marks	Always
Max Voltage Drop For Branch Circuit Wire Sizing	2.00%
Max Voltage Drop For Feeder Circuit Wire Sizing	3.00%
Arrow for Multi-Circuits Home Run	Multiple Arrows
Home Run Arrow Style	Arrow Filled 15 Degree

Ambient Temperature	Specifies the temperature of the environment in which the wiring will exist.
Gap of Wiring Crossing	Specifies the width of the gap used to display non-connected wires that cross, as shown.
Hot Wire Tick Mark	Specifies the style of tick mark that displays for Hot Conductor, Ground Conductor, and Neutral Conductor.
Ground Wire Tick Mark	Revit has four tick mark styles:
Neutral Wire Tick Mark	 Short Wire Tick Mark Circle Wire Tick Mark Hook Wire Tick Mark Long Wire Tick Mark

Slanted Line across Tick Marks	Specifies whether to display the tick mark for the ground conductor as a diagonal line that crosses the tick marks for the other conductors, as shown.
Show Tick Marks	Specifies whether to always hide tick marks, always show them, or show them for home runs only.
Max Voltage Drop for Branch Circuit Wire Sizing	Specifies the percentage for the maximum voltage drop allowed for branch circuits.
Max Voltage Drop for Feeder Circuit Wire Sizing	Specifies the percentage for the maximum voltage drop allowed for feeder circuits.
Arrow for Multi-Circuits Home Run	Specifies whether a single arrow or multiple arrows display on all circuit wires or the end wire only.
Home Run Arrow Style	Specifies the style for the home run arrow, including the arrow angle and size.

To load a tick mark family

1. Click Insert tab→Load from Library panel→ (Load Family).
2. In the Open dialog, navigate to Annotations→Electrical→Tick Marks.
3. Select one or more tick mark family files and click Open.

You can assign a different style to each conductor.

Click the Value column, click , and select a tick mark style.

You can use the Family Editor to customize an existing tick mark or create additional tick marks.

Wire Sizes

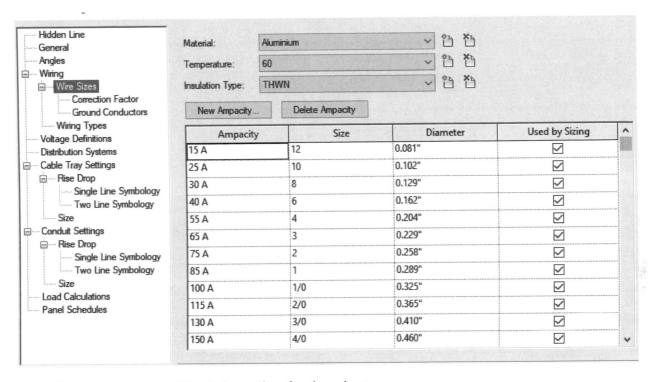

The wire sizes pane provides information for the wire types.

The right pane lists wire types, based on material, temperature ratings (60 degrees C, 75 degrees C, and 90 degrees C) and insulation type. Correction Factor and Ground Conductor branches list correction factors (based on ambient temperature), and ampacity for ground conductor sizes.

The wire size table lists the wire sizes currently available for a given combination of wire material, temperature rating, and insulation type. Each combination of material, temperature rating, and insulation type is associated with a separate table of wire sizes. Whenever a new attribute (material, temperature rating, or insulation type) is created, a new table of sizes is also created for the attribute. Each size shows ampacity, AWG (American Wire Gauge) size, and diameter of wires available in a project. Revit calculates the wire size for circuits (based on the current rating of the circuit) to maintain a voltage drop of less than 3 percent.

- Material: (default values are Aluminum and Copper)

Click ⌦Delete to remove a selected material from the table. A material cannot be deleted if it is in use in a project or if it is the only one specified in a project.

Click ⧉Add to open the New Material dialog, where you specify values for a new wire material. Material names must be unique with a project.

- Temperature: The temperature (Celsius) determines which insulation types are available for a particular material. There is a greater selection of insulation types available for lower temperatures than for high temperatures. For example, type UF insulation is available for Aluminum wire at 60C, but when you select 90C for Aluminum wire, type UF insulation is not available.

  Click ⌖Delete to remove a setting. A temperature cannot be deleted if it is in use in a project or if it is the only one specified in a project.

  Click ⌗Add to open the New Temperature dialog where you specify values for a new temperature. Temperature names must be unique for a specified material.

- Insulation: The default insulation values depend on the selected material and temperature.

  Click ⌖Delete to remove a setting. An insulation type cannot be deleted if it is in use in a project or if it is the only one specified in a project.

  Click ⌗Add to open the New Insulation dialog where you specify values for a new wire insulation type. Insulation type names must be unique for a specified material.

- Used by Sizing:

  When selected for a specific wire size, that wire size is made available for use in circuits where Revit calculates the wire size. When cleared, the size is not available for use with the sizing feature.

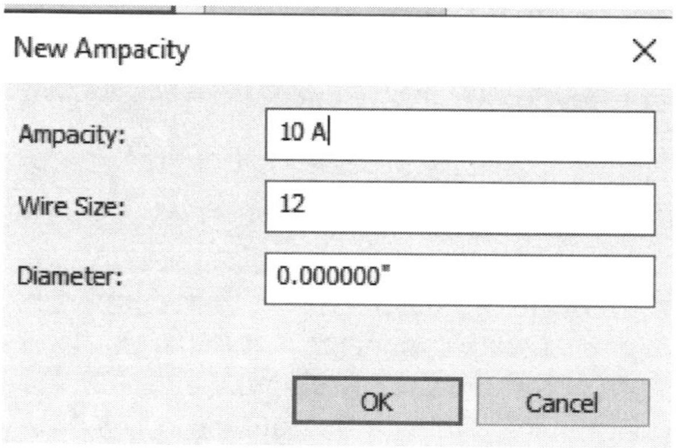

New Ampacity Dialog

New Ampacity ✕

Ampacity:	10 A
Wire Size:	12
Diameter:	0.000000"

OK   Cancel

Use this dialog to specify a new ampacity (wire size) to add to the currently selected wire sizes table.

Specify the ampacity, wire size (AWG), and diameter for the new wire.

Click OK.

The ampacity is added to the table and available for use.

Correction Factor

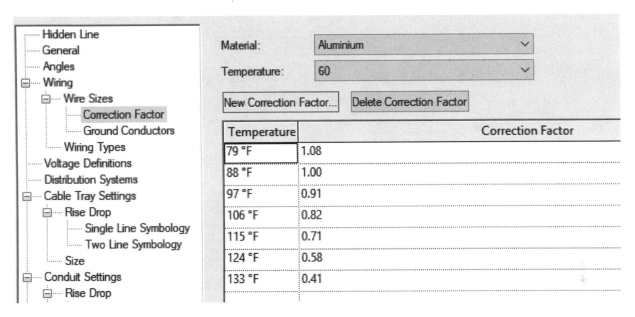

The correction factor is used when calculating wire sizes in the project.

Choosing the right cable with the right capacity isn't as simple as looking at conductor size. When reviewing cable ampacity, consider the application and the environment. Depending upon the application and/or environment, a correction factor may need to be applied to obtain the correct, safe ampacity rating. A correction factor is a multiplier applied to the ampacity rating to adjust the value based on a specific condition. The multiplier may be less than, equal to, or greater than one.

Ambient temperature affects the current carrying capability of wire. This effect is specified as a value for the wire material at specific ambient temperatures. The correction factor is used when calculating wire sizes in the project. You can click New Temperature to add custom correction factors to a project.

As the application temperature rises above ambient, the ampacity value is reduced, and vice-versa.

Ground Conductors

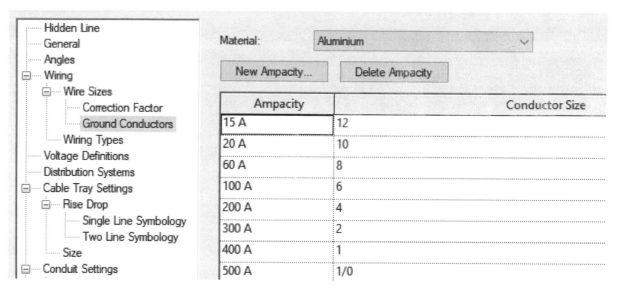

Number of Conductors

As more conductors are energized in a cable, more heat is generated. Therefore, a correction factor is applied when a cable holds more than three current-carrying conductors. Ampacity values are reduced to compensate for the added heat generated within the conductors.

Reeling Applications

Cable on a reel with multiple layers of cable does not have the ability to radiate heat away from the lower layers. When multiple layers of cable are installed on an energized cable reel, you must apply correction factors to compensate for the inability of the cable to radiate heat away, or risk de-grading or damaging the cable.

Duty-Cycle

In certain applications, especially resistance welders, a correction factor based on the duty-cycle is applied to gain the correct ampacity value.

In summary, the first step in applying correction factors in regards to the ampacity value of a cable is to define and understand the application, and to consult the standards or authority having jurisdiction (National Electrical Code, UL, CSA, ICEA, etc.). With this understanding, choosing the correct conductor size for your application can reduce the time and costs associated with unnecessary repairs or replacement of materials.

Wiring Types

	Name	Material	Temperature Rating (°C)	Insulation	Max Size	Neutral Multiplier	Neutral Required	Neutral Size	Conduit Type
1	THWN	Copper	60	THWN	2000	1.00	☑	Hot Conductor Size	Non-Magnetic
2	XHHW	Copper	60	XHHW	2000	1.00	☑	Hot Conductor Size	Non-Magnetic
3	Default	Aluminium	60	TW	2000	1.00	☑	Hot Conductor Size	Non-Magnetic

Use the Wiring Types table to specify the wire types that can be used in your project.

You can add or remove wire types as needed. Multiple wire types can be specified for a project. The first entry specified in the wire types table dialog is the default wire type used for circuits created in the project. This should be the wire type used for the majority of the wiring in a project. You can select a different wire type for a circuit on its Properties palette.

Name	User-defined alphanumeric string used to identify the wire type.
Material	Copper or Aluminum are available by default or the user can create a custom material.
Temperature Rating	30C, 60C, or 95C, or a project specific temperature rating as defined in the New Temperature dialog.
Insulation	Depending on the material selected, different insulation types can be specified, including a project specific insulation as defined in the New Insulation dialog. By default, you have three insulation types available: THWN/XHHW/TW. THWN- Thermoplastic Heat and Water-resistant Nylon-coated. XHHW -XLPE (cross-linked polyethylene) High Heat-resistant Water-resistant. TW - TW/THW is a solid or stranded, soft annealed copper conductor insulated with Polyvinylchloride (PVC).
Max Size	This is the maximum conductor size to be used when sizing wires of this type from 14 to 2000 MCM (thousand circular mils). This parameter lets you control when wires start being sized in parallel runs rather than by simply increasing the wire size until 2000 MCM is reached.
Neutral Multiplier	Using this field, in combination with the next two fields, you can specify how the neutral conductor of a system will be sized. The value specified here is used to increase or decrease the calculated size of the neutral conductor based on a multiplier of the conductor's size. Similar to the way that ground conductors are oversized for voltage drop situations, the neutral conductor can be sized to be larger than the calculated size. The neutral multiplier is applied to the neutral conductor based on cross-sectional area rather than ampacity. It is intended to handle the current increase that results from harmonic loads. Harmonic loads are

	caused by switching the power supplies found in many types of electronic equipment. These switching power supplies create harmonic distortion in the current waveform and cause current to flow at a higher value than would be expected in an electrical system. The Neutral Multiplier is applied after the neutral size is calculated, either by sizing the same as the hot conductors or according to unbalanced current.
Neutral Required	If selected, all wiring runs using this wire type will include a neutral, even in the case of a balanced 3-phase load, where a neutral may not be required by the load itself. If not selected, a neutral will be omitted for balanced loads and will be included for unbalanced loads.
Neutral Size	In this field you can specify whether the neutral is sized by Hot Conductor Size (the baseline for the size of the neutral will be the same as the hot conductor) or by Unbalanced Current (the neutral will be sized based on the amount of current flowing in the neutral).
Conduit Type	Conduit material affects the impedance of wire and determines what portion of the wire impedance table is used for the voltage drop calculations. Click the value, then select either Steel or Non-Magnetic.

Exercise 4-1:

Place Wiring Manually

Drawing Name: *wiring_manual.rvt*
Estimated Time: 15 minutes

This exercise reinforces the following skills:

    ❏   Add Wires Manually

1.   Open *wiring_manual.rvt*.

2.   Activate the **Insert** ribbon.

3.   Select the **Manage Links** tool.

      Note that there is a floor plan file linked to the project.

      Use RELOAD FROM to reload the file from the current folder.

      Click **OK**.

4.   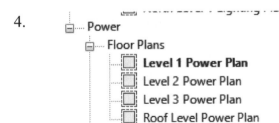   Verify that the **Level 1 Power Plan** floor plan is the active view.

5.

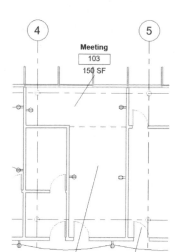

Zoom into Room 103 – between Grids 4 &5 and Grids 19-21.

6.

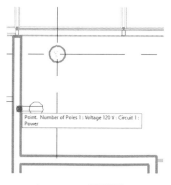

Activate the **Systems** ribbon.

7.

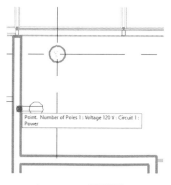

Select the **Arc Wire** tool under Wire.

8.

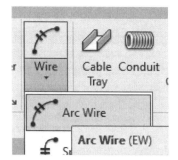

Verify that the **THWN** wire type is active in the Properties palette.

9.

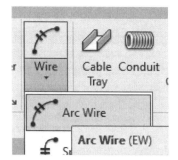

Move the cursor over the receptacle on the left wall until the wiring connection point appears.

Left click on the connection point to set the first point for the arced wiring.

10.

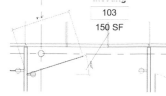

Left click to place a second point for the arc around the midpoint of the grid line in the room.

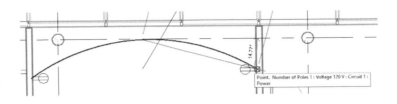

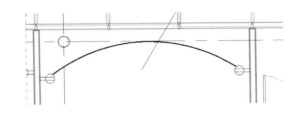

11. Move the cursor over the receptacle on the right wall until the wiring connection point appears.

   Left click on the connection point to set the end point for the arced wiring.

12. The wire is placed.

13. Save as *4-1.rvt*.

Exercise 4-2:

Display Wire Tick Marks

Drawing Name: *wire_tick_marks.rvt*
Estimated Time: 10 minutes

This exercise reinforces the following skills:

- ❏ Display Wire Tick Marks
- ❏ Load Family
- ❏ Electrical Settings

1. 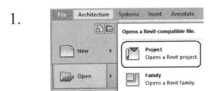 Open *wire_tick_marks.rvt*.

2. Insert   Activate the **Insert** ribbon.

3.

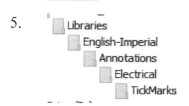

Select the **Manage Links** tool.

Note that there is a floor plan file linked to the project.

You may need to use RELOAD FROM to update the link to the file.

Click **OK**.

4.

Select the **Load Family** tool.

5.

Browse to the *TickMarks* folder located under the *English- Imperial/Annotations/Electrical* library.

6.

Hold down the Control key and select:

- *Long Wire Tick Mark*

- *Short Wire Tick Mark*

- *Circle Wire Tick Mark*

Click **Open**.

7.

Verify that the **Level 1 Power Plan** floor plan is the active view.

8.

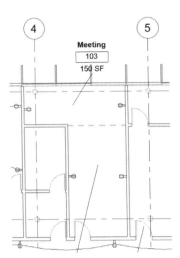

Zoom into Room 103 – between Grids 4 &5 and Grids 19-21.

9.

Systems

Activate the **Systems** ribbon.

10.

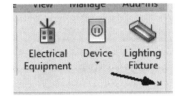

Launch the **Electrical Settings** dialog.

11. Highlight **Wiring** in the left pane.

Set the Hot Wire Tick Mark to **Long Wire Tick Mark.**

Set the Ground Wire Tick Mark to **Circle Wire Tick Mark.**

Set the Neutral Wire Tick Mark to **Short Wire Tick Mark**.

Setting	
Ambient Temperature	86 °F
Gap of Wiring Crossing	0' - 0 1/16"
Hot Wire Tick Mark	Long Wire Tick Mark
Ground Wire Tick Mark	Circle Wire Tick Mark
Neutral Wire Tick Mark	Short Wire Tick Mark
Slanted Line across Tick Marks	No
Show Tick Marks	Always
Max Voltage Drop For Branch ...	2.00%

Verify that Show Tick Marks is set to **Always**.

Click **OK** to close the dialog.

12. Tick marks are now displayed on all the wires.

Save as *ex4-2.rvt*.

Exercise 4-3:

Create a Home Run Wire

Drawing Name: *wire_home_run.rvt*
Estimated Time: 5 minutes

This exercise reinforces the following skills:

- ❑ Place a home run wire
- ❑ Load Family
- ❑ Electrical Settings

1. 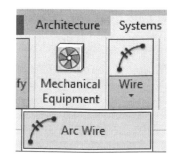 Open *wire_home_run.rvt*.

2. 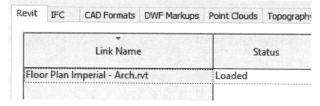 Select the **Manage Links** tool.

   Note that there is a floor plan file linked to the project.

   You may need to use RELOAD FROM to update the link to the file.

   Click **OK**.

3. 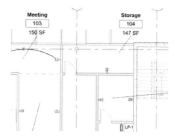 Zoom into Room 103 & Room 104 – between Grids 4 & 6 and Grids 19-21.

4. Activate the Systems ribbon.

   Select the **Arc Wire** tool.

5. Left click on the connection point for the receptacle on the right wall in Room 103.

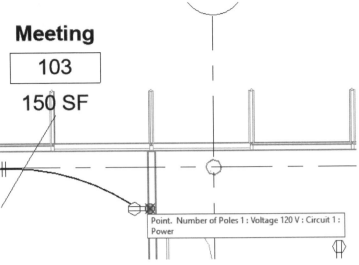

Point. Number of Poles 1 : Voltage 120 V : Circuit 1 : Power

6. Left click beneath the mid-point of the grid line in Room 104 to place the second point for the arc.

    Left click slightly below and to the right of the second point to create the home run wire.

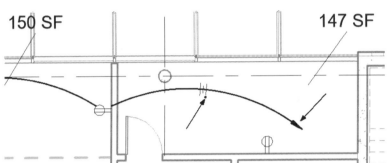

7.  Click the **Modify** tool on the ribbon to exit the command.

8. Save as *ex4-3.rvt*.

Exercise 4-4:

Create a Multiple Circuit Home Run Wire

Drawing Name: *multiple_home_run.rvt*
Estimated Time: 10 minutes

This exercise reinforces the following skills:

❑ Place a home run wire
❑ Add wires

1. Open *multiple_home_run.rvt*.

2. Zoom into Room 103 & Room 107 – between Grids 4 & 6 and Grids 19-21.

3. Select the Home Run wire located in Room 104.

   Select the grip located at the arrowhead.

4. Drag the grip to place it at the connection point for the receptacle located on the south wall of Room 104.

5. Activate the Systems ribbon.

   Select the **Arc Wire** tool.

6.

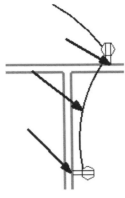

Select the south wall receptacle connection point as the start point for the wire.

Left click below and to the left of the south wall receptacle for the second arc wire point.

Left click to select the connection point on the receptacle located on the west wall of Room 107.

You are still in Place Wire mode.

7.

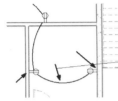

Select the west wall receptacle connection point on the west wall of Room 107 as the start point for the wire.

Left click below and to the right of the west wall receptacle for the second arc wire point.

Left click to select the connection point on the receptacle located on the east wall of Room 107.

You are still in Place Wire mode.

8.

Left click to select the connection point on the receptacle located on the east wall of Room 107.

Left click below and to the left of the east wall receptacle for the second arc wire point.

Left click below and to the left of the east wall receptacle for the end arc wire point.

Click ESC to exit the command.

9.

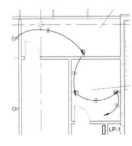

Note the home run arrows on the wires.

Save as *ex4-4.rvt*.

Exercise 4-5:

Create a Circuit

Drawing Name: *circuits.rvt*
Estimated Time: 20 minutes

This exercise reinforces the following skills:

- ❏ Manage Links
- ❏ Electrical Circuits
- ❏ Arc Wires

1.

Open *circuits.rvt*.

2.
Activate the Insert ribbon.
Select **Manage Links**.

3.
Manage Links

Reload From...

Use the **Reload From** button to reload the *Architecture_Model.rvt* file from the location where your exercise files are stored.
Click **OK**.

4.
Verify that the active view is
POWER – First Floor floor plan.

5.

Zoom into **OFFICE 104**.

Hold down the **CTL** key and select the four receptacles in the room.

Select the **Power** button on the ribbon.

6. 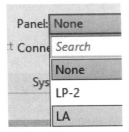 On the ribbon:
Assign the circuit to Panel **LA**.

7. 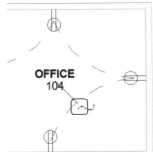 A preview of the circuit will appear in the display window.

 Select **Arc Wire** from the ribbon.

Alternatively, you can select the Arc Wire icon on the preview to convert the preview to place arc wires.

This places wires based on the preview.

8.  Click anywhere in the window to release the selection. Zoom into **OFFICE 103**.

Hold down the **CTL** key and select the three duplex receptacles in the room. *Do not select the receptacle on the east wall.*

 Select the **Power** button on the ribbon.

9. A preview of the circuit will appear in the display window. Select **Arc Wire** from the ribbon or use the convert to wire icon.
This places wires based on the preview.

10. Locate the quadruple receptacle in OFFICE 103.

Click on the bottom connector point.

11.

| Create Power Circuit |
| Add to Circuit |
| Remove from Circuit |
| Hide in View |
| Override Graphics in View |
| Create Similar |
| Edit Family |
| Select Previous |
| Select All Instances |
| Delete |
| Find Referring Views |
| Zoom In Region |
| Zoom Out (2x) |
| Zoom To Fit |
| Previous Pan/Zoom |
| Next Pan/Zoom |
| Browsers |
| Properties |

#1

Right click and select **Add to Circuit**.

Select one of the duplex receptacles in the room.

Select the **Arc Wire** tool or use the convert to wire icon to place a wire.

Click ESC to exit the command.

This is what has been done so far.

12. Power
  Ceiling Plans
    EL - Level 1

Open the **EL – Level 1** Ceiling Plan under Power.

13. 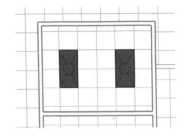 Select the two lighting fixtures in OFFICE 104.

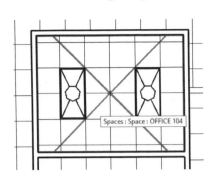

If you aren't sure which room is OFFICE 104, hover over the space and Revit will display the information.

Select the **Power** button on the ribbon.

14. Panel: HA   Set the Panel to **HA** on the ribbon.

15.   A preview of the circuit will appear in the display window. Select **Arc Wire** from the ribbon.

16. 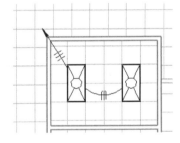 The circuit is placed.

17.

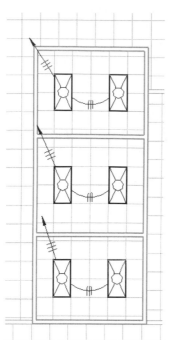

Repeat to place circuits in OFFICE 103 and 102.

18.

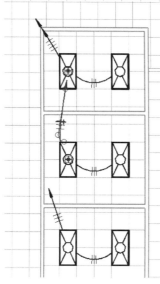

Select the home run wire in OFFICE 103 and drag to connect to the left light fixture in OFFICE 104.

Save as *ex4-5.rvt.*

Exercise 4-6:

Defining Switch Legs

Drawing Name: *lighting_switch_legs.rvt*
Estimated Time: 25 minutes

This exercise reinforces the following skills:

- ❑ Switch Systems
- ❑ Wires
- ❑ Filters
- ❑ View Overrides

1.  Activate the Insert ribbon.
Select **Manage Links**.

2. 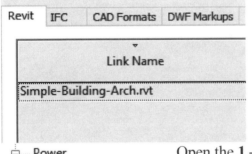 Use the **Reload From** button to reload the *Simple-Building-Arch.rvt* file from the location where your exercise files are stored.
Click **OK**.

3. 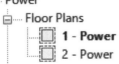 Open the **1 – Power** floor plan view.

4.

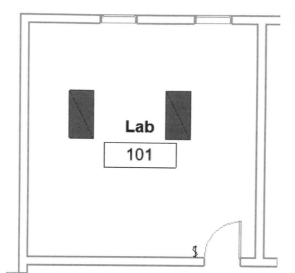

Hold down the CTL key or use a window to select the lighting fixtures in Lab – Room 101.

5.

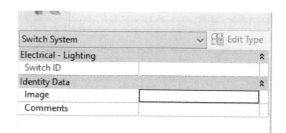

Select the **Switch Systems** tab on the ribbon.

6.

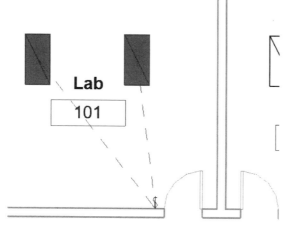

The circuit displays showing that the light fixtures are connected to the Switch.

7.

On the Properties palette, notice that the Switch ID is blank.

Click **ESC** to release the selection.

8.

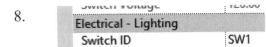

Select the switch in the Room 101.

Type **SW1** as the Switch ID in the Properties palette.

9.

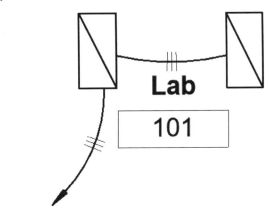

Activate the Systems ribbon.

Select the **Arc Wire** tool.

10.

Draw a wire between the two light fixtures.

Draw a home run wire from the left light fixture pointing to the lower left corner of the room.

ESC out of the command.

11.

Select one of the arc wires you just placed.

On the Properties palette, select **Edit Type**.

12.

Select **Duplicate.**

13.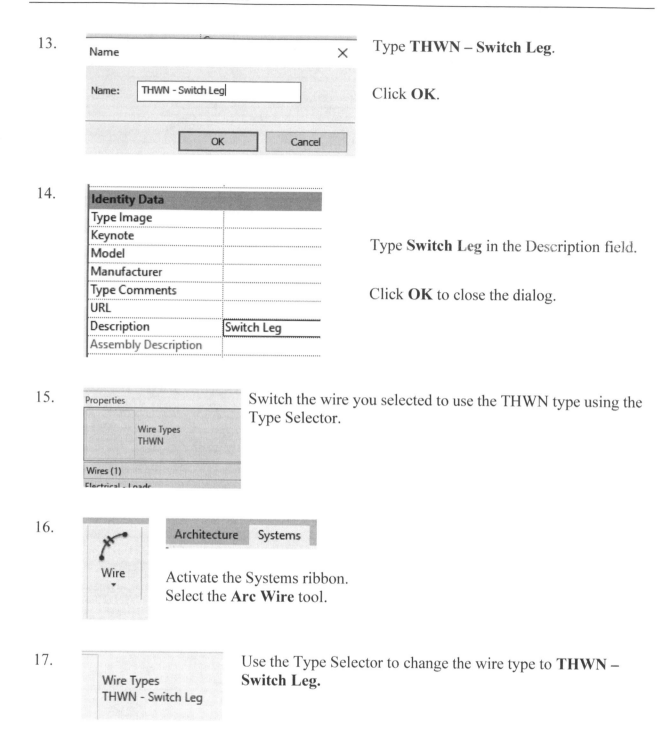

Type **THWN – Switch Leg**.

Click **OK**.

14.

Type **Switch Leg** in the Description field.

Click **OK** to close the dialog.

15.

Switch the wire you selected to use the THWN type using the Type Selector.

16.

Activate the Systems ribbon.
Select the **Arc Wire** tool.

17.

Use the Type Selector to change the wire type to **THWN – Switch Leg.**

18.

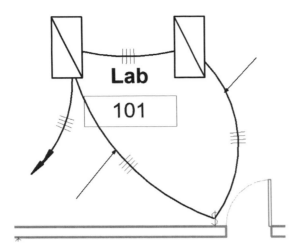

Draw two arc wires from each light fixture to the switch.

19.

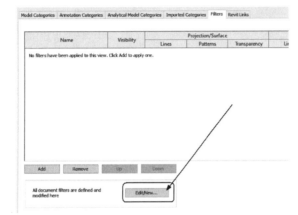

Type **VV** to open the Visibility/Graphics dialog.

Click on the **Filters** tab.

Select **Edit/New** at the bottom of the dialog to access the **Filters** tool.

20.

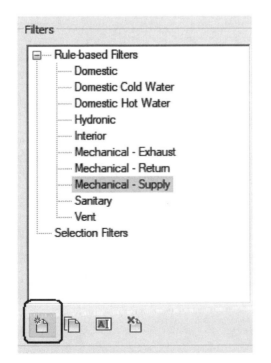

Select **New**.

21.

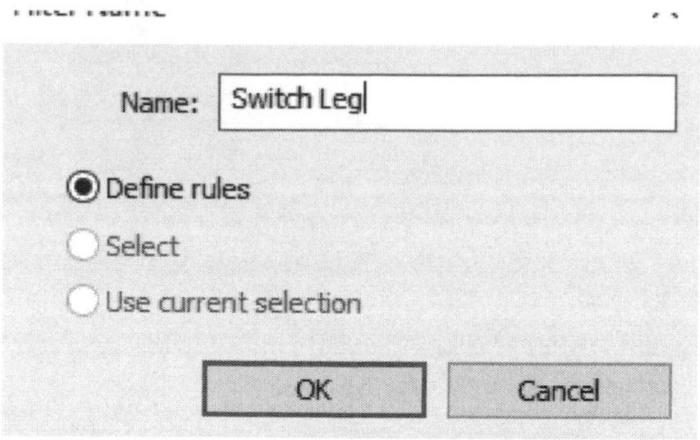

Name: Switch Leg

⦿ Define rules
○ Select
○ Use current selection

OK   Cancel

Type **Switch Leg** for the Name.

Click **OK**.

22.

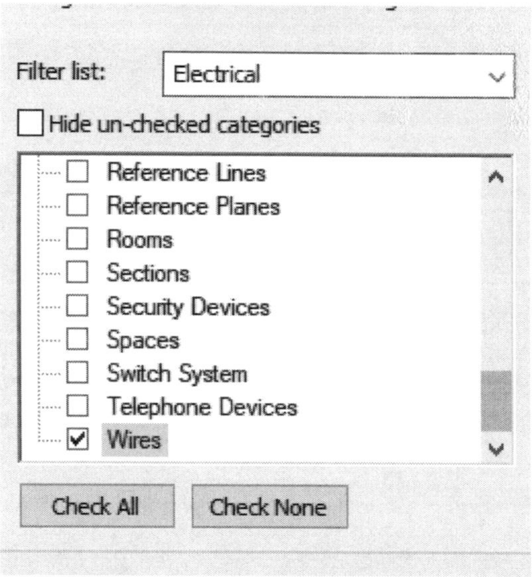

Filter list: Electrical

☐ Hide un-checked categories

- ☐ Reference Lines
- ☐ Reference Planes
- ☐ Rooms
- ☐ Sections
- ☐ Security Devices
- ☐ Spaces
- ☐ Switch System
- ☐ Telephone Devices
- ☑ Wires

Check All   Check None

Enable **Wires** in the category list.

23.

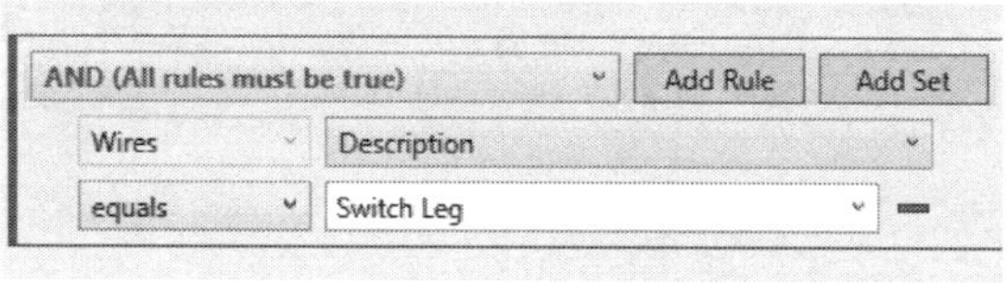

AND (All rules must be true)   Add Rule   Add Set

Wires   Description
equals   Switch Leg

Select **Description**.

Select **equals**.

Select **Switch Leg**.

Click **OK** to create the filter.

24.

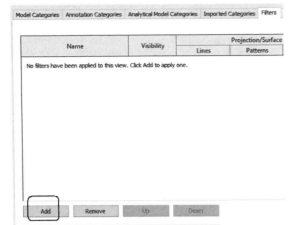

Select **Add** from the bottom of the dialog.

25.

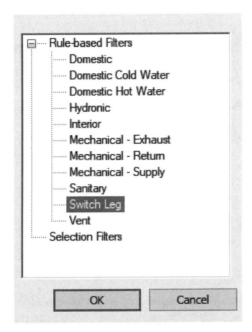

Highlight **Switch Leg**.

Click **OK**.

26.

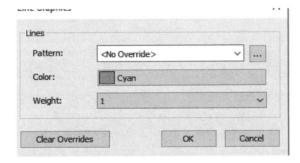

Left click in the Lines column.

Set the Color to **Cyan**.

Set the Weight to **1.**

Click **OK**.

Close the dialog box.

27.

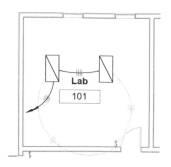

The wire display updates.

28.   Save the file as *ex4-6.rvt*.

Exercise 4-7:

Wiring to a Junction Box

Drawing Name: *junction_box_wiring.rvt*
Estimated Time: 30 minutes

This exercise reinforces the following skills:

- ❑ Electrical Fixtures
- ❑ Wires
- ❑ Circuits

1.   Open the **East – Elec** elevation view.

Elevations (Building Elevation)
     East - Elec
     North - Elec
     South - Elec
     West - Elec

2.   Set the display to **Wireframe**.

1/8" = 1'-0"

3.   Notice that there is a floor placed on Level 2 and a ceiling placed on Level 1.
Notice that ceilings go up and floors go down.

Level 2
10' - 0"

Ceiling
8' - 0"

Level 1
0' - 0"

4.  Select the **Measure** tool located on the QAT.

5. Measure between the floor and the ceiling to determine how much space you have for electrical equipment.
There is a 1' 3 ¾" air space between the floor and the ceiling.

6. Measure the thickness of the ceiling.
To verify, select the ceiling.

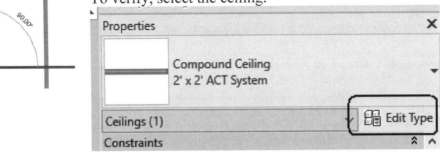

Select **Edit Type** on the Properties palette.

7. Locate the thickness value of the ceiling.

The thickness is 2 ¼".

This means that any electrical equipment should be placed at 8' 2 ¼" elevation from Level 1.

Click **OK** to close the Type Properties dialog.
Press **ESC** to release the selection.

8. 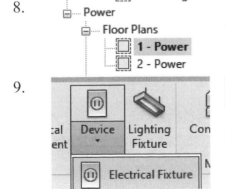 Open the **1 – Power** floor plan.

9. Switch to the **Systems** ribbon.

Select **Electrical Fixture** on the Device fly-out.

10.

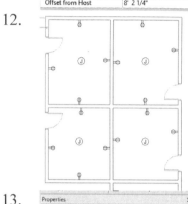

Locate the **4" Square 120 -1 Junction box** using the Type Selector.

11.

On the Properties palette:
Set the Elevation from level to **8' 2 ¼"**.

12.

Place a junction box in the center of each of the rooms shown.

13.

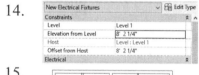

Locate the **12" Square 120 Junction box** using the Type Selector.

14.

On the Properties palette:
Set the Elevation from level to **8' 2 ¼"**.

15.

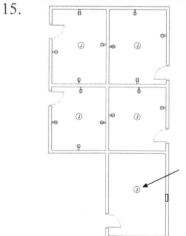

Place the 12" square junction box in the center of the lower room.

16.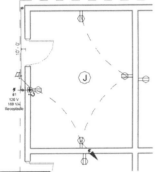

Select the receptacle on the west wall in the upper left room.

17. Electrical Circuits

Select the **Electrical Circuits** tab on the ribbon.

18.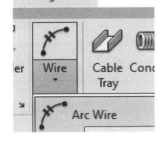

A preview will display showing the receptacles are connected.

Edit Circuit

Select **Edit Circuit** on the ribbon.

19. Add to Circuit

Verify that **Add to Circuit** is enabled on the ribbon.

Notice that the receptacles are in the foreground and all the other devices are grayed out. Anything that is not part of the circuit is faded.

Select the junction box in the center of the room to add it to the circuit.

20. Finish Editing Circuit

Select **Finish Editing Circuit** on the ribbon.

21.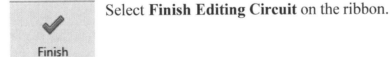

Select **Arc Wire** from the ribbon.

22.

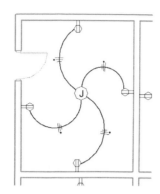

Draw a wire from each receptacle to the junction box in the middle of the room.

23.

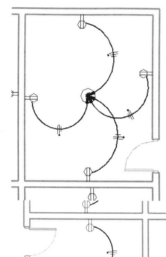

Draw a wire from each receptacle to the junction box in the middle of the upper right room.

24.

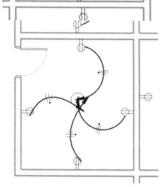

Draw a wire from each receptacle to the junction box in the middle of the lower left room.

25. Draw a wire from each receptacle to the junction box in the middle of the lower right room.

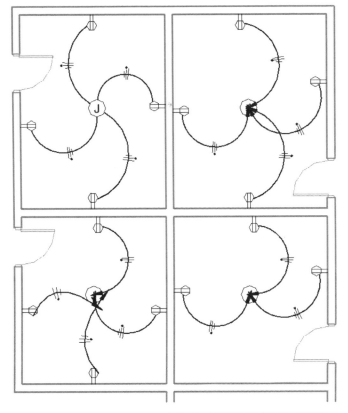

26. Draw a wire between the lower left room junction box and the upper left room junction box.

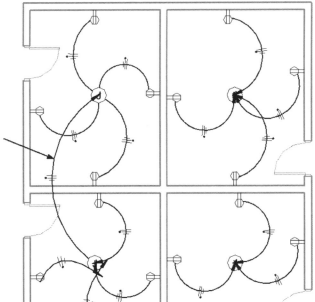

27. Draw a wire between the lower right room junction box and the upper right room junction box.

28.

Draw a wire from the lower left room to the 12" junction box.

Draw a wire from the lower right room to the 12" junction box.

29.

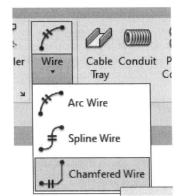

Select the **Chamfered Wire** tool from the ribbon.

30.

Select the 12" junction box as the starting point for the chamfered wire.

Left click to the right of the junction box.

Move the mouse down and left click again to place the home run wire.

31. *Your view should look similar to this.*

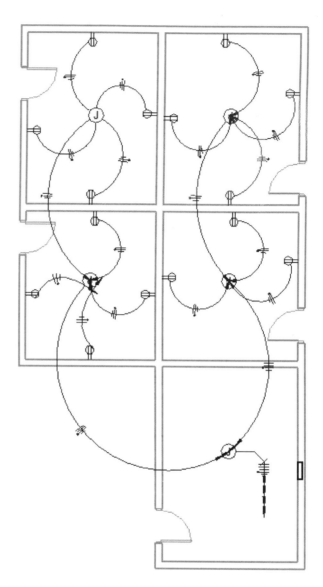

32.

Elevations (Building Elevation)
- East - Elec
- North - Elec
- South - Elec
- West - Elec

Open the **East – Elec** elevation view.

33. You see the junction boxes placed on the ceilings of each room.

Notice that you don't see the wires. This is because wires are view-specific. They are only visible in the view where they are placed.

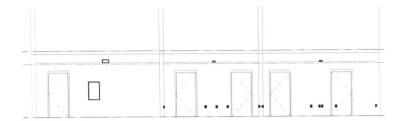

Save as *ex4-7.rvt*.

Lab Exercise

Open *wiring_lab.rvt*.

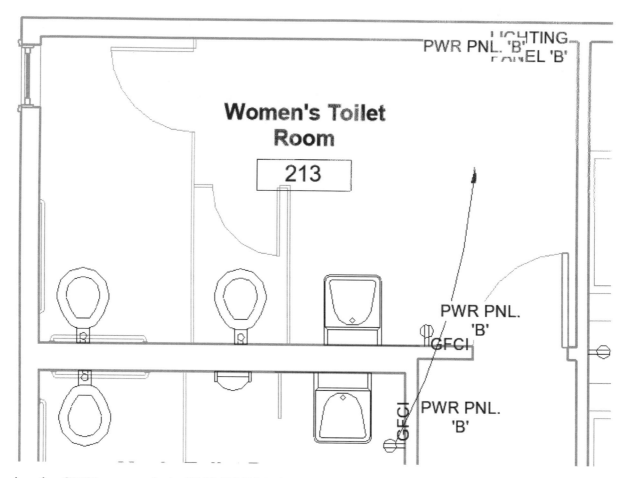

Assign the GFCI receptacle in WOMEN'S ROOM 213 & MEN'S ROOM 212 to PWR PNL B.

Add an arc wire.

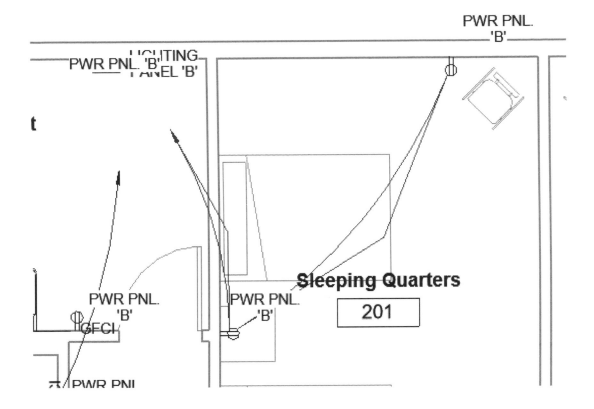

Assign the receptacle in SLEEPING QUARTERS 201 to PWR PNL B.

Add an arc wire.

Continue through the Main Floor floor plan assigning receptacles to panels and adding wires.

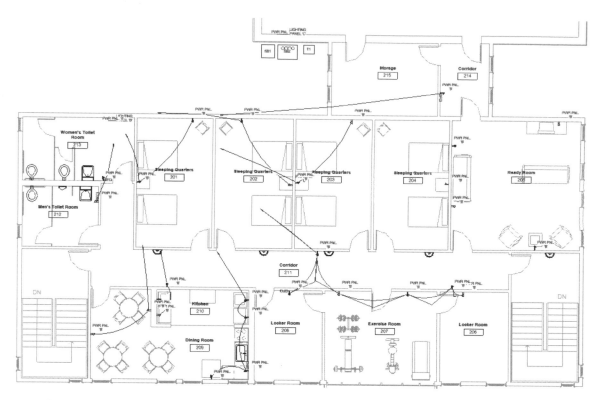

Create a duplicate view of the Main Floor Lighting Ceiling Plan.

Rename the view **Main Floor Lighting**.

Turn off the visibility of elevations, sections, roofs, ceilings, and reference planes.

Add circuits to the lighting fixtures. Connect the lighting fixtures to LIGHTING PANEL B.

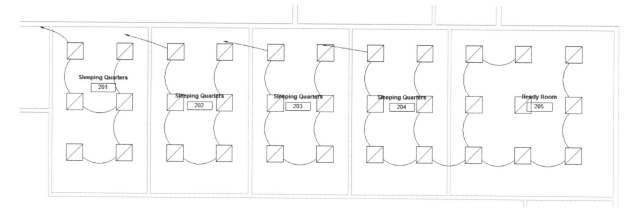

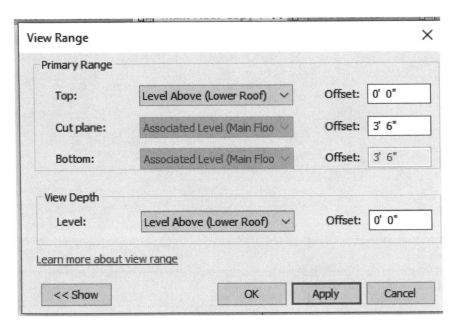

Change the View Range for the Main Floor Lighting Ceiling Plan so that the Cut Plane is set to 3' 6". *This allows you to see any switches which are placed.*

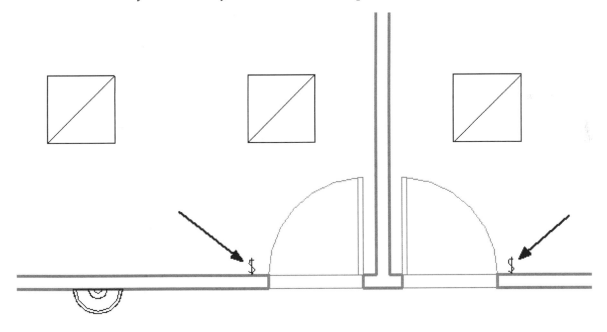

Add single pole light switches to each room.

Connect the light fixtures to the switches.

Create switch leg wires to differentiate between the wires going from the light fixture to the switch and the wiring between the light fixtures.

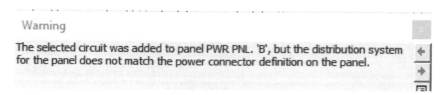

How can you fix this error?

Hint: Add a new distribution system that will work with the panel and assign the panel to the correct distribution system.

Lesson

05

Conduits

An electrical conduit is a tube used to protect and route electrical wiring in a building or structure. Most conduits use PVC piping, which is rigid. Revit allows you to define different conduit types and sizes.

Revit allows you to apply EMT (electrical metallic tubing), IMC (intermediate metal conduit), RMC (rigid metal conduit), RNC Schedule 40 (rigid nonmetallic conduit) or RNC Schedule80 (rigid nonmetallic conduit) to your conduit families. RNC Schedule 40 is Heavy Wall PVC piping standard. RNC Schedule 80 is an Extra Heavy Wall standard. RNC Schedule 40 is listed for underground applications encased in concrete and can also be used in exposed or concealed areas above ground. RNC Schedule 80 is used in areas where it may be subjected to physical damage. You have the ability to create and apply your own conduit standards. For example, if you wanted to use GRC (galvanized rigid conduit), you would have to create that standard before you can apply it to a conduit family.

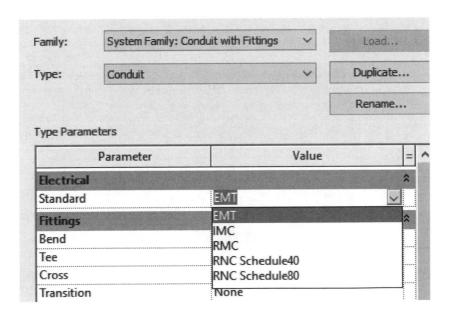

Conduits are system families, so any conduit definitions you create are local to the active project only. I recommend that you define your conduit families for your projects and place them in your templates, so you don't have to redefine them for each project.

You can change the elevation of a conduit as you are placing it by changing the off-set value in the drop-down. Revit will automatically insert the proper fittings and vertical conduits to transition to the new elevation.

To connect a conduit to equipment or to a device, the object must have a conduit connector available. When you select the object, you can right click the conduit connector and choose the option to draw conduit from the connector. You can connect to equipment in plan view, elevation view or 3D view.

In most projects, you will want to assign colors to your conduits to make them easier to identify. You need to create a family type for each conduit color you want to use. In order to see the different colors, you need to define a View Filter and then apply it to your view. You may want to create a view template to make it easier for you to assign the view filter to your views displaying conduits.

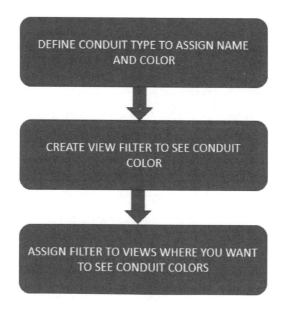

Conduit Settings

	Setting	
Hidden Line	Use Annot. Scale for Single Line Fittings	☐
General	Conduit Fitting Annotation Size	0' 0 1/8"
Angles	Conduit Size Prefix	
Wiring	Conduit Size Suffix	ø
Wire Sizes	Conduit Connector Separator	-
Correction Factor		
Ground Conductors		
Wiring Types		
Voltage Definitions		
Distribution Systems		
Cable Tray Settings		
Rise Drop		
Single Line Symbology		
Two Line Symbology		
Size		
Conduit Settings		

Use Annot. Scale for Single Line Fittings	Specifies whether conduit fittings are drawn at the size specified by the Conduit Fitting Annotation Size parameter. Changing this setting does not change the plotted size of components already placed in a project.
Conduit Fitting Annotation Size	Specifies the plotted size of fittings drawn in single-line views. This size is maintained regardless of the drawing scale.
Conduit Size Prefix	Specifies the symbol preceding the conduit size.
Conduit Size Suffix	Specifies the symbol following the conduit size.
Conduit Connector Separator	Specifies the symbol used to separate information between 2 different connectors.

Access conduit settings on the Electrical Settings dialog. To open the dialog, type ES.

Conduit Settings – Rise Drop

Setting		Specifies the plotted size of rise/drop symbols drawn in single-line views. This size is maintained regardless of the drawing scale.
Conduit Rise/Drop Annotation Size	1/8"	

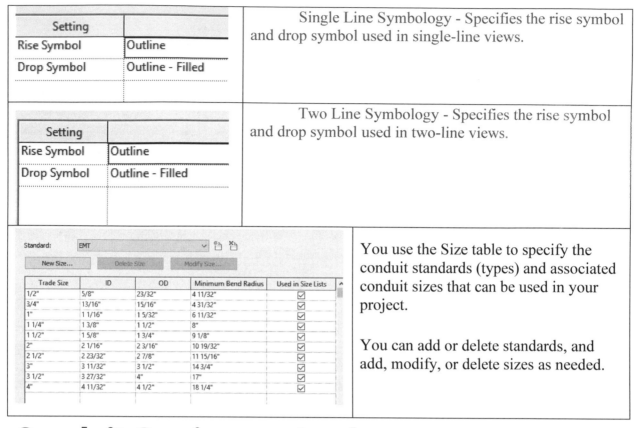

Setting	
Rise Symbol	Outline
Drop Symbol	Outline - Filled

Single Line Symbology - Specifies the rise symbol and drop symbol used in single-line views.

Setting	
Rise Symbol	Outline
Drop Symbol	Outline - Filled

Two Line Symbology - Specifies the rise symbol and drop symbol used in two-line views.

Standard: EMT

Trade Size	ID	OD	Minimum Bend Radius	Used in Size Lists
1/2"	5/8"	23/32"	4 11/32"	☑
3/4"	13/16"	15/16"	4 31/32"	☑
1"	1 1/16"	1 5/32"	6 11/32"	☑
1 1/4"	1 3/8"	1 1/2"	8"	☑
1 1/2"	1 5/8"	1 3/4"	9 1/8"	☑
2"	2 1/16"	2 3/16"	10 19/32"	☑
2 1/2"	2 23/32"	2 7/8"	11 15/16"	☑
3"	3 11/32"	3 1/2"	14 3/4"	☑
3 1/2"	3 27/32"	4"	17"	☑
4"	4 11/32"	4 1/2"	18 1/4"	☑

You use the Size table to specify the conduit standards (types) and associated conduit sizes that can be used in your project.

You can add or delete standards, and add, modify, or delete sizes as needed.

Conduit Settings – Angles

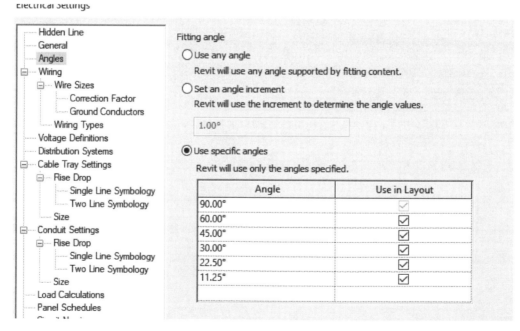

Electrical Settings

- Hidden Line
- General
- **Angles**
- Wiring
  - Wire Sizes
    - Correction Factor
    - Ground Conductors
  - Wiring Types
- Voltage Definitions
- Distribution Systems
- Cable Tray Settings
  - Rise Drop
    - Single Line Symbology
    - Two Line Symbology
  - Size
- Conduit Settings
  - Rise Drop
    - Single Line Symbology
    - Two Line Symbology
  - Size
- Load Calculations
- Panel Schedules

Fitting angle

○ Use any angle
   Revit will use any angle supported by fitting content.

○ Set an angle increment
   Revit will use the increment to determine the angle values.
   1.00°

● Use specific angles
   Revit will use only the angles specified.

Angle	Use in Layout
90.00°	☑
60.00°	☑
45.00°	☑
30.00°	☑
22.50°	☑
11.25°	☑

Specify the allowable angles to be used for conduits, cable trays, etc. based on the equipment to be installed.

Exercise 5-1:

Creating a Conduit Standard

Drawing Name: *conduit_GRC.rvt*
Estimated Time: 5 minutes

This exercise reinforces the following skills:
- ❑ Electrical Settings
- ❑ Conduits

1.   Open **Level 1- Rooms** floor plan.

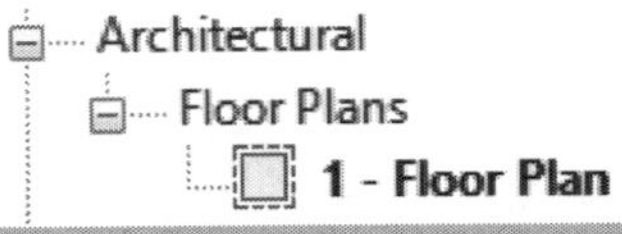

```
⊟··· Architectural
     ⊟··· Floor Plans
          ···□ 1 - Floor Plan
```

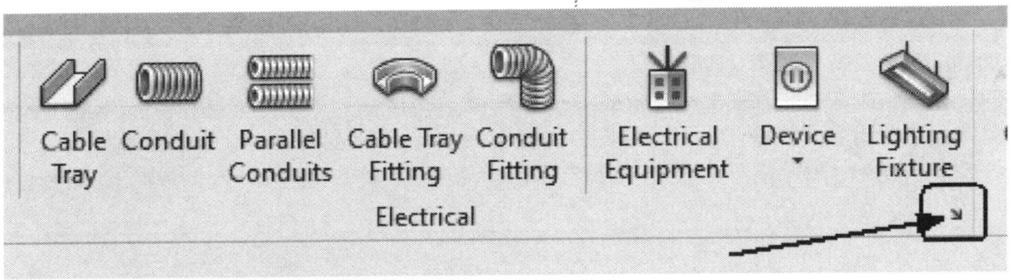

2.   Switch to the Systems ribbon.
Open the Electrical Settings panel.

3.

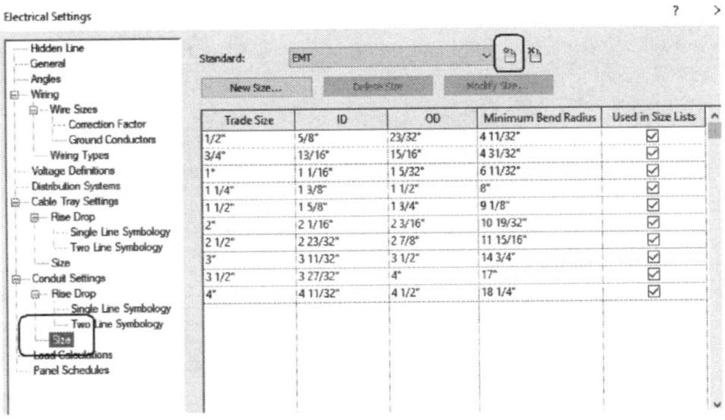

Highlight Size in the left panel.

Select the **New** icon next to Standard.

4.

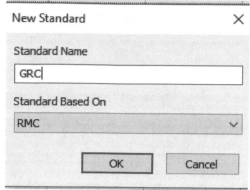

Type **GRC** for the Standard Name.

Select **RMC** (Rigid Metal Conduit) for the Standard Based On value.

Click **OK**.

5. Close the dialog.

Exercise 5-2:

Creating a Conduit Family

Drawing Name: *Conduit_family.rvt*
Estimated Time: 15 minutes

This exercise reinforces the following skills:
- Electrical Settings
- Conduits
- System Families

1. Open **1- Power** floor plan.

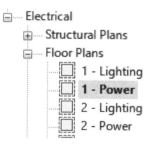

2.

Switch to the **Systems** ribbon.
Select the **Conduit** tool on the Electrical panel.

3.

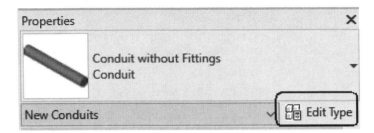

Select the **Conduit without Fittings** using the Type Selector.

Click **Edit Type**.

4.

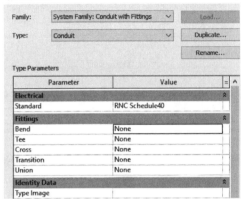

Under Standard:
Select **RNC Schedule40**.

This is a Rigid Non-metal conduit, usually PVC pipe.

Notice that no fittings are loaded for this family.

5.

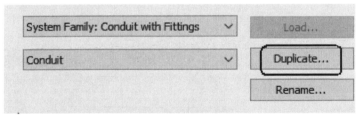

Select **Duplicate.**

6.

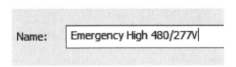

Type **Emergency High 480/277V** in the Name field.

Click **OK**.

7.

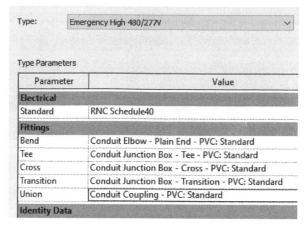

Using the drop-down next to each field:

Assign **the Conduit-Elbow – Plain End- PVC: Standard** to the Bend fitting.

Assign the **Conduit Junction Box – Tee -PVC: Standard** to the Tee fitting.

Assign the **Conduit Junction Box – Cross -PVC: Standard** to the Cross fitting.

Assign the **Conduit Junction Box – Transition -PVC: Standard** to the Transition fitting.

Assign the **Conduit Coupling -PVC: Standard** to the Union fitting.

8.

Union	None
Identity Data	
Type Image	
Keynote	
Model	
Manufacturer	
Type Comments	
URL	
Description	Red
Assembly Description	
Assembly Code	
Type Mark	
Cost	

In the Description field, type **Red**.

9.

Family:	System Family: Conduit with Fittings	Load...
Type:	Emergency High 120/208V	Duplicate...
		Rename...

Type Parameters

Select **Duplicate.**

10.

Type: Emergency Low 120/208V

Type **Emergency Low 120/208V** in the Name field.

Click **OK**.

11.

| Type: | Emergency Low 120/208V |

Type Parameters

Parameter	Value
Electrical	
Standard	RNC Schedule40
Fittings	
Bend	Conduit Elbow - Plain End - PVC: Standard
Tee	Conduit Junction Box - Tee - PVC: Standard
Cross	Conduit Junction Box - Cross - PVC: Standard
Transition	Conduit Junction Box - Transition - PVC: Standard
Union	Conduit Coupling - PVC: Standard
Identity Data	
Type Image	
Keynote	
Model	
Manufacturer	
Type Comments	
URL	
Description	Magenta
Assembly Descrip	

In the Description field, type **Magenta**.

12.

Family:	System Family: Conduit without Fittings	Load...
Type:	Emergency Low 480/277V	Duplicate...
		Rename...

Type Parameters

Select **Duplicate.**

13.

| Name: | Normal High 480/277V |

Type **Normal High 480/277V** in the Name field.

Click **OK**.

14.

| Type: | Normal High 480/277V | ⌄ |

Type Parameters

Parameter	Value
Electrical	
Standard	RNC Schedule40
Fittings	
Bend	Conduit Elbow - Plain End - PVC: Standard
Tee	Conduit Junction Box - Tee - PVC: Standard
Cross	Conduit Junction Box - Cross - PVC: Standard
Transition	Conduit Junction Box - Transition - PVC: Standard
Union	Conduit Coupling - PVC: Standard
Identity Data	
Type Image	
Keynote	
Model	
Manufacturer	
Type Comments	
URL	
Description	Cyan
Assembly Descrip	

In the Description field, type **Cyan**.

Click **Apply**.

15.

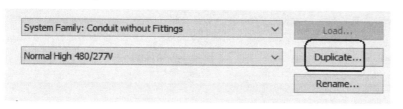

Select **Duplicate.**

16.

| Name: | Normal Low 120/208V |

Type **Normal Low 120/208V** in the Name field.

Click **OK**.

17.

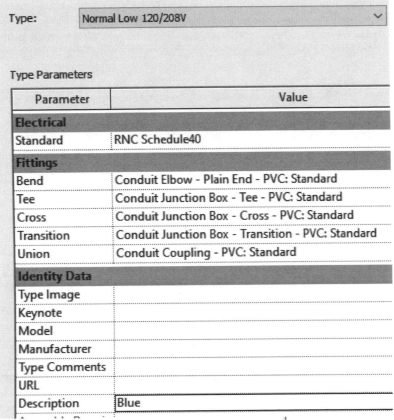

In the Description field, type **Blue**.

Click **OK** to close the dialog.

We have created four conduit types:
- Emergency High (Red)
- Emergency Low (Magenta)
- Normal High (Cyan)
- Normal Low (Blue)

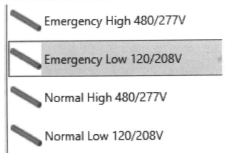

18.   Save as *ex5-2.rvt*.

Filters

Filters provide a way to override the graphic display and control the visibility of elements that share common properties in a view.

For example, if you need to change the line style and color for different conduit types, you can create a filter that selects all conduits in the view that have the color 'red' in the description parameter. You can then select the filter, define the visibility and graphic display settings (such as line style and color), and apply the filter to the view. When you do this, all conduits that meet the criteria defined in the filter update with the appropriate visibility and graphics settings. You need to set up the view filters and then apply those filters to each view in order to display the conduits with the correct colors and linetypes.

Exercise 5-3:

Defining View Filters

Drawing Name: *Conduit_views.rvt*
Estimated Time: 20 minutes

This exercise reinforces the following skills:
- ❑ View Filters
- ❑ Conduits

1. Open **Vault Level** floor plan.

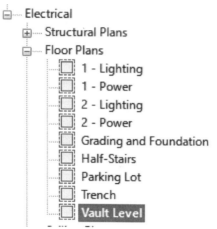

Activate the View ribbon.
Select **Filters**.

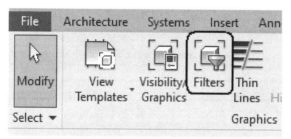

2.

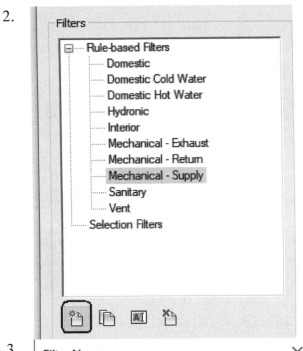

Click the **New** filter icon located at the lower left of the Filters palette.

3.

Type **Emergency High UG** in the Name field.

This filter will control the display properties for the underground emergency high conduit.

Click **OK**.

4.

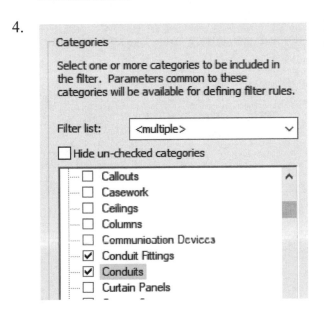

Place a check next to **Conduits** and **Conduit Fittings** in the Category panel.

5.

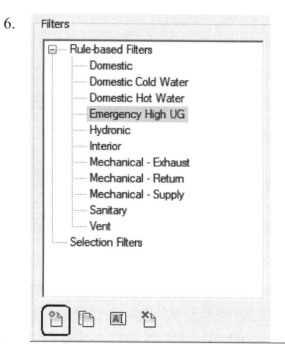

Select **Description** from the parameter list.

Select **equals**.

You can use the drop-down list to determine which criteria to use.

Select **Red-Dashed**.

Click **Apply**.

6.

Click the **New** filter icon located at the lower left of the Filters palette.

7.

Type **Emergency Low UG** in the Name field.

This filter will control the display properties for the underground emergency low conduit.

Click **OK**.

8.

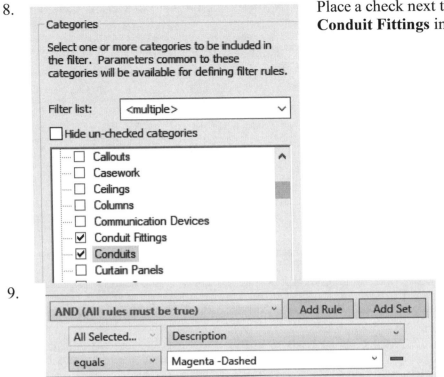

Place a check next to **Conduits** and **Conduit Fittings** in the Category panel.

9.

Select **Description** from the parameter list.

Select **equals**.

You can use the drop-down list to determine which criteria to use.

Select **Magenta-Dashed**.

Click **Apply**.

10.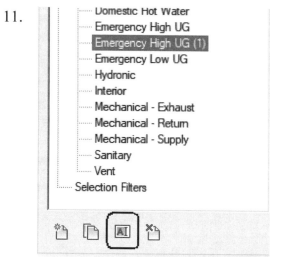

Highlight the **Emergency High UG** filter.

Select **Duplicate**.

11.

Highlight the copied filter.

Select **Rename**.

12.

Type **Normal Low UG** in the new Name field.

This filter will control the display properties for the underground normal low conduit.

Click **OK**.

13.

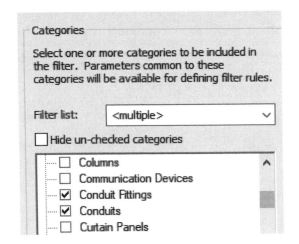

Because you used DUPLICATE, the conduits and conduit fittings categories are already selected.

14.

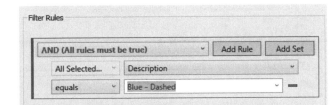

Because you used DUPLICATE, the filter rules are already set up.

Change the color and linetype to **Blue-Dashed**.

Click **Apply**.

15.

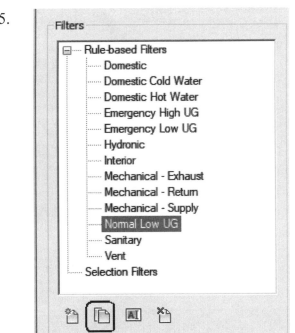

Highlight the Normal Low UG filter.

Select **Duplicate**.

16.

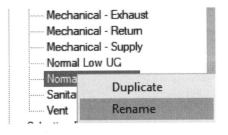

Highlight the copied filter.

Right click and select **Rename**.

17.

Previous: Normal Low UG (1)

New: Normal High|UG

OK

Type **Normal High UG** in the new Name field.

This filter will control the display properties for the underground normal high conduit.

Click **OK**.

18.

Categories

Select one or more categories to be included in the filter. Parameters common to these categories will be available for defining filter rules.

Filter list: <multiple>

☐ Hide un-checked categories

☐ Columns
☐ Communication Devices
☑ Conduit Fittings
☑ Conduits
☐ Curtain Panels

Verify that Conduit Fittings and Conduits are checked.

19.

AND (All rules must be true)   Add Rule

All Selected...   Description

equals   Cyan - Dashed

Because you used DUPLICATE, the filter rules are already set up.

Change the color and linetype to **Cyan-Dashed**.

Click **Apply**.

20.

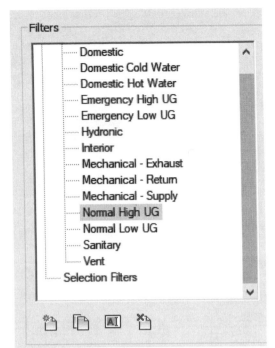

Check the filters list and verify that you have defined four filters:

- Emergency High UG
- Emergency Low UG
- Normal High UG
- Normal Low UG

Click **OK** to close the Filters palette.

21. Save as *ex5-3.rvt*.

Exercise 5-4:

Applying View Filters to a View

Drawing Name: *conduit_view_setting.rvt*
Estimated Time: 10 minutes

This exercise reinforces the following skills:
- ❏ View Filters
- ❏ Conduits

1.   Open **Vault Level** floor plan.

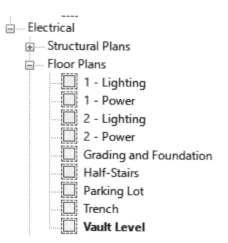

2.

Switch to the View ribbon.

Select the **Visibility/Graphics** tool.
You can also access the Visibility/Graphics palette by typing VV.

3.   Select the **Filters** tab.

4.   Click **Add.**

5.

Select one or more filters to insert.

- Rule-based Filters
  - Domestic
  - Domestic Cold Water
  - Domestic Hot Water
  - Emergency High UG
  - Emergency Low UG
  - Hydronic
  - Interior
  - Mechanical - Exhaust
  - Mechanical - Return
  - Mechanical - Supply
  - Normal High UG
  - Normal Low UG
  - Sanitary
  - Vent

Edit/New...

OK   Cancel   Help

Hold down the CTL key and select the four filters used for conduits:

- Emergency High UG
- Emergency Low UG
- Normal High UG
- Normal Low UG

Click **OK**.

6.

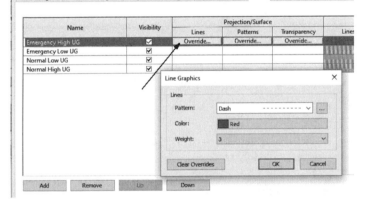

Highlight the **Emergency High UG** conduit filter.

Click in the cell in the Lines column.
Set the Pattern to **Dash**.
Set the Color to **Red**.
Set the Weight to **3**.

Click **OK**.

7.

Name	Visibility	Lines
Emergency High UG	☑	------------------
Emergency Low UG	☑	
Normal Low UG	☑	
Normal High UG	☑	

A preview of how the conduit will be displayed is shown.

8.

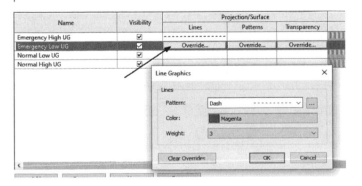

Highlight the **Emergency Low UG** conduit filter.

Click in the cell in the Lines column.
Set the Pattern to **Dash**.
Set the Color to **Magenta**.
Set the Weight to **3**.

Click **OK**.

9.

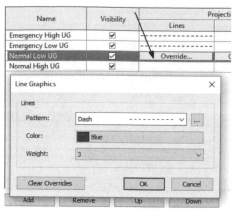

Highlight the **Normal Low UG** conduit filter.

Click in the cell in the Lines column.
Set the Pattern to **Dash**.
Set the Color to **Blue**.
Set the Weight to **3**.

Click **OK**.

10.

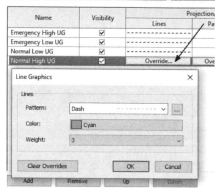

Highlight the **Normal High UG** conduit filter.

Click in the cell in the Lines column.
Set the Pattern to **Dash**.
Set the Color to **Cyan**.
Set the Weight to **3**.

Click **OK**.

11.

Name	Visibility	Lines
Emergency High UG	☑	- - - - - - - - - - - -
Emergency Low UG	☑	- - - - - - - - - - - -
Normal Low UG	☑	- - - - - - - - - - - -
Normal High UG	☑	- - - - - - - - - - - -

Verify that all four filters have been defined correctly.

Click **OK**.

12. Save as *ex5-4.rvt*.

Exercise 5-5:

Placing Conduits

Drawing Name: *add_conduit_2.rvt*
Estimated Time: 15 minutes

This exercise reinforces the following skills:
- ❑ Conduit Fittings
- ❑ Conduits

1. Open the **EL2A Panel** section view.

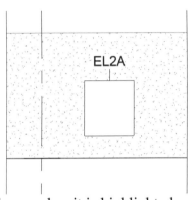

2.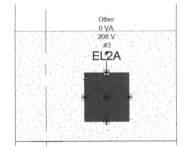

Select the panel so it is highlighted.

3.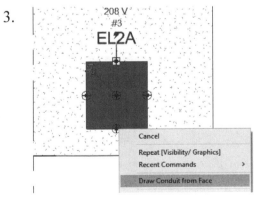

Select the bottom connector.
Right click and select **Draw Conduit from Face**.

4. 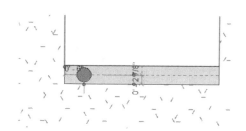 Adjust the position of the connector so it is 6" from the left side.

5. 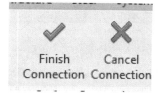 Click on **Finish Connection** on the ribbon.

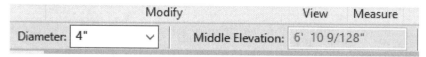

6. On the Options bar: Set the diameter to **4"**.

7. 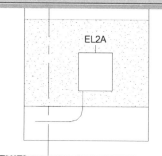 Using the Type Selector on the Properties palette: Set the Conduit Type to **Emergency High 480/277V OH**.

8. Draw the conduit down 3' 6" and to the left.

    Notice the conduit is the correct color, but not the fitting.

    Select the fitting.

9. 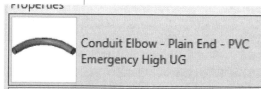 Use the Type Selector to set the fitting to **Conduit Elbow – Plain End – PVC Emergency High UG.**

10. 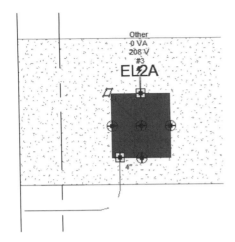 Select the panel so it is highlighted.

11. 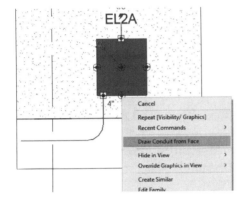 Select the bottom center connector.

Notice there is a new connector available on the bottom face.

Right click and select **Draw Conduit from Face**.

12. 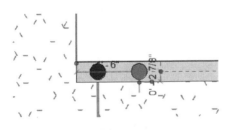 Adjust the position of the connector so it is 1' 6" from the left side.

13. 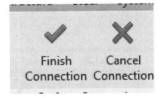 Click on **Finish Connection** on the ribbon.

14.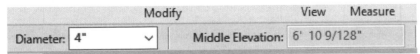

15. On the Options bar: Set the diameter to **4"**.

16.

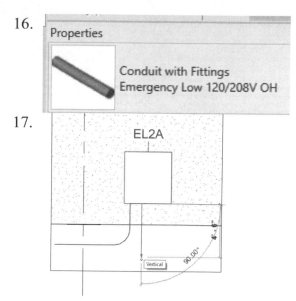

Using the Type Selector on the Properties palette:
Set the Conduit Type to **Emergency Low 120/208V OH**.

17.

Draw the conduit down 4' 6" and to the left.

Notice the conduit is the correct color, but not the fitting.

Select the fitting.

18.

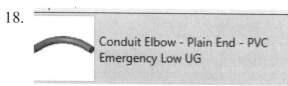

Use the Type Selector to change the fitting to **Emergency Low UG.**

The color should update.

19.

Select the panel so it is highlighted.

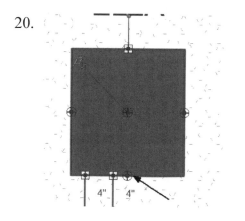

20.

Select the bottom connector that doesn't have a conduit attached.

Notice there is a new connector available on the bottom face.

Right click and select **Draw Conduit from Face**.

21.

Adjust the position of the connector so it is 2' 6" from the left side.

22. 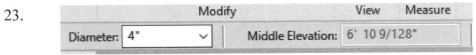 Click on **Finish Connection** on the ribbon.

23.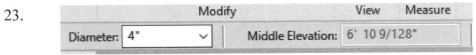

24. On the Options bar: Set the diameter to **4"**.

25. 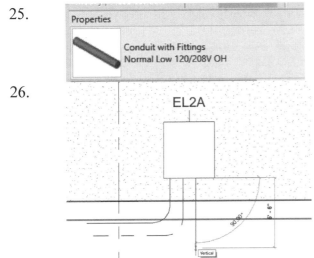 Using the Type Selector on the Properties palette:
Set the Conduit Type to **Normal Low 120/208V OH**.

26. Draw the conduit down 5' 6" and to the left.

Notice the conduit is the correct color, but not the fitting.

Select the fitting.

27. 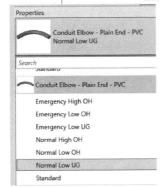 Use the Type Selector on the Properties panel to change the fitting to use the **Normal Low UG** fitting type.

28.

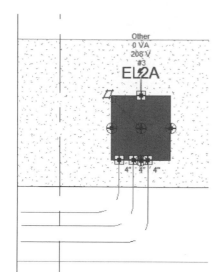

Select the panel so it is highlighted.

29.

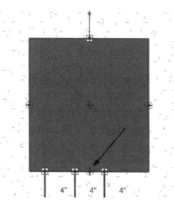

Select the bottom connector that doesn't have a conduit attached.

Notice there is a new connector available on the bottom face.

Right click and select **Draw Conduit from Face**.

30.

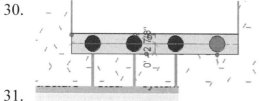

Adjust the position of the connector so it is 3' 6" from the left side.

31.

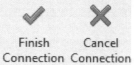

Click on **Finish Connection** on the ribbon.

32.

	Modify		View	Measure
Diameter:	4"	v	Middle Elevation: 6' 10 9/128"	

33. On the Options bar: Set the diameter to **4"**.

34.

Using the Type Selector on the Properties palette:
Set the Conduit Type to **Normal High 480/277V OH.**

35.

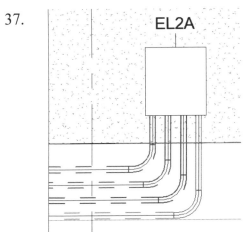

Draw the conduit down 6' 6" and to the left.

Notice the conduit is the correct color, but not the fitting.

Select the fitting.

36.

Use the Type Selector on the Properties panel to change the fitting to use the **Normal High UG** fitting type.

37.

You should have four conduits from the panel.

Each conduit should display a different color.

You can use the Type Selector to change the conduits to display as underground or overhead.

Use the Display Bar to change the Detail Level to Fine to see how the conduit display changes.

38. Save as *ex5-5.rvt*.

Exercise 5-6:

Assigning Conduit Fittings to Conduit Families

Drawing Name: *conduit_types.rvt*
Estimated Time: 15 minutes

This exercise reinforces the following skills:
- ❏ Conduit Fittings
- ❏ Conduits

1. Open the **EL2A Panel** section view.

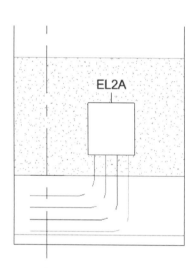

2.

In the Project Browser, navigate to the Families folder.

Locate and expand the *Conduit with Fittings* category.

3.

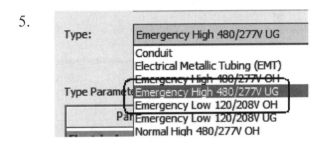

Highlight the **Emergency High 480/277V OH** family.

Right click and select **Type Properties**.

4.

Parameter	Value
Electrical	
Standard	RNC Schedule40
Fittings	
Bend	Conduit Elbow - Plain End - PVC: Emergency High OH
Tee	Conduit Junction Box - Tee - PVC: Standard
Cross	Conduit Junction Box - Cross - PVC: Standard
Transition	Conduit Junction Box - Transition - PVC: Standard
Union	Conduit Coupling - PVC: Standard
Identity Data	

Type Parameters

For the Bend parameter, use the drop-down list to select the **Emergency High OH** Conduit Elbow.

Click **Apply**.

5.

Type: Emergency High 480/277V UG

Conduit
Electrical Metallic Tubing (EMT)
Emergency High 480/277V OH
Emergency High 480/277V UG
Emergency Low 120/208V OH
Emergency Low 120/208V UG
Normal High 480/277V OH

Type Parameters

Select **Emergency High 480/277V UG** from the Type drop-down list.

6.

Type: Emergency High 480/277V UG

Type Parameters

Parameter	Value
Electrical	
Standard	RNC Schedule40
Fittings	
Bend	Conduit Elbow - Plain End - PVC: Emergency High UG
Tee	Conduit Junction Box - Tee - PVC: Standard
Cross	Conduit Junction Box - Cross - PVC: Standard
Transition	Conduit Junction Box - Transition - PVC: Standard
Union	Conduit Coupling - PVC: Standard
Identity Data	

For the Bend parameter, use the drop-down list to select the **Emergency High UG** Conduit Elbow.

Click **Apply**.

7.

Family: System Family: Conduit with Fittings

Type: Emergency Low 120/208V OH

Select **Emergency Low 120/208V OH** from the Type drop-down list.

8.

Parameter	Value
Type:	Emergency Low 120/208V OH

Type Parameters

Parameter	Value
Electrical	
Standard	RNC Schedule40
Fittings	
Bend	Conduit Elbow - Plain End - PVC: Emergency Low OH
Tee	Conduit Junction Box - Tee - PVC: Standard
Cross	Conduit Junction Box - Cross - PVC: Standard
Transition	Conduit Junction Box - Transition - PVC: Standard
Union	Conduit Coupling - PVC: Standard

For the Bend parameter, use the drop-down list to select the **Emergency Low OH** Conduit Elbow.

Click **Apply**.

9.

Family:	System Family: Conduit with Fittings
Type:	Emergency Low 120/208V UG

Select **Emergency Low 120/208V UG** from the Type drop-down list.

10.

Type:	Emergency Low 120/208V UG

Type Parameters

Parameter	Value
Electrical	
Standard	RNC Schedule40
Fittings	
Bend	Conduit Elbow - Plain End - PVC: Emergency Low UG
Tee	Conduit Junction Box - Tee - PVC: Standard

For the Bend parameter, use the drop-down list to select the **Emergency Low UG** Conduit Elbow.

Click **Apply**.

Click **OK**.

11.   Save as ex5-6.rvt.

Exercise 5-7:

Adding a Conduit

Drawing Name: *add_conduit.rvt*

Estimated Time: 30 minutes

This exercise reinforces the following skills:
- ❑ Conduit Fittings
- ❑ Conduits

1. 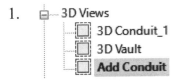   Open the **Add Conduit** view.

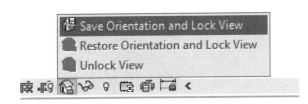

Click on the Display Bar to **Save Orientation and Lock View.**

You see that the 3D view is saved and locked. This means you can't modify this view.

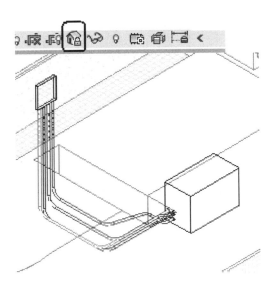

We want to place conduits to connect from the panel to the utility vault located below ground.

This is what it should look like when we are done.

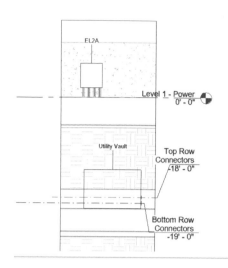

Switch to the **EL2A Panel to Utility Vault** Section view.

I have added levels on the EL2A Panel to Utility Vault section view to help you see how far down you need to bring the conduits to align them with the connectors located at the utility vault.

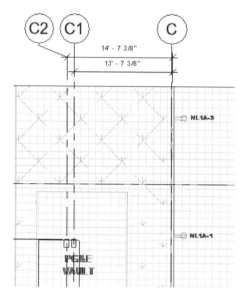

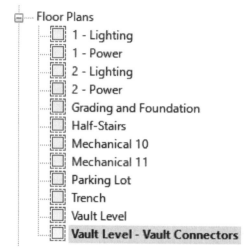

*If you activate the **Vault Level- Vault Connectors** Floor plan view, you see how far from the wall center each set of connectors is located in relation to the Vault.*

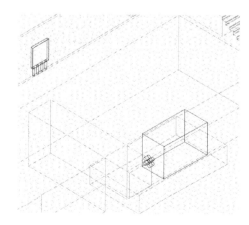

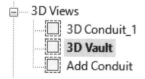

*If you open the **Add Conduit** view, you can see that the vault has two rows of connectors with two connectors on each row – a total of four connectors for the four conduits.*

2.

Why don't you see the colors that were defined for each conduit type?

In order to see the colors, a view filter must be applied to the view.

3.

Conduits : Conduit with Fittings : Normal High 120/208V OH

If you hover your mouse over each conduit, you will see the conduit type displayed.

4.

Sections (Building Section)
  EL2A Panel
  EL2A Panel to Utility Vault
  Main Electrical Room 1

Switch to the **EL2A Panel to Utility Vault** Section view.

5.

Cancel

Repeat [Move]
Recent Commands   >
Draw Conduit

Select the **Normal High** cyan conduit.

To continue an existing conduit:

Select the end point of the **Normal High** cyan conduit, right click and select **Draw Conduit**.

6.

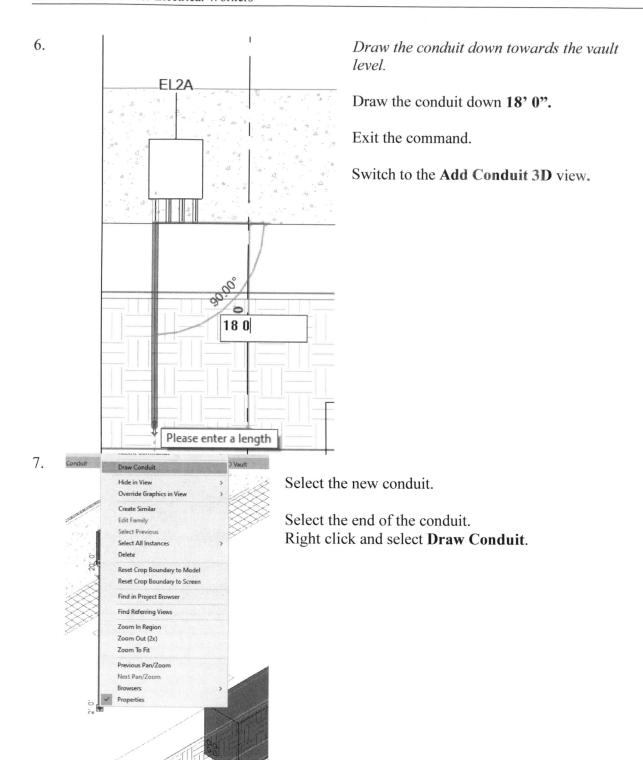

Draw the conduit down towards the vault level.

Draw the conduit down **18' 0"**.

Exit the command.

Switch to the **Add Conduit 3D** view.

7.

Select the new conduit.

Select the end of the conduit.
Right click and select **Draw Conduit**.

8.

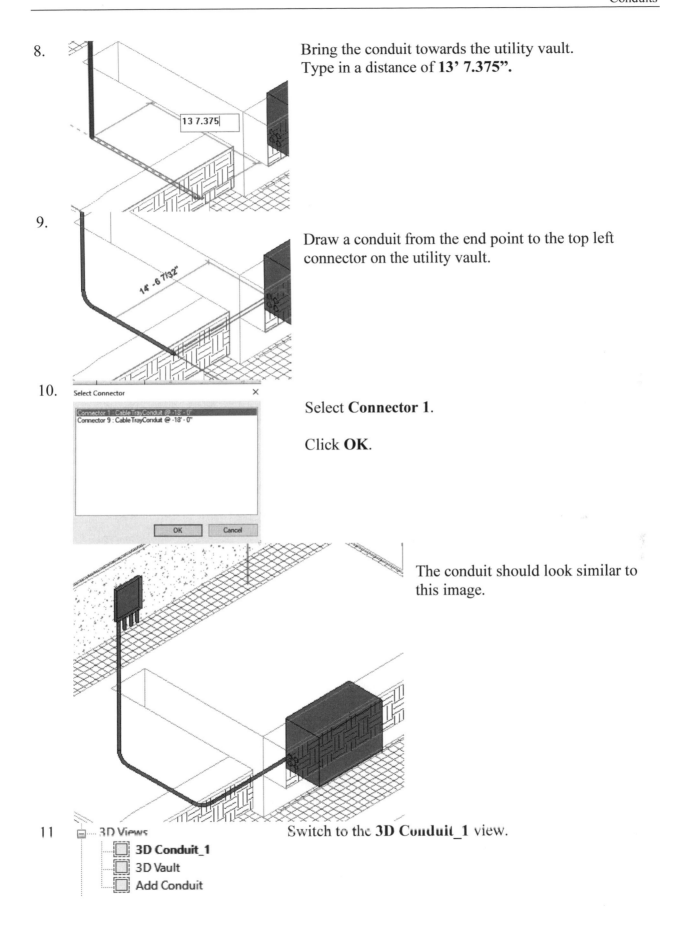

Bring the conduit towards the utility vault.
Type in a distance of **13' 7.375"**.

9.

Draw a conduit from the end point to the top left connector on the utility vault.

10.

Select **Connector 1**.

Click **OK**.

The conduit should look similar to this image.

11   Switch to the **3D Conduit_1** view.

12.

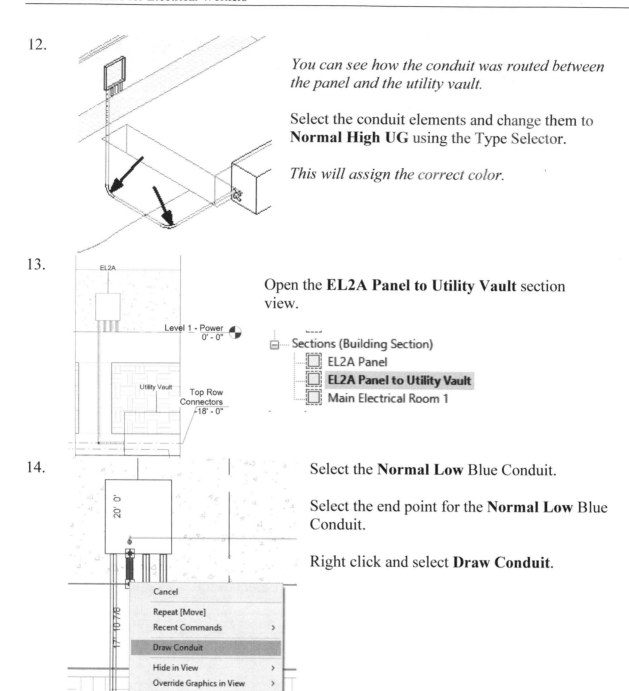

You can see how the conduit was routed between the panel and the utility vault.

Select the conduit elements and change them to **Normal High UG** using the Type Selector.

This will assign the correct color.

13.

Open the **EL2A Panel to Utility Vault** section view.

14.

Select the **Normal Low** Blue Conduit.

Select the end point for the **Normal Low** Blue Conduit.

Right click and select **Draw Conduit**.

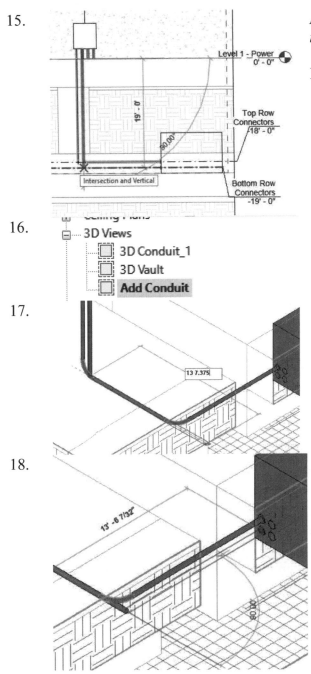

15. *Draw the conduit down towards the vault level.*

    Draw the conduit down **19' 0"**.

16. Switch to the **Add Conduit** 3D view.

17. Select the end of the conduit. Right click and select **Draw Conduit.**
    Draw the conduit **13' 7 3/8"** towards the utility vault.

18. Select the end of the conduit.
    Right click and select **Draw Conduit.**
    Draw the conduit to the connector on the left side bottom row on the utility vault.

    Exit the command.

19.

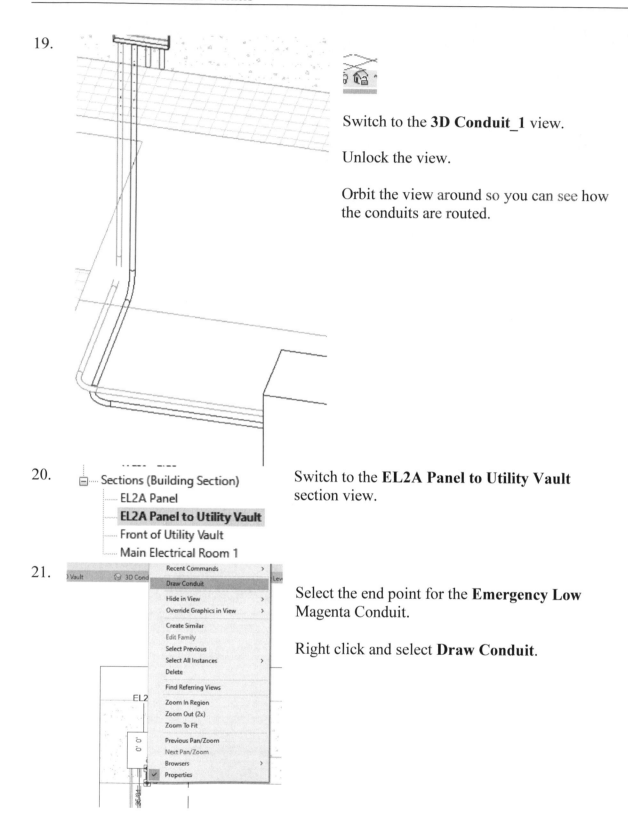

Switch to the **3D Conduit_1** view.

Unlock the view.

Orbit the view around so you can see how the conduits are routed.

20.

Switch to the **EL2A Panel to Utility Vault** section view.

21.

Select the end point for the **Emergency Low Magenta Conduit**.

Right click and select **Draw Conduit**.

22.

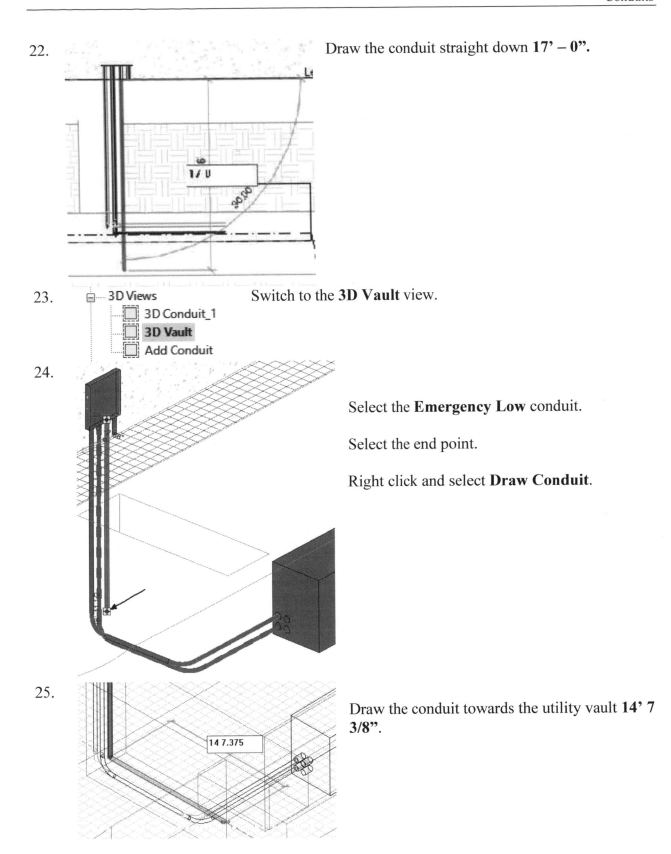

Draw the conduit straight down **17' – 0"**.

23.

Switch to the **3D Vault** view.

3D Views
- 3D Conduit_1
- **3D Vault**
- Add Conduit

24.

Select the **Emergency Low** conduit.

Select the end point.

Right click and select **Draw Conduit**.

25.

Draw the conduit towards the utility vault **14' 7 3/8"**.

14 7.375

26.

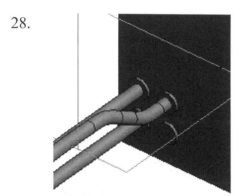

Select the upper right connector on the vault.

27.

Select Connector

Connector 5 : CableTrayConduit @ -19' - 0"
Connector 9 : CableTrayConduit @ -18' - 0"

A dialog should appear allowing you to select which connector to use.
Select **Connector 9**.

Click **OK**.
If you see an error, just dismiss it.

28.

Revit will auto-route the connection.

29.

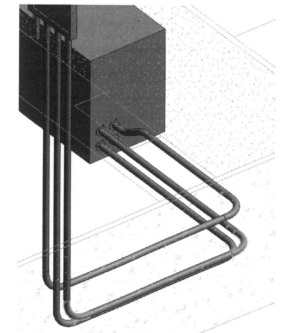

Orbit around the view to inspect the conduit connections.

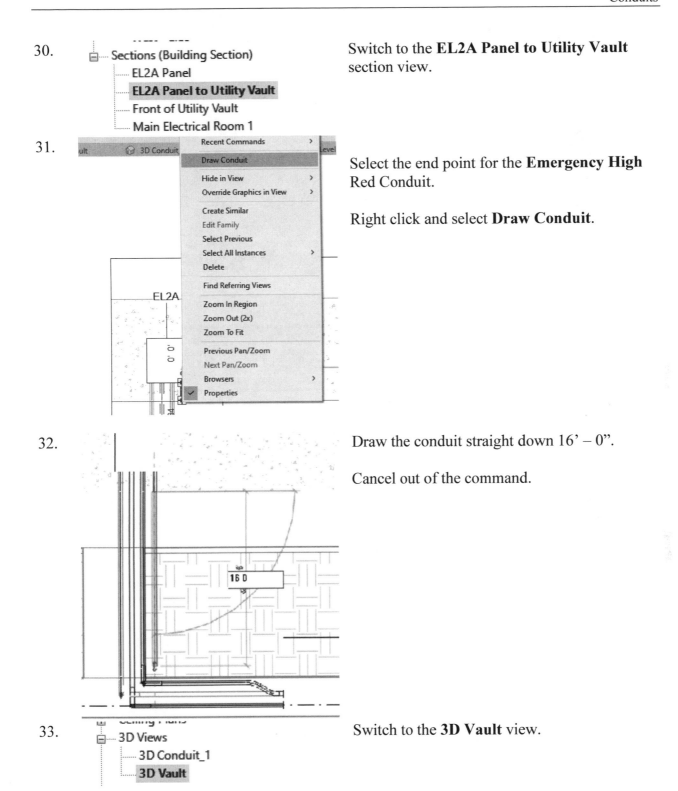

30. Sections (Building Section)
     ── EL2A Panel
     ── **EL2A Panel to Utility Vault**
     ── Front of Utility Vault
     ── Main Electrical Room 1

Switch to the **EL2A Panel to Utility Vault** section view.

31.

Select the end point for the **Emergency High Red Conduit**.

Right click and select **Draw Conduit**.

32.

Draw the conduit straight down 16' – 0".

Cancel out of the command.

33. 3D Views
     ── 3D Conduit_1
     ── **3D Vault**

Switch to the **3D Vault** view.

34.

Select the Emergency High conduit.

Select the end point.

Right click and select **Draw Conduit**.

35.

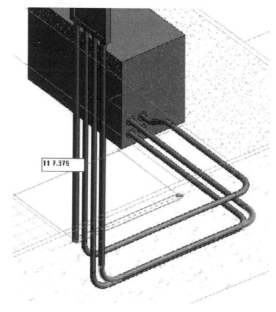

Draw the conduit towards the utility vault **11' 7 3/8"**.

36.

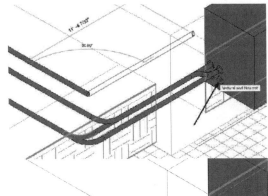

Continue drawing the conduit.
Select the lower right connector on the utility vault.

37.

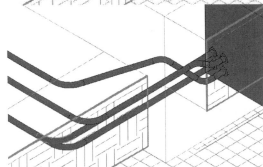

Revit will auto-route the connection.

Cancel out of the command.

38.

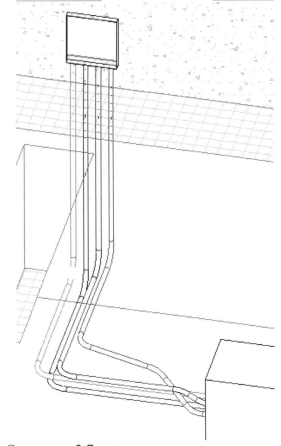

3D Views
　　3D Conduit_1
　　3D Vault
　　Add Conduit

Switch to the **3D Conduit_1** view.

Can you change the types of conduits applied to the elbows to the correct type of conduit using the Type Selector?

39. Save as *ex5.7.rvt*.

Exercise 5-8:

Adding Parallel Conduits

Drawing Name: *parallel_conduits.rvt*

Estimated Time: 15 minutes

This exercise reinforces the following skills:
- ❑ Conduit Fittings
- ❑ Conduits

1.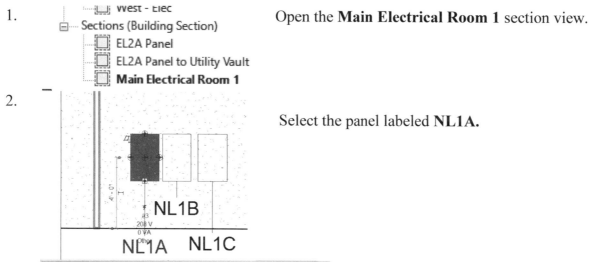
    Open the **Main Electrical Room 1** section view.

2. Select the panel labeled **NL1A.**

3. Select **Edit Type** on the Properties panel.

4. Change the Width to **4' 0".**

    Click **OK**.

    The panel size adjusts.

5.

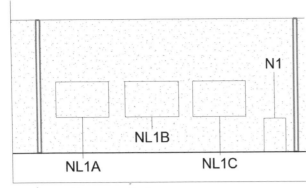

Adjust the position of the panels so they fit inside the room properly.

Select the panel labeled **NL1A.**

6.

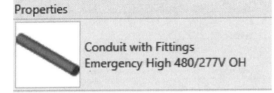

Select the top connector on the panel.
Right click and select **Draw Conduit**.
Adjust the position of the connector 6" from the left side.

7.

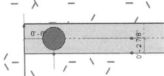

Select **Finish Connection** on the ribbon.

8.

Select the Conduit to **Emergency High 480/277V OH** using the Type Selector on the Properties panel.

9.

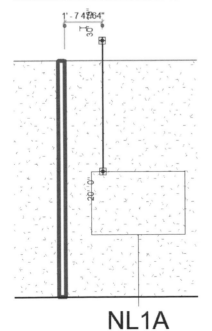

Draw the conduit straight up to the ceiling.

Click Cancel to exit the command.

10.

Switch to the Systems tab on the ribbon.

Select **Parallel Conduits**.

11.
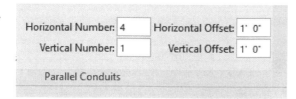
On the ribbon:
Set the Horizontal Number to 4.
Set the Horizontal Offset to 1' 0".
Set the Vertical Number to 1.
Set the Vertical Offset to 1' 0".

12.
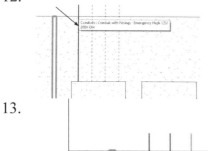
Select the conduit that was placed and check the preview placement of the new conduits.

Left click to place the new conduits.

13.
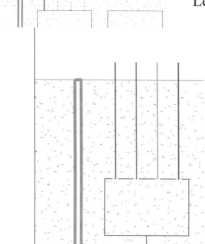
The conduits that were placed are all using the Emergency High conduit type.

Use the Type Selector to change the conduits so that they are from left to right:

- Emergency High (Red)
- Emergency Low (Magenta)
- Normal High (Cyan)
- Normal Low (Blue)

14.
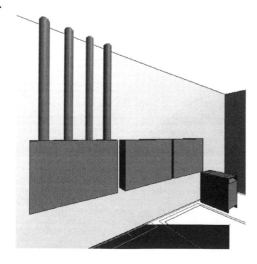

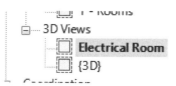

Switch to the **Electrical Room** 3D view to inspect your conduits.

Save as *ex5-8.rvt*.

View Templates

A view template is a collection of view properties, such as view scale, discipline, detail level, and visibility settings.

Use view templates to apply standard settings to views. View templates can help to ensure adherence to office standards and achieve consistency across construction document sets.

Before creating view templates, first think about how you use views. For each type of view (floor plan, elevation, section, 3D view, and so on), what styles do you use? For example, an architect may use many styles of floor plan views, such as power and signal, partition, demolition, furniture, and enlarged.

You can create a view template for each style to control settings for the visibility/graphics overrides of categories, view scales, detail levels, graphic display options, and more.

Exercise 5-9:

Using View Templates

Drawing Name: *conduit_view_templates.rvt*

Estimated Time: 15 minutes

This exercise reinforces the following skills:
- ❑ Visibility Graphics Overrides
- ❑ Visibility Graphics Filters
- ❑ View Templates

1.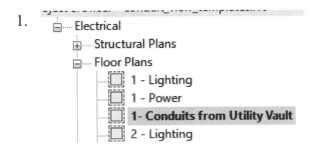

Open the **Conduits from Utility Vault** floor plan.

2.

You can see the conduits, but you don't see any of the colors applied.

The colors are controlled by filters in the Visibility/Graphics dialog. Each view needs to be set up to have the overrides applied or you can save the view settings to a view template and apply it to any view.

3.

Sections (Building Section)
- EL2A Panel
- **EL2A Panel to Utility Vault**
- Main Electrical Room 1

Open the **El2A Panel to Utility Vault** section view.

The conduits in this view are displaying colors.

4.

Scope Box	None
Identity Data	
View Template	Conduit Overrides
View Name	EL2A Panel to Utility Vault
Dependency	Independent

Look under Identity Data in the Properties panel.

A View Template called Conduit Overrides has been applied to this view.

5.

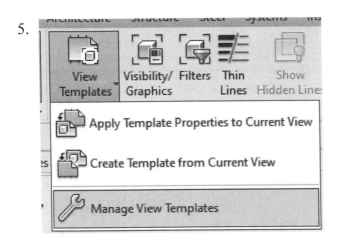

Switch to the **View** ribbon.

Under View Templates, select **Manage View Templates**.

6.

Parts Visibility	Show Original	☑
V/G Overrides Model	Edit...	☑
V/G Overrides Annotation	Edit...	☐
V/G Overrides Analytical Mod	Edit...	☑
V/G Overrides Import	Edit...	☑
V/G Overrides Filters	Edit...	☑

Highlight Conduit Overrides.

Notice that there is no check next to the Annotations category.

Notice the Detail Level is set to Fine.

Click on **Edit** next to **Filters**.

7.

Name	Enable Filter	Visibility	Proje Lines
Emergency High UG	☑	☑	--------
Emergency Low UG	☑	☑	--------
Normal Low UG	☑	☑	--------
Normal High UG	☑	☑	--------
Emergency High OH	☑	☑	————
Emergency Low OH	☑	☑	————
Normal High OH	☑	☑	————
Normal Low OH	☑	☑	————

Notice the filters that have been applied in this view template.

Click **OK**.

Click **OK** to close the dialog.

8.

- Electrical
  - Structural Plans
  - Floor Plans
    - 1 - Lighting
    - 1 - Power
    - 1- Conduits from Utility Vault
    - 2 - Lighting

Open the **Conduits from Utility Vault** floor plan.

9.

Locate the **View Template** parameter under Identity Data.

Click on the **<None>** button.

10.

Set the View type filter to all.

Highlight the **Conduit Overrides** template.

11.

Uncheck all the parameters to include.

Place a check next to **V/G Overrides Filters**.

Click **Apply**.

Click **OK** to close the dialog.

12.

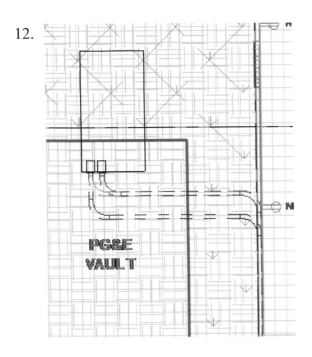

The conduits now display the assigned colors.

Save as *ex5-9.rvt*.

Exercise 5-10:

Create a Conduit Run Schedule

Note: You cannot create a conduit run schedule using Conduits with Fittings. You must use Conduits without Fittings families. Fittings will still be automatically added to the conduit runs.

Drawing Name: *conduit_schedules..rvt*

Estimated Time: 15 minutes

This exercise reinforces the following skills:
- ❑ Parameters
- ❑ Schedules
- ❑ Conduits

1.

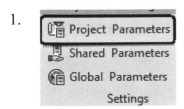

Activate the Manage ribbon.

Select **Project Parameters**.

2.

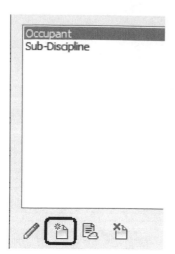

Select **New.**

3.

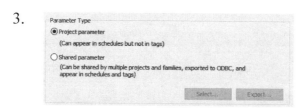

Verify that Project parameter is enabled.

Note that project parameters are only available in this file.

4.

Type **Conduit Run Name** for the Name.
Set the Type of Parameter to **Text.**
Group the parameter under **Text.**
Enable **Instance**.

5.

☐ Columns
☐ Communication Devices
☑ Conduit Fittings
☑ Conduit Runs
☑ Conduits
☐ Curtain Panels

Enable **Conduit Fittings, Conduit Runs**, and **Conduits** in the Categories list.

Click **OK**.

6.

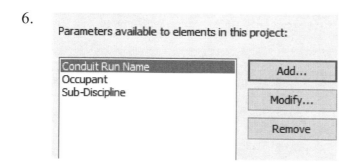

The parameter is listed and will now be added to the enabled categories.

Click **OK.**

7. 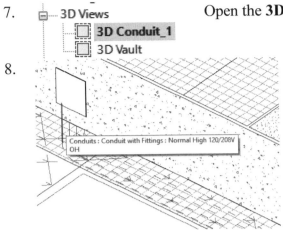 Open the **3D Conduit_1** view.

8. Select the start of the Normal High (Cyan) Conduit run.

9. 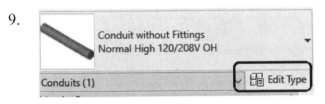 Click **Edit Type** on the Type Selector.

10. 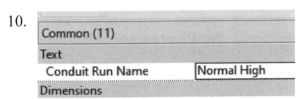 Type **Normal High** for the Conduit Run Name in the Properties palette.

Click **Apply**.

Make sure you Click ENTER after you type in the name or the value may not "stick".
Click **OK** to release your selection.

Click **ESC** to release your selection.
In the Project Browser:

11. 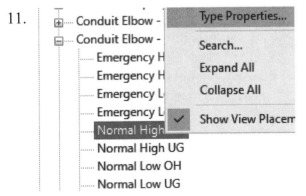 Locate the **Normal High OH** type for the Conduit Elbow – Plain End- PVC.
Right click and select **Type Properties**.

12.

Family:	Conduit Elbow - Plain End - PVC
Type:	Normal High OH

Type Parameters

Parameter	
Constraints	
Default Elevation	0' 0"
Text	
Conduit Run Name	Normal High

Type **Normal High** for the Conduit Run Name in the Properties palette.

Click **OK**.

13.

Select the start of the Normal Low (Blue) Conduit run.

Click **Edit Type** on the Type Selector.

14.

Common (11)	
Text	
Conduit Run Name	Normal Low
Dimensions	
Size	

Type **Normal Low** for the Conduit Run Name in the Properties palette.
Click.
OK
Click **ESC** to release your selection.

15.

Select the start of the **Emergency Low** (Magenta) Conduit run.

Click **Edit Type** on the Type Selector.

16.

Common (11)	
Text	
Conduit Run Name	Emergency Low
Dimensions	

Type **Emergency Low** for the Conduit Run Name in the Properties palette.

Click **OK**.

Click **ESC** to release your selection.

17.

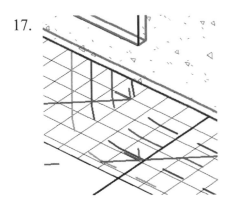

Select the start of the Emergency High (Red) Conduit run.

Click **Edit Type** on the Type Selector.

18.

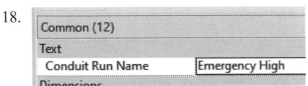

Type **Emergency High** for the Conduit Run Name in the Properties palette.
Click **OK**.

Click **ESC** to release your selection.

19. Switch to the **View** ribbon.

Select **Schedules/Quantities**.

20.

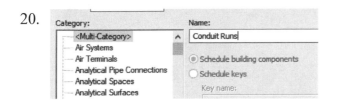

Highlight **Multi-Category**.in the Category pane.

Type **Conduit Runs** in the Name field.

Click **OK.**

21. Select the following fields:

Scheduled fields (in order):

Conduit Run Name
Family and Type
Length

- Conduit Run Name
- Family and Type

22.

Fields Filter Sorting/Grouping Formatting Appearance

Filter by: Conduit Run Name   has no value

And: (none)

Select the **Filter** tab.

Filter by **Conduit Run Name**
Has no value.

This means any elements with a Conduit Run Name assigned won't be included in the schedule.

23.

Select the Sorting/Grouping tab.

Set Sort by: **Conduit Run Name**.

Click **OK**.

24.

Scroll down to the bottom of the schedule.
Locate the conduit elements.

The Conduit Runs will be listed.
Type in the values for the Conduit Run Names using the values shown in the Family and Type Column.

As you assign run names, those rows will disappear because they no longer meet the filter setting.

25.

A dialog will appear explaining that if you assign the new Conduit Run Name to the type of conduit, it will be applied to all conduits of the same type.

Click **OK**.

26.

Switch to the View ribbon.

Select **Schedules/Quantities**.

27.

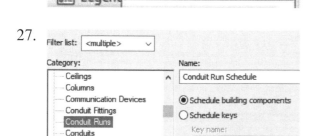

Highlight **Conduit Runs**.

Change the Name to **Conduit Run Schedule**.

Click **OK**.

28.

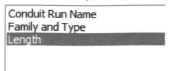

On the Fields tab:
Select:
- Conduit Run Name
- Family and Type
- Length

29.

Select the Filter tab.

Change the Filter by: to **None**.

Click OK.

30.

Select the Formatting tab.

Highlight **Length.**

Select **Calculate totals** from the drop-down list.

31.

Select the Sorting/Grouping tab.
Sort by: **Conduit Run Name**
Enable **Footer**.
Select **Totals only** from the drop-down list.

Click **OK**.

32.

The total conduit run length is displayed for each conduit type.

33. Save as *ex5-10.rvt*

Exercise 5-11:

Create a Conduit Saddle

Drawing Name: *conduit_saddle.rvt*

Estimated Time: 25 minutes

This exercise reinforces the following skills:
- ❑ Conduits
- ❑ Electrical Fixtures

1.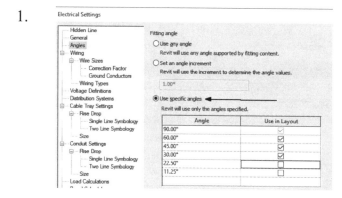

Type **ES** to open the Electrical Settings palette.

Highlight **Angles**.

Enable **Use specific angles**.

90° is enabled by default.
Enable 60°, 45°, and 30°.

Click **OK.**

Open the **Lab 101** section view.

2.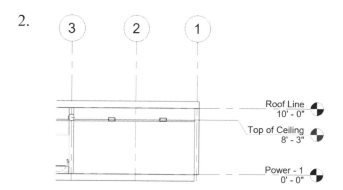

Notice that the top of the ceiling is located at 8' 3" above Level 1.

3.

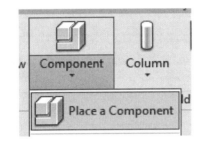

Ceiling Plans
1 - Top of Ceiling

Open the **1 – Top of Ceiling** ceiling plan under Power.

Click on the view in the Project Browser. *The level in the section view is a reference level.*

Locate the intersection of Grids A and 2.

We are going to place a junction box at this intersection.

4.

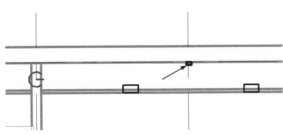

Select the **Place a Component** tool from the Architecture ribbon.

It is useful to use this tool if you don't know how an element is defined.

5.

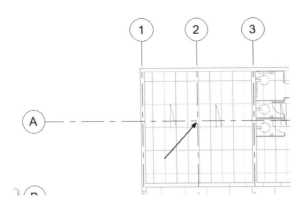

Select the **Load w Conduit Connectors 4" Square 120 Junction Box**.

Set the Elevation from Level to **9' 9 7/8"**.

This places it above the piping.

Place the component at A2.

West - Elec
Sections (Building Section)
Conduit over Pipes
Lab 101

6.

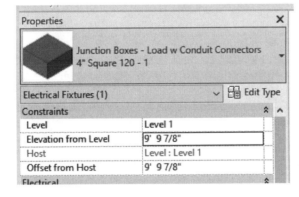

Switch to the Lab 101 Section view and verify the placement of the junction box.

It should be attached to the roof joist.

7.

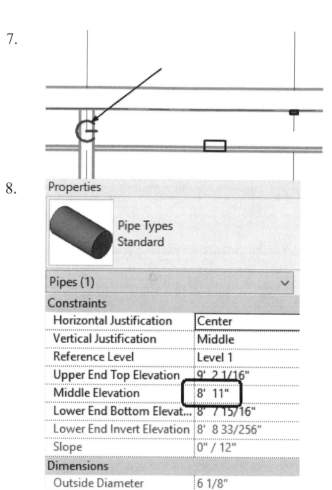

The green circle on the left is the pipe.

Select the pipe.

8.

On the Properties palette:

Verify that the pipe has a diameter of 6" and has a middle elevation of 8' 11".

Our conduit needs to go below this pipe.

Properties	
Pipe Types Standard	
Pipes (1)	
Constraints	
Horizontal Justification	Center
Vertical Justification	Middle
Reference Level	Level 1
Upper End Top Elevation	9' 2 1/16"
Middle Elevation	8' 11"
Lower End Bottom Elevat...	8' 7 15/16"
Lower End Invert Elevation	8' 8 33/256"
Slope	0" / 12"
Dimensions	
Outside Diameter	6 1/8"
Inside Diameter	5 95/128"
Size	6"ø
Length	13' 9 11/16"

9.

If you look on the Properties palette, you see the bottom of the pipe is located at 8' 7 15/16".

The top of the ceiling is at 8' 3".

This means we can place the bottom part of the conduit saddle at 8' 5".

10.

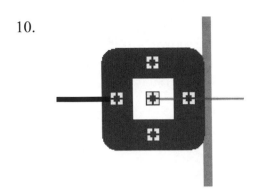

Switch back to the **1 – Top of Ceiling** view.

Select the junction box to activate the connectors.

You should see five connectors – one at the top and one on each side.

11.

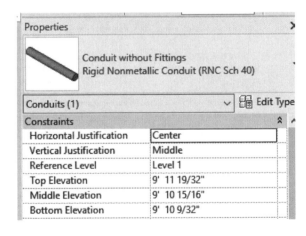

Select the south/bottom connector.

Right click and select **Draw Conduit**.

12.

On the Option bar:
Set the conduit diameter to **1"**.
Set the Middle Elevation to **9' 10 15/16"**.

13.

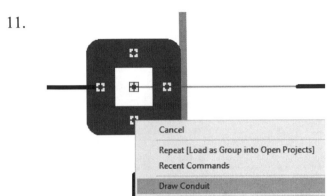

Use the Type Selector to place a **Conduit without Fittings (RNC Sch 40)**.

14.

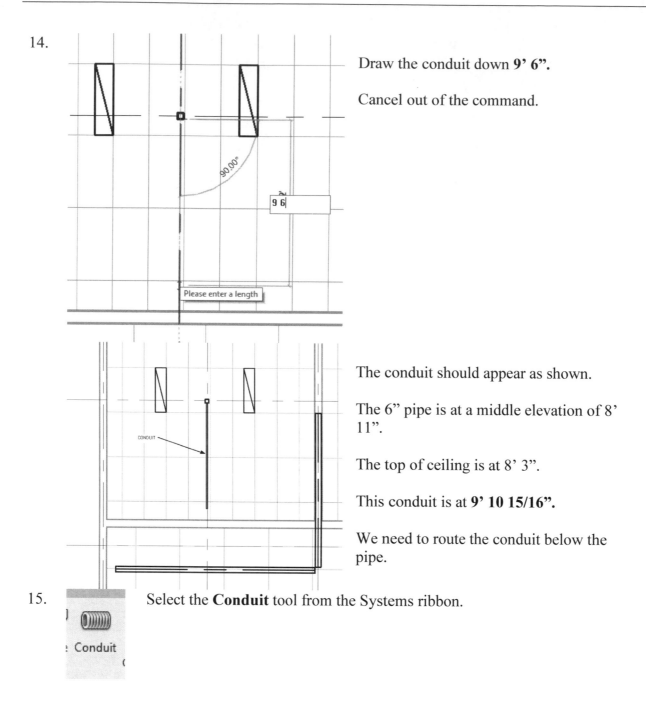

Draw the conduit down **9' 6"**.

Cancel out of the command.

The conduit should appear as shown.

The 6" pipe is at a middle elevation of 8' 11".

The top of ceiling is at 8' 3".

This conduit is at **9' 10 15/16"**.

We need to route the conduit below the pipe.

15.　　　　Select the **Conduit** tool from the Systems ribbon.

16.

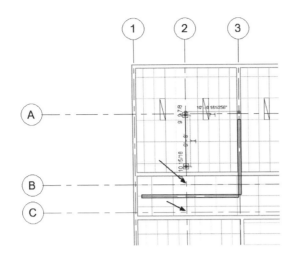

Draw the conduit from B2 to C2.

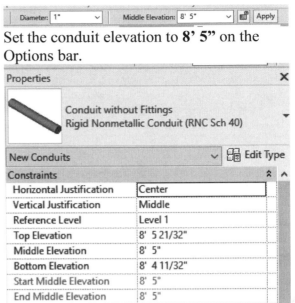

Set the conduit elevation to **8' 5"** on the Options bar.

If you look on the Properties palette, you can confirm the elevation.

17.

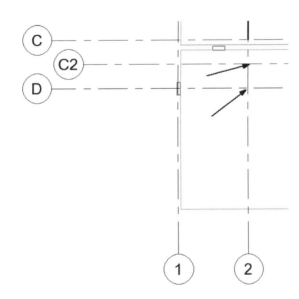

Draw a conduit from C2-2 to D2 at an elevation of **9' 10 15/16"**.

Type **CN** to start the Conduit command.

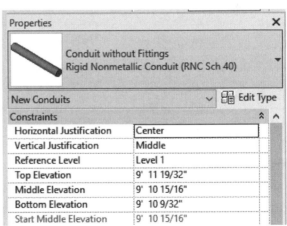

18.

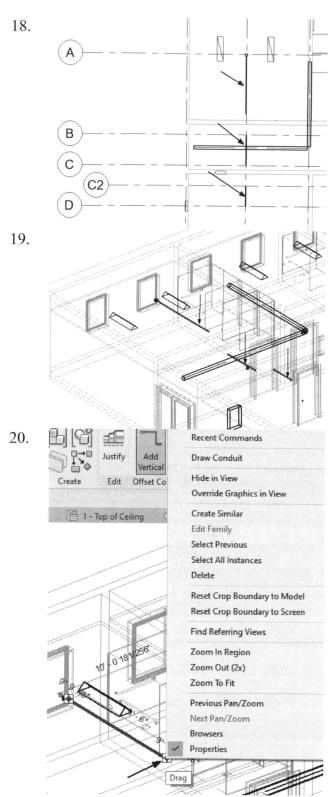

At this point, you should have three sections of conduit placed.

Two of the conduits are at 9' 10 15/16" and one which is located beneath the pipe at 8' 5".

Switch to the **3D Elec** view.

19.

Locate the three conduit sections in the view.

20.

Select the end of the first conduit section.

Right click and select **Draw Conduit**.

21.

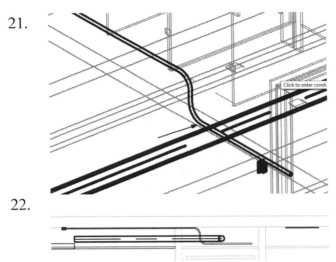

Select the end point of the lower conduit.

Revit will automatically create the bend and add the appropriate fittings based on the angles which were enabled in Electrical Settings.

Exit the command.

22.

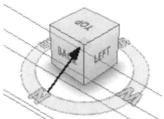

Use the Viewcube to switch to a Left view and you can see that the conduit is routed below the pipe.

23.

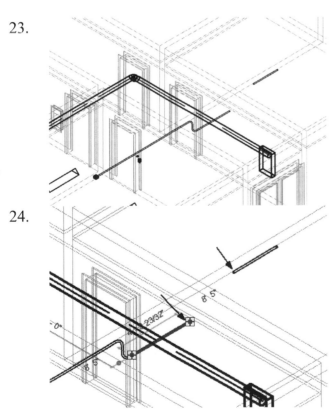

Click on the upper corner of the Viewcube to restore the 3D view.

24.

Select the end point of the lower conduit. Right click and select **Draw Conduit**.

Select the end point of the third conduit.

25.

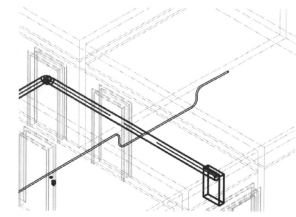

The saddle is created.

Use **Orbit** and/or the Viewcube to inspect the conduit.

Save as *ex5-11.rvt*.

Challenge Exercise 1:

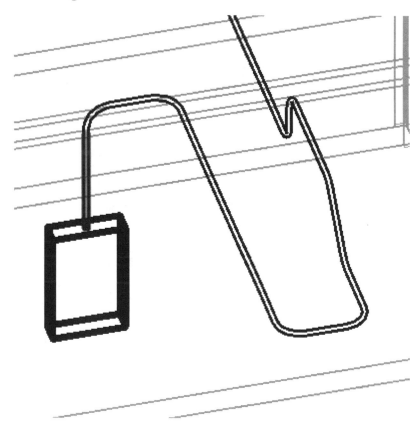

See if you can connect the top connector on the LP1-A panel to the conduit you created.

Hint: *Make sure you provide plenty of space for fittings to be added.*

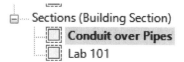

Switch to the **Conduit over Pipes** section view and draw a section of conduit straight up 3'3".

Then try to connect the two ends of conduit.

Exercise 5-12:

Create a Conduit Roll

Drawing Name: *conduit_roll.rvt*

Estimated Time: 20 minutes

This exercise reinforces the following skills:
- ❑ Conduits
- ❑ Electrical Fixtures

1.

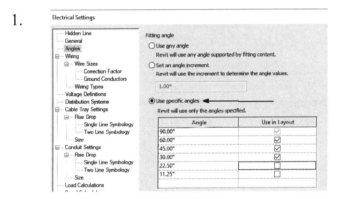

Type **ES** to open the Electrical Settings palette.

Highlight **Angles**.

Enable **Use specific angles**.

90° is enabled by default.
Enable 60°, 45°, and 30°.

Click **OK.**

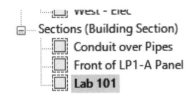

Open the **Lab 101** section view.

Notice that the top of the ceiling is located at 8' 3" above Level 1.

2.

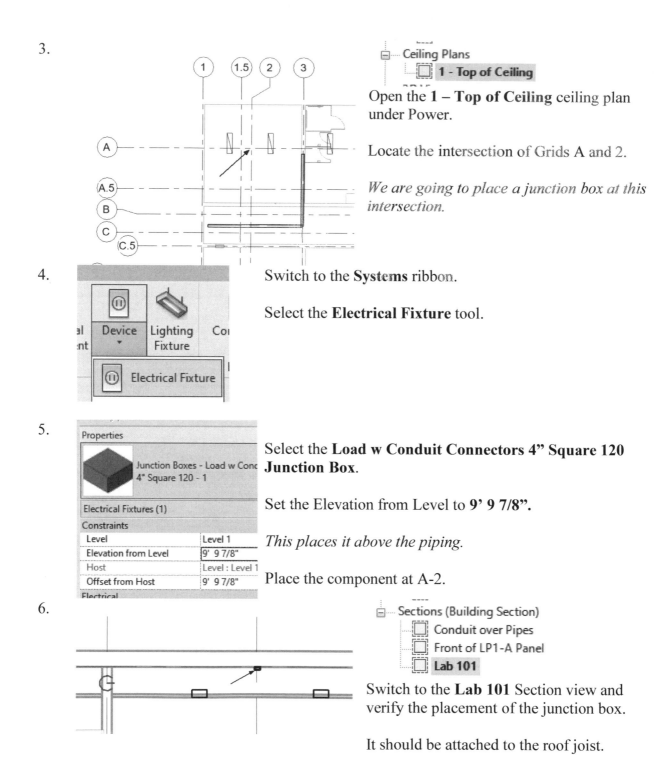

3.

Open the **1 – Top of Ceiling** ceiling plan under Power.

Locate the intersection of Grids A and 2.

We are going to place a junction box at this intersection.

4.

Switch to the **Systems** ribbon.

Select the **Electrical Fixture** tool.

5.

Select the **Load w Conduit Connectors 4" Square 120 Junction Box**.

Set the Elevation from Level to **9' 9 7/8"**.

This places it above the piping.

Place the component at A-2.

6.

Switch to the **Lab 101** Section view and verify the placement of the junction box.

It should be attached to the roof joist.

7.

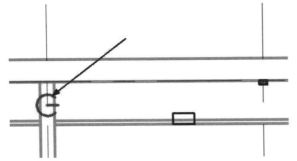

The green circle on the left is the pipe.

Select the pipe.

8.

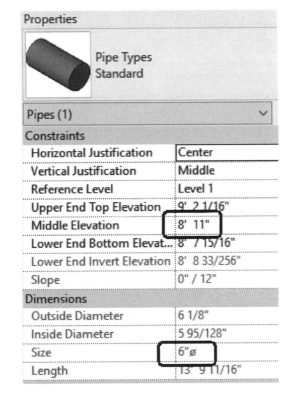

On the Properties palette:
See that the pipe has a diameter of 6" and has a middle elevation of 8' 11".

Our conduit needs to go below this pipe.

9.

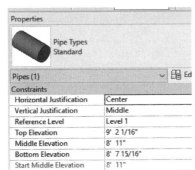

If you look on the Properties palette, you see the bottom of the pipe is located at 8' 7 15/16".

The top of the ceiling is at 8' 3".

This means we can place the bottom part of the conduit saddle at 8' 5".

10.

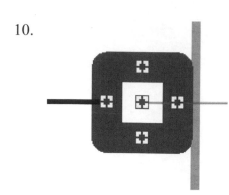
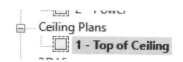

Switch back to the **1 – Top of Ceiling** view.

Select the junction box to activate the connectors.

You should see five connectors – one at the top and one on each side.

11.

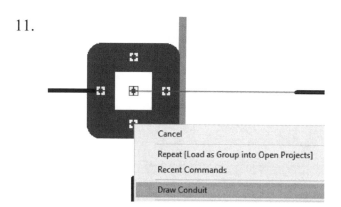

Select the south/bottom connector.

Right click and select **Draw Conduit.**

12.

On the Option bar:
Set the conduit diameter to **1"**.
Set the Middle Elevation to **9' 10 15/16".**

13.

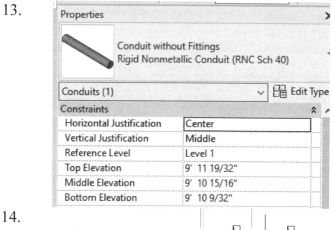

Use the Type Selector to place a **Conduit without Fittings (RNC Sch 40).**

14.

Draw the conduit straight down to the A.5 grid lines.

Left click on the A.5 grid line to end this conduit section.

Click to end this conduit section.

15.

Diameter: 1"   Middle Elevation: 8' 5"

On the Options bar:

Change the conduit elevation to **8' 5"**.

16.

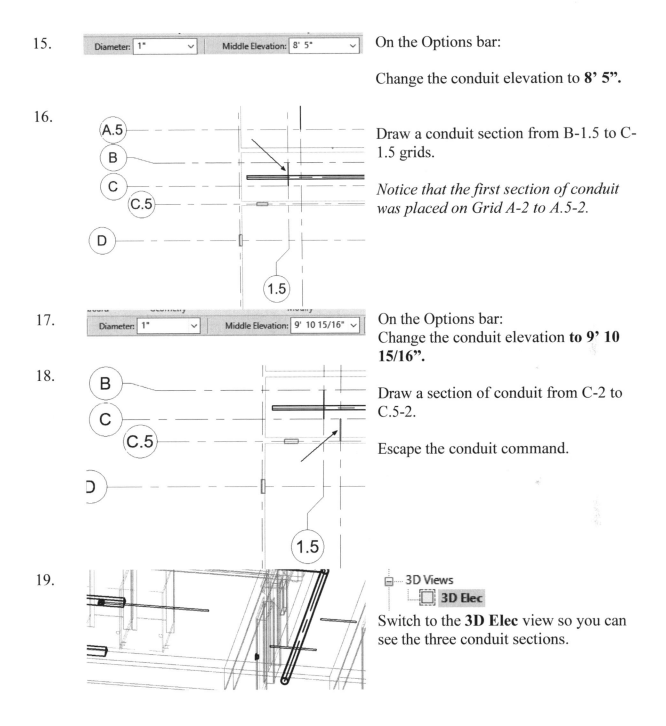

Draw a conduit section from B-1.5 to C-1.5 grids.

Notice that the first section of conduit was placed on Grid A-2 to A.5-2.

17.

Diameter: 1"   Middle Elevation: 9' 10 15/16"

On the Options bar:
Change the conduit elevation **to 9' 10 15/16"**.

18.

Draw a section of conduit from C-2 to C.5-2.

Escape the conduit command.

19.

⊟ 3D Views
    ☐ **3D Elec**

Switch to the **3D Elec** view so you can see the three conduit sections.

20.

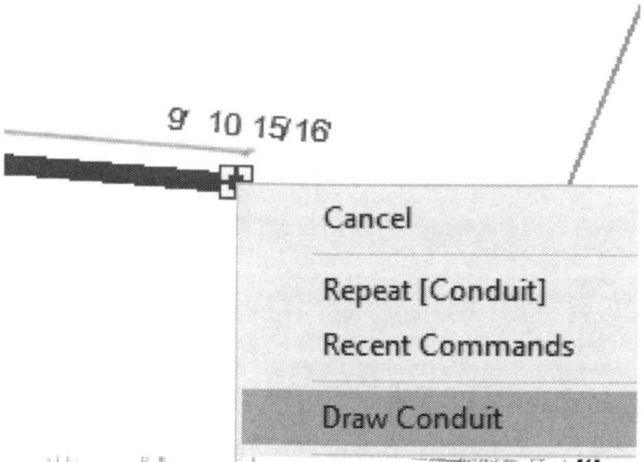

9 10 15 16

Cancel

Repeat [Conduit]
Recent Commands

Draw Conduit

Select the end point of the first conduit section.

Right click and select **Draw Conduit**.

21.

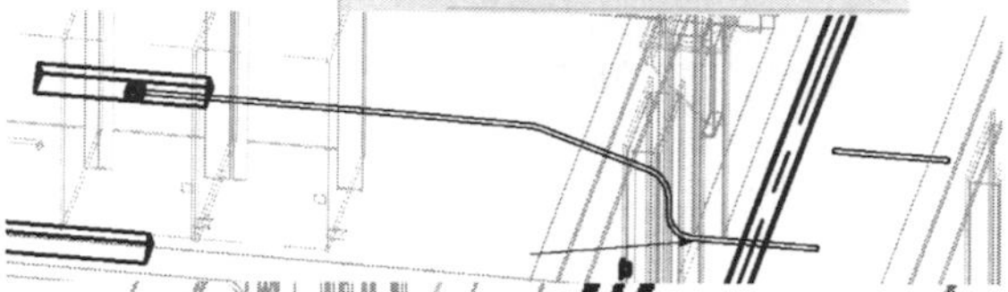

Select the end point of the conduit located below the pipe.

22.

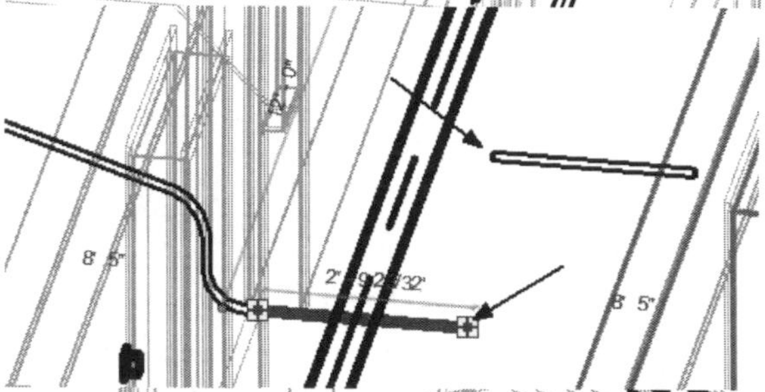

8 5"

2' 9½" 32"

3 5"

Select the end point of the lower conduit section.

Right click and select **Draw Conduit**.

Select the endpoint of the third conduit section.

23.

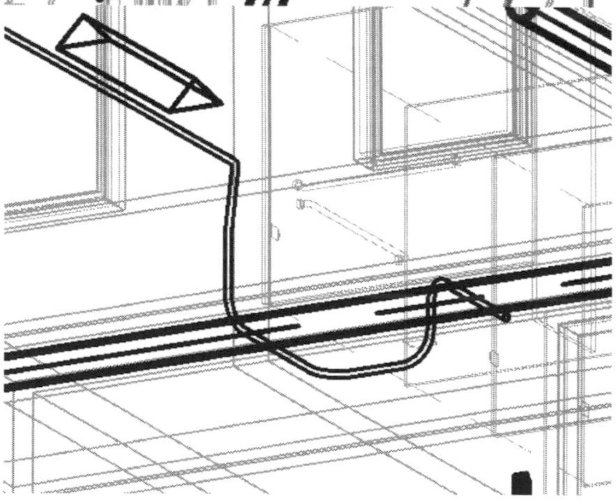

Orbit around to inspect how the conduit was routed.

Save as *ex5-12.rvt*.

Challenge Exercise 2:

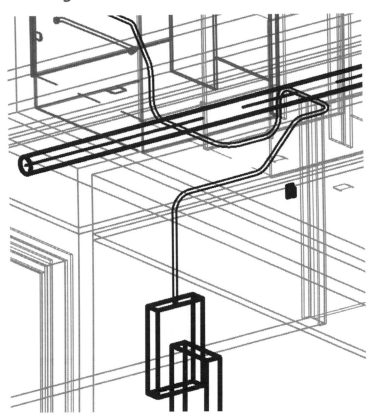

See if you can connect the top connector on the LP1-A panel to the conduit you created.

Hint: *Make sure you provide plenty of space for fittings to be added.*

Switch to the Front of LP1-A Panel section view and draw a section of conduit straight up 3'3".

Then try to connect the two ends of conduit.

Exercise 5-13:

Place a Conduit through a Pipe

Drawing Name: *pipe and conduit.rvt*

Estimated Time: 10 minutes

This exercise reinforces the following skills:
- Conduits
- Electrical Fixtures

1.

Open the 3D view.

There is a 6" pipe and a 6" thick concrete slab.

We want to run a conduit through the pipe.

2.
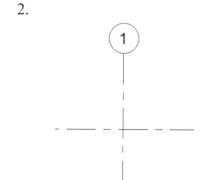

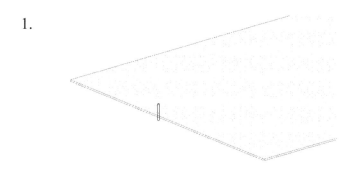

Switch to an **East – Elec** elevation.

The pipe is aligned with Grid 1.

3.
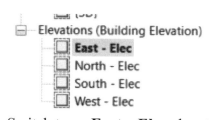

Set the Detail Level to **Fine**.
Set the Display to **Wireframe**.

This will allow you to see the pipe.

4.

    Conduit

    Select the **Conduit** tool from the Systems ribbon.

5. | Diameter: 4" ⌄ | Middle Elevation: 9' 0" |

    Set the conduit diameter to 4" on the Options bar.

6.

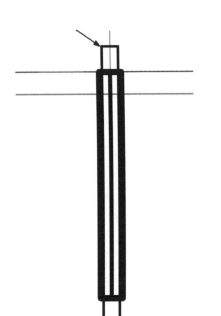

Draw the conduit centered on the pipe.

Draw the conduit so it goes through the pipe.

You can use the grips to extend the conduit so it is outside the pipe.

7.

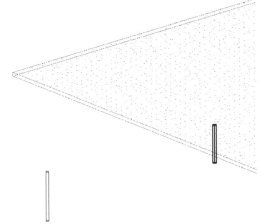

⊟ 3D Views
　　🗔 (3D)

Switch to a 3D view.

The conduit is actually located away from the pipe.

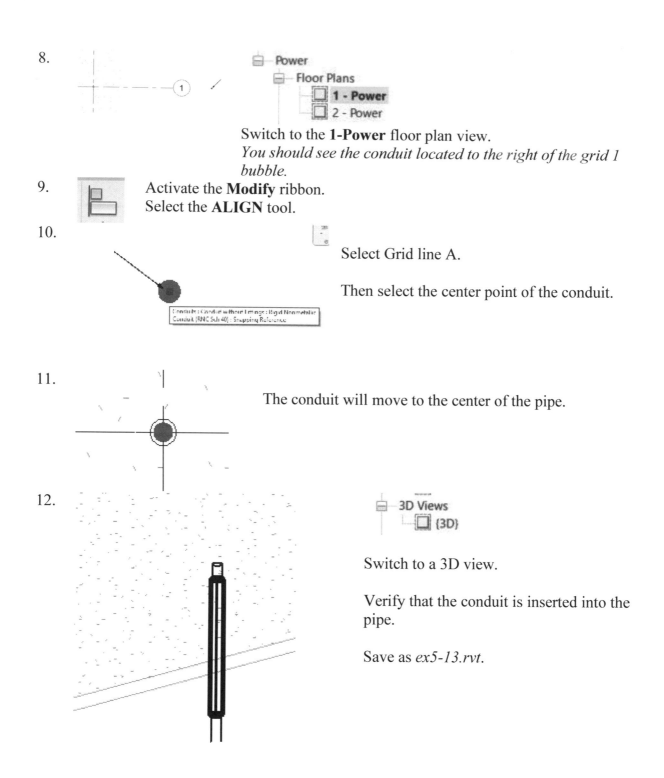

8.

Switch to the **1-Power** floor plan view.
You should see the conduit located to the right of the grid 1 bubble.

9. Activate the **Modify** ribbon.
Select the **ALIGN** tool.

10. Select Grid line A.

Then select the center point of the conduit.

11. The conduit will move to the center of the pipe.

12. Switch to a 3D view.

Verify that the conduit is inserted into the pipe.

Save as *ex5-13.rvt*.

Lab Exercises

Open *Electrical Project_conduit.rvt*.

Open the Manage Links dialog.

Reload the linked files: Architectural.rvt and Structural.rvt.

Open the Electrical Power, Floor Plan Level 1.

Zoom in on the ELECTRICAL Room 107.

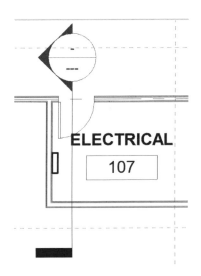

Create a section view to see an elevation of the electrical panel.

Open the new section.

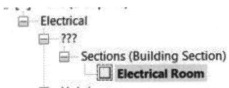

Rename it **Electrical Room** elevation.

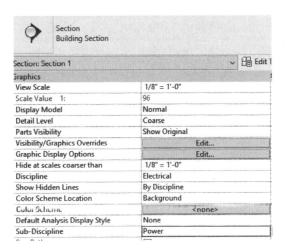

Change the sub-discipline of the view to Power so it sorts correctly in the project browser.

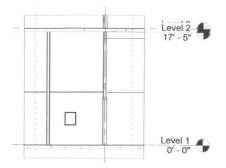

Adjust the crop region for the view.

Set the Detail Level to **Fine**.

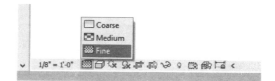

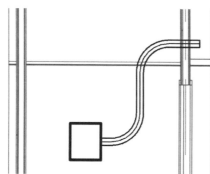

Select the panel so you can see the connectors.

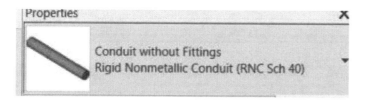

Locate the connector using the right side connector.

Change the conduit diameter to 4".

Draw a conduit as shown.

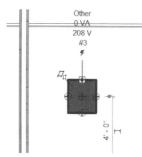

Select the conduit that was just placed.

Define four new conduit types called

- **Emergency High – OH**
- **Emergency High – UG**
- **Emergency Low – OH**
- **Emergency Low- UG**

The descriptions for each type should be:

- Emergency High OH – Red Solid
- Emergency High UG – Red Dashed
- Energency Low OH – Magenta Solid
- Emergency Low UG – Magenta Dashed

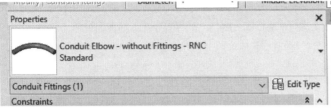

Select the conduit elbow that was just placed.

Define four new conduit fitting types called
- **Emergency High – OH**
- **Emergency High – UG**
- **Emergency Low – OH**
- **Emergency Low- UG**

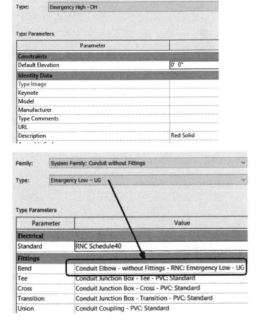

The descriptions for each type should be:

- Emergency High OH – Red Solid
- Emergency High UG – Red Dashed
- Energency Low OH – Magenta Solid
- Emergency Low UG – Magenta Dashed

Change the family definitions for the Emergency High and Emergency Low conduits to use the correct fittings/conduit elbows.

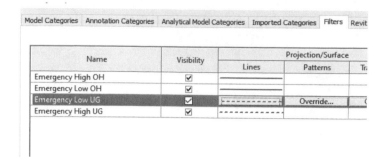

Launch the Visibility/Graphics dialog.

Define filters to display the conduits using the appropriate colors, linetypes, and lineweights.

*Hint: You can also use **Transfer Project Standards** from the Manage ribbon to copy filters from other Revit project files.*

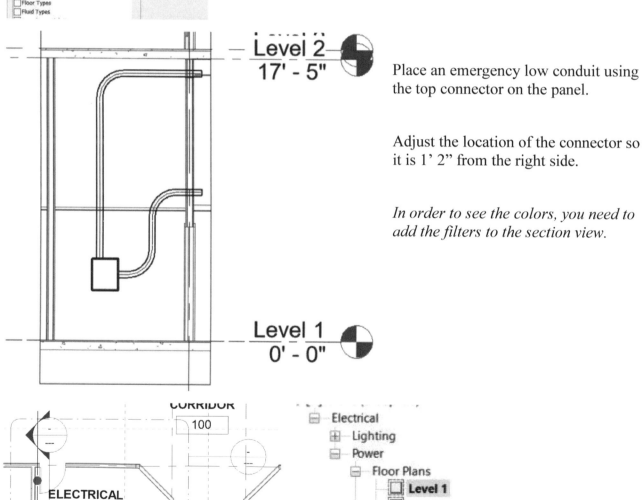

Place an emergency low conduit using the top connector on the panel.

Adjust the location of the connector so it is 1' 2" from the right side.

In order to see the colors, you need to add the filters to the section view.

Add a callout view to Level 1 Power floor plan around the electrical room 107.

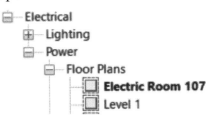

Open the Callout View.

Rename the CalloutView **Electrical Room 107**.

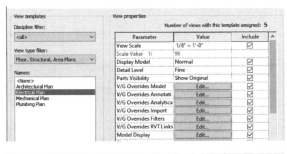

Select the Electrical Plan view template used on the Callout view.

Change the Detail Level to Fine.

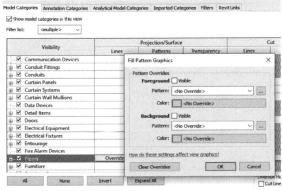

Select the V/G Overrides Model button.

Change the Floors under Model Categories by selecting override under Patterns.

Disable the visibility of the foreground and background patterns.

Edit the Filters to add the filters for the conduits.

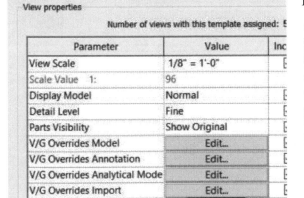

Name	Enable Filter	Visibility	Proje
			Lines
Emergency High UG	☑	☑	- - - - - - - - - -
Emergency Low UG	☑	☑	- - - - - - - - -
Emergency High OH	☑	☑	————
Emergency Low OH	☑	☑	————

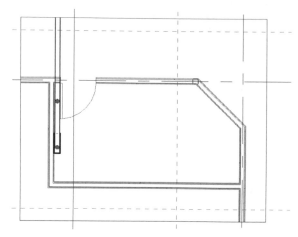

The callout view appearance should no longer display the hatch patterns for the floors.

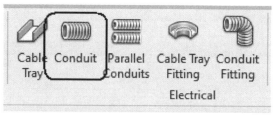

Go to the Systems tab.

Select the **Conduit** tool.

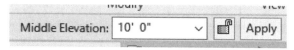

Set the Middle Elevation to 10' 0" above the level.

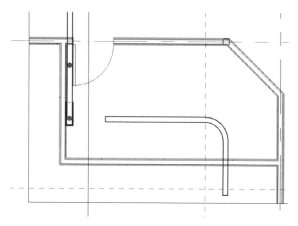

Select the **Emergency High OH** conduit family from the Type Selector.

Align the start of the conduit with the end of the lower connector.

Start the conduit about 5' to the right of the panel, drawing the conduit horizontally 9' and then draw a vertical conduit down.

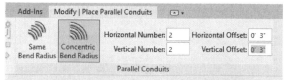

Use the Place parallel conduits tool to place two horizontal and two vertical conduits.

Enable **Concentric Bend Radius**.

Set the Horizontal Offset to 3".

Set the Vertical Offset to 3".

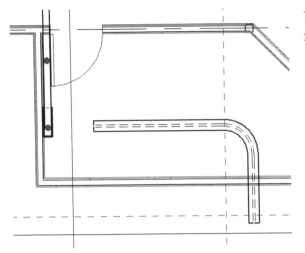

Use the TAB key to select the entire conduit run.

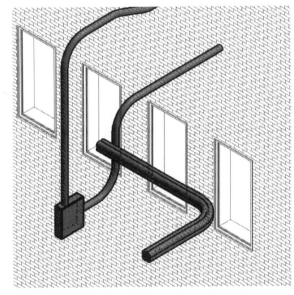

Switch to a 3D view to see your conduits.

Create a view template to apply filters to your conduits.

Apply the view template to a 3D view.

If you don't see the colors, verify that dashes and spaces are exactly the same for definitions, filters, and overrides.

Notes:

Lesson
06

Schedules

With equipment, devices, and lights in place, creating schedules helps leverage the database Revit creates to the needs of the designer. MA Lighting fixture schedule can be created the same way the space schedule was created. Panels schedules are a slightly different creature. Select a panel, then pick the Create Panel Schedule tool on the contextual tab. There is a standard template, but custom templates can be created to serve the needs of the individual firms. Once created, Panel schedules become available under their own heading in the Project Browser, similar to other schedules.

Schedules provide information about building elements, such as lighting fixtures, that can be exported to other applications, like Excel, for cost lists, estimates and other quantity tallies. Schedules update automatically when the building model is changed, eliminating the possibility of errors.

A schedule is a formatted view of a building model in tabular format. It is considered a view-element. Each property of an element is represented as a field in the schedule. Schedules can list every instance of an element type in different rows or condense the information to multiple instances of an element in a single row.

Guidelines for working with Schedules

- Create schedules that display only important or critical fields so that the schedules are easy to understand.
- Use the Hidden Field check box on the Formatting tab to hide fields that you want to use but not show in the schedule.
- Use Sorting/Grouping to organize your schedule.
- You can locate elements by using the schedule. Simply highlight the element in the schedule and select Show.

Exercise 6-1:

Creating a Lighting Fixture Schedule

Drawing Name: *schedules.rvt*
Estimated Time: 15 minutes

This exercise reinforces the following skills:
- User Interface
- Ribbon
- System Browser

1. Go to **File**.

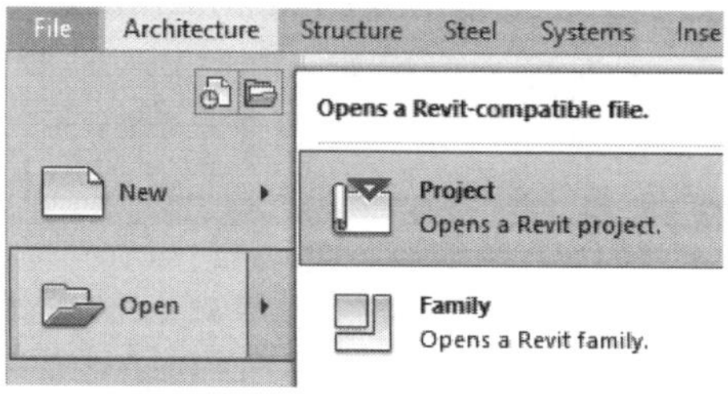

    Select **Open→Project**.

    Open *schedules.rvt*.

2. Activate the **View** tab.

3. Select **Schedules→Schedule/Quantities**.

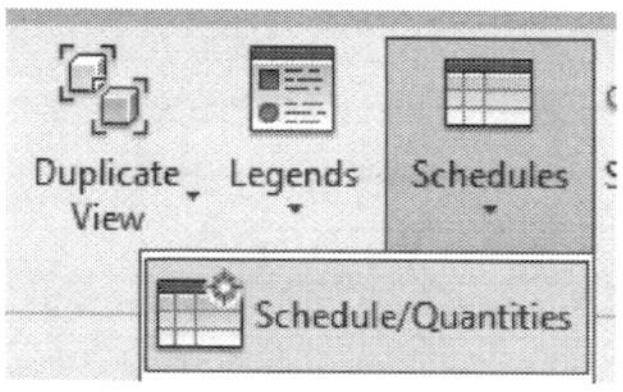

4. Highlight **Lighting Fixtures** in the Category pane.

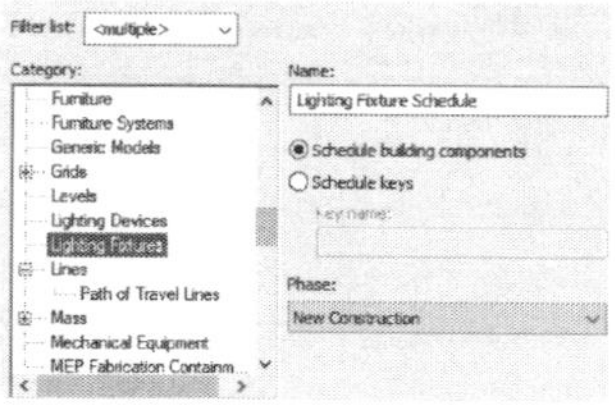

    Enable **Schedule building components**.

    Click **OK**.

5.  Use the Add tool to add the fields to the schedule. The order of the fields determines their column position in the schedule.

Add the following fields:

- Type Mark

- Manufacturer

- Model

- Lamp

- Electrical Data

- Description

6.  Select the **Sorting/Grouping** tab.

Set Sort by: to **Type Mark**.

Enable **Itemize every instance** at the bottom of the dialog.

Click **OK** to create the schedule.

7. 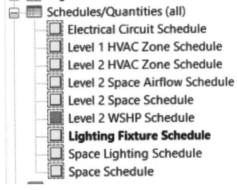 The Lighting Fixture Schedule appears in the browser.

Right click on the Lighting Fixture Schedule and select **Rename**.

8.  Rename **Level 2 – Lighting Fixture Schedule.**

9.

On the Properties palette:

Select **Edit** next to Fields.

10.

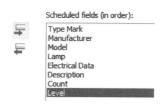

Add **Count** and Level to the selected Fields.

11. Activate the **Filter** tab.

Set Filter by: **Level.**

Select **equals** from the drop-down list.

Set the Level to **Level 2**.

12. Select the **Formatting** tab.

Highlight **Level.**

Enable **Hidden field**.

This means the field will not appear in the schedule, but will be used by the Filter to determine which elements appear in the schedule.

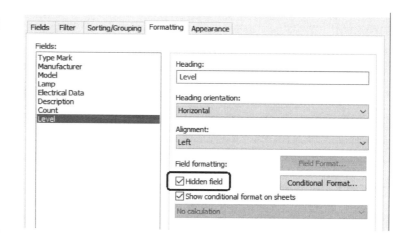

13. Select the **Sorting/Grouping** tab.

    Disable **Itemize every instance** at the bottom of the dialog.

    Click **OK** to close the dialog.

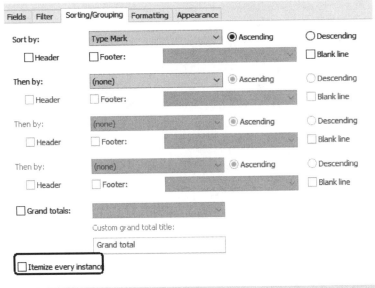

14. The schedule updates.

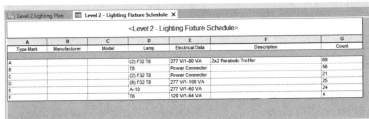

15. Save as *ex6-1.rvt*.

Exercise 6-2:

Creating a Lighting and Power Usage Schedule

Drawing Name: *power_usage.rvt*
Estimated Time: 10 minutes

This exercise reinforces the following skills:
- ❑ Schedules
- ❑ Conditional Format
- ❑ Energy Settings

1. 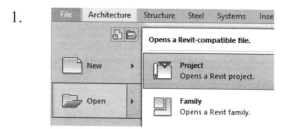 Go to **File**.

   Select **Open→Project**.

   Open *power_usage.rvt*.

2. Activate the **View** tab.

3. Select **Schedules→Schedule/Quantities**.

4. Highlight **Spaces** in the Category pane.

5. Change the Name to **Lighting & Power Usage**.

   Click **OK**.

6.

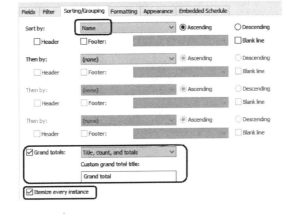

Select the following fields in this order:

- Name

- Number

- Area

- Actual Lighting Load

- Actual Lighting Load per Area

- Actual Power Load

- Actual Power Load per Area

7.

Select the Sorting/Grouping tab.

Sort by: **Name**.

Enable **Grand totals**.

Set to **Title, count, and totals**

Enable **Itemize every instance**.

8.

Select the **Filter** tab.

Select **Actual Power Load per Area**

Is Greater Than

Use the drop-down list to select **1.97 W/ft^2**.

Click **OK**.

9.

The schedule view opens.

10. Save as *ex6-2.rvt*.

Sheet Lists

One of the types of schedules available in Revit is sheet lists. A sheet list schedule is built the same way as a component model schedule, but all the relevant parameters are used by sheets. Creating a sheet list is useful for managing your construction documentation and organizing any documentation packages you sent out as it operates as a table of contents.

If you are required to submit a list of all drawings in a submittal package, you can use a Sheet List schedule.

As with most schedules that appear on construction documents, it is good practice to have two different versions of each schedule – one that has all the parameters necessary for tracking the sheets and any revisions and another that is used for the actual documentation package.

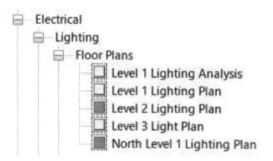

The Project Browser has icons to indicate if a view has been placed on a sheet.

If the square next to the view name is filled in/colored, this indicates the view has been placed on a sheet.

Views can only be used once. If you need to use a view on more than one sheet, you need to duplicate the view.

Legends are the only view type that can be placed on more than one sheet.

Exercise 6-3:

Creating a Sheet List

Drawing Name: *sheet_lists.rvt*
Estimated Time: 5 minutes

This exercise reinforces the following skills:
- ❑ Schedules
- ❑ Sheets

1.    Go to **File**.

   Select **Open→Project**.

   Open *sheet_lists.rvt.*

2. Activate the **View** ribbon.

3. 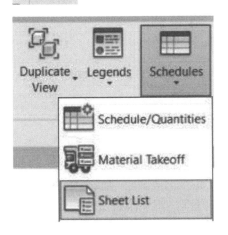   Go to **Schedules→Sheet List.**

4.

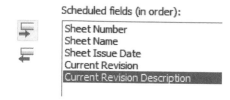

Scheduled fields (in order):

- Sheet Number
- Sheet Name
- Sheet Issue Date
- Current Revision
- Current Revision Description

Select the following fields in order:

- Sheet Number

- Sheet Name

- Sheet Issue Date

- Current Revision

- Current Revision Description

Click **OK**.

5.

Level 2	Sheet List	X		
		<Sheet List>		
A	B	C	D	E
Sheet Number	Sheet Name	Sheet Issue Date	Current Revision	Current Revision D
E601	Panel Schedules	02/11/08		
E201	Second Floor Power Plan	02/11/08		
M201	1ST FLOOR NORTH - HVAC	03/05/08		
E301	NORTH LEVEL 1 LIGHTING PLAN	03/05/08		
M601	DUCT SECTIONS	03/05/08		
M701	MECHANICAL SCHEDULES	03/05/08		

The schedule view opens.

Save as *ex6-3.rvt*.

Note Blocks

A note block is a schedule of an annotation family that is used in the project. Note Block schedules are useful in managing the plan notes on your construction documents as an alternative to keynotes. As note annotations are placed in a view, they can be given a description. Those descriptions are gathered into the Note Block schedule. A usage parameter can determine on which sheet the note has been placed.

A benefit of creating a Note Block schedule is that if you need to update or delete a note, you can make the desired changes in the schedule and the notes will update across all the sheets.

Exercise 6-4:

Creating a Note Block

Drawing Name: *note_block.rvt*
Estimated Time: 45 minutes

This exercise reinforces the following skills:
- ❑ Families
- ❑ Symbols
- ❑ Schedules
- ❑ Sheets

1. 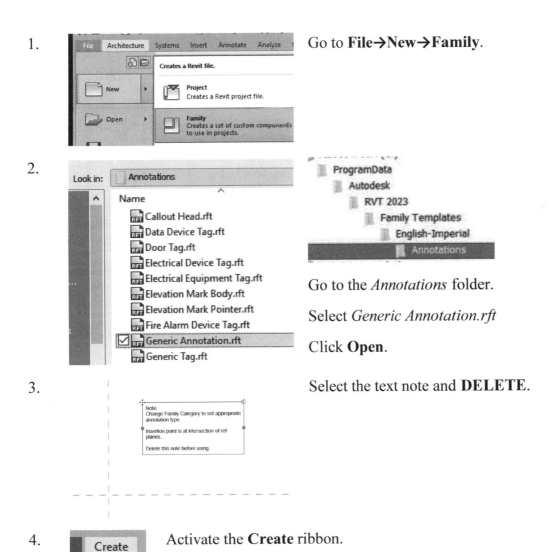 Go to **File→New→Family**.

2. Go to the *Annotations* folder.

    Select *Generic Annotation.rft*

    Click **Open**.

3. Select the text note and **DELETE**.

4. Activate the **Create** ribbon.

5. Select the **Label** tool.

Left click on the intersection of the two reference planes for the insertion point.

6.  Select **New** located at the bottom left of the dialog.

7.  Type **Note Number** for the Name.

Enable **Instance**.

Set the Type of Parameter to **Text**.

Group Parameter under **Text**.

Click **OK**.

8.  Add to the right panel.

Set the Sample Value to **00**.

Click **OK**.

9. Select the **Label** tool.

Left click below the first label.

10.

Select **New** located at the bottom left of the dialog.

11.

Type **Note Description** for the Name.
Enable **Instance**.
Set the Type of Parameter to **Text**.
Group Parameter under **Text**.

Exit out of the Label command.

12.

Add **Note Description** to the Label Parameters.

Click **OK**.
We are not placing this label, but it will still be available in the family.
Right click and select **Cancel**

13.

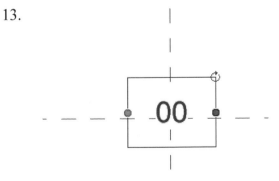

Use the grips to reduce the size of the label.

14.

Create

Activate the **Create** ribbon.

15. 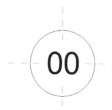 Select the **Line** tool.

16. 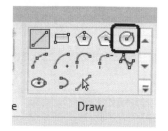 Select the **Circle** tool from the Draw panel.

17. 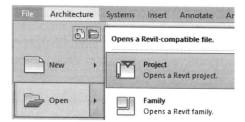 Draw a circle around the label.

Position the label so it is centered in the circle on the intersection.

Save as *Note Block.rfa*.

18. 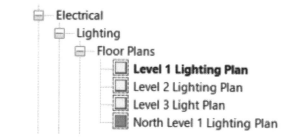 Go to **File**.

Select **Open→Project**.

Open *note_block.rvt*.

19. 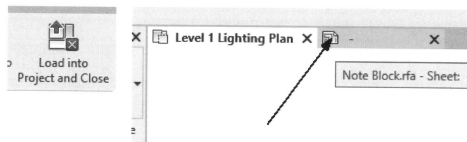 Open **Level 1 Lighting Plan** floor plan.

20.

Select the tab for the Note Block.

Select **Load into Project and Close**.

Select the Note_Block project file if a dialog appears.

21. Activate the **Annotate** ribbon.

22. Select the **Symbol** tool.

23. 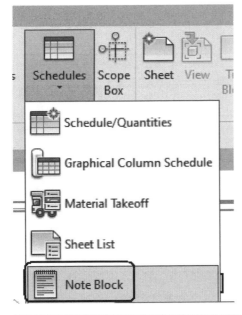 Left click to place in **MEETING 105**.

Cancel out of the command.

24. Activate the View ribbon.

Select **Schedules→Note Block**.

25. 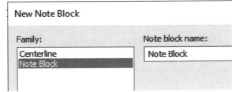 *The Note Block that was just placed is listed.*

Highlight **Note Block**.

Click **OK**.

26.

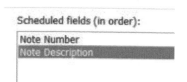

Highlight **Note Number** and **Note Description**.

You can select them together.

Add to the **Scheduled fields**.

Click **OK**.

27.

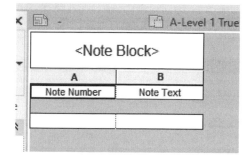

The schedule opens.

The cells are blank because you haven't added anything to the instance properties.

28.

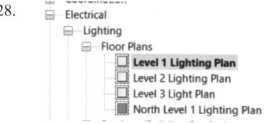

Open Level 1 Lighting Plan floor plan.

29.

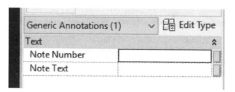

Select the Note Block in MEETING 105.

On the Properties palette:

Note the two instance properties that were created.

30.

Type **01** in the Note Number field.

In the Note Description field type:

LIGHTING FIXTURES WITH MORE THAN TWO LAMPS SHALL HAVE THE TWO OUTER LAMPS CONTROLLED BY ONE SWITCH AND INNER LAMP(S) CONTROLLED BY A SECOND SWITCH

31.

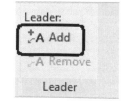

With the note symbol selected, select **Add** leader from the ribbon.

32.

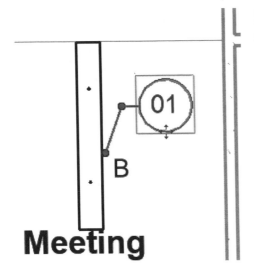

Position the leader so it is attached to the light fixture as shown.

33.

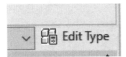

Select **Edit Type** on the Properties palette.

34.

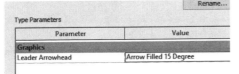

Set the Leader Arrowhead to **Arrow Filled 15 Degree.**

Click **OK.**

35.

Verify the note block is still selected.

Select the **Copy** tool from the Modify panel.

36.

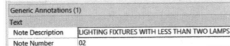

Left click to select a base point.

Left click to place in **MEETING 103.**

Use the grip to adjust the position of the leader to attach to lighting fixture.

37.

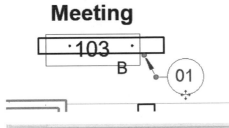

In the Properties palette:

Type **02** in the Note Number field.

In the Note Description field type:

LIGHTING FIXTURES WITH LESS THAN TWO LAMPS SHALL BE CONTROLLED BY ONE SWITCH

Release the selection by left clicking in the display window.

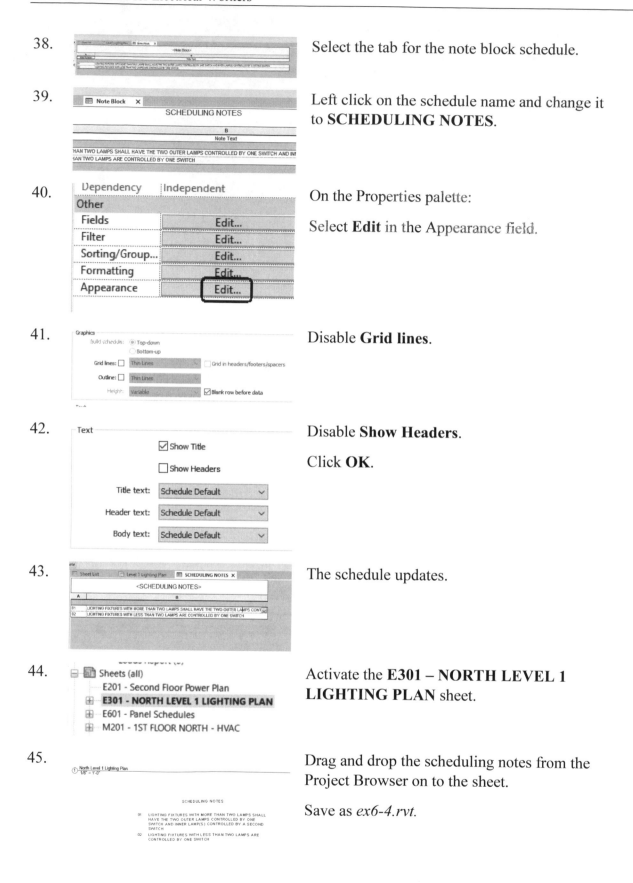

38. Select the tab for the note block schedule.

39. Left click on the schedule name and change it to **SCHEDULING NOTES**.

40. On the Properties palette:

Select **Edit** in the Appearance field.

41. Disable **Grid lines**.

42. Disable **Show Headers**.

Click **OK**.

43. The schedule updates.

44. Activate the **E301 – NORTH LEVEL 1 LIGHTING PLAN** sheet.

45. Drag and drop the scheduling notes from the Project Browser on to the sheet.

Save as *ex6-4.rvt*.

Schedule Keys

You can assign a target lighting level for all the spaces to be analyzed. Create a project parameter to be used for your targeted lighting level. This should be an instance parameter as it will be unique to each space element. Set the Discipline of the parameter to Electrical and the type to Illuminance. Group the parameter in the Electrical-Lighting group so that it can be easily located. Name the parameter Required Lighting Level, so the intended use of the parameter is clear. Project parameters can be used in schedules, but they cannot be tagged.

Once you have created the project parameter, you can create another type of schedule to associate lighting levels with types of spaces. This will not be a schedule of building components, but a schedule key.

Creating a schedule key makes it easy to assign target lighting levels to spaces.

Exercise 6-5:

Creating a Schedule Key

Drawing Name: *schedule_key.rvt*
Estimated Time: 20 minutes

This exercise reinforces the following skills:
- ❑ Schedules
- ❑ Project Parameters
- ❑ Spaces

1. 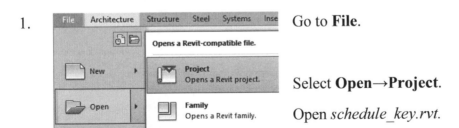 Go to **File**.

   Select **Open→Project**.

   Open *schedule_key.rvt*.

2. Activate the Manage ribbon.

   Select the **Project Parameters** tool.

3.

Select **New**.

4.

In the Name field:

Type **Required Lighting Level**.

Enable **Instance**.

Under Discipline:

Select **Electrical**.

Under Type of Parameter:

Select **Illuminance**.

Group parameter under **Electrical -Lighting**.

Select **Spaces** in the right Categories pane.

Click **OK**.

5.

The new parameter is now listed in the dialog.

Click **OK** to close the dialog.

6. View   Activate the **View** ribbon.

7.

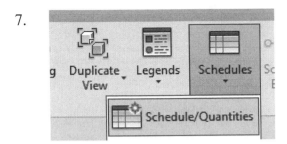

Select **Schedule/Quantities**.

8.

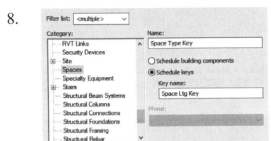

Highlight **Spaces** in the left pane.

In the Name field, type **Space Type Key**.

Enable **Schedule Keys**

Type **Space Ltg Key** in the Key Name field.

Click **OK**.

9.

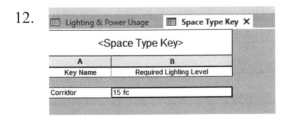

Add **Required Lighting Level** to the scheduled fields.

This is the project parameter you just added.

Click **OK** to create the schedule.

10.

The schedule will not contain any data rows because no values have been assigned to any of the spaces.

11.

Select **Insert Data Row** from the ribbon.

12.

In the new row:

Type **Corridor** for the Key Name.

Type **15 fc** for the Required Lighting Level.

13.

A	B
Key Name	Required Lighting Level
Admin	50 fc
Cafeteria	50 fc
Classroom	30 fc
Conference	30 fc
Copy	30 fc
Corridor	15 fc
Elev	15 fc
Hall	15 fc

<Space Type Key>

Use the Insert Data Row tool to add more data rows and their values.

This key schedule can now be used with the Lighting Analysis schedule.

14.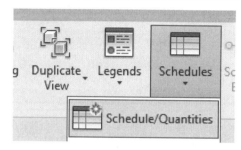

Select **Schedule/Quantities** from the View ribbon.

15.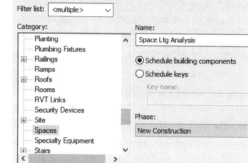

Highlight **Spaces** in the left Category pane.

Type **Space Ltg Analysis** in the Name field.

Enable **Schedule Building Components**.

Click **OK**.

16.

Scheduled fields (in order):

Number
Name
Space Ltg Key
Required Lighting Level
Average Estimated Illumination
Ceiling Reflectance
Lighting Calculation Workplane

Select the following fields in order for the schedule:

- Number

- Name

- Space Ltg Key

- Required Lighting Level

- Average Estimated Illumination

- Ceiling Reflectance

- Lighting Calculation Workplane

Click **OK**.

17.

109	WC	(none)
103	Meeting	(none)
102	Archive	(none)
101	Cafeteria	(none)
111	Air Lock	(none)
114	Workshop	Admin
115	Room	Cafeteria
116	Room	Classroom
118	Room	Conference
121	Manager	Copy
122	Manager	Corridor
		(none)

In the Space Ltg Key column, use the drop-down list to assign the correct key to spaces where keys have been provided.

Once the key is assigned, the associated value will appear in the Required Lighting Level column.

18.

	A	B	C	D	E	F	
	Number	Name	Space Ltg Key	Required Lighting Level	Average Estimated Illumination	Ceiling Reflectance	Lighting Ca
	104	Storage	(none)		28 fc	75.00%	2' - 6"
	107	Mech	(none)		37 fc	75.00%	2' - 6"
	108	Janitor	(none)		24 fc	75.00%	2' - 6"
	105	Meeting	(none)		42 fc	75.00%	2' - 6"
	109	WC	(none)		38 fc	75.00%	2' - 6"
	103	Meeting	(none)		28 fc	75.00%	2' - 6"
	102	Archive	(none)		22 fc	75.00%	2' - 6"
	101	Cafeteria	Cafeteria	50 fc	49 fc	75.00%	2' - 6"
	111	Air Lock	(none)		10 fc	75.00%	2' - 6"

Reviewing the schedule you can quickly see where you need to make changes in the model.

19. Save as *ex6-5.rvt*.

Exercise 6-6:

Creating a Panel Schedule

Drawing Name: *load_schedule.rvt*
Estimated Time: 10 minutes

This exercise reinforces the following skills:

- ❑ Load Classifications
- ❑ Families
- ❑ Schedules

1. 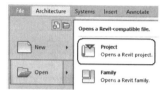 Open *load_schedule.rvt*.

2. 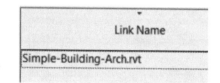 Go to the Insert ribbon.

   Use **Manage Links** to reload the *Simple-Building-Arch.rvt* file.

3. 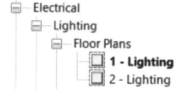 Open the **1 – Lighting** floor plan view.

4. 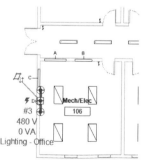 Select **Panel D** in the Mech/Elec Room 106.

5. On the ribbon:

Select **Create Panel Schedules→Choose a Template**.

6. There is only one template available called Branch Panel.

Highlight this template.

Click **OK**.

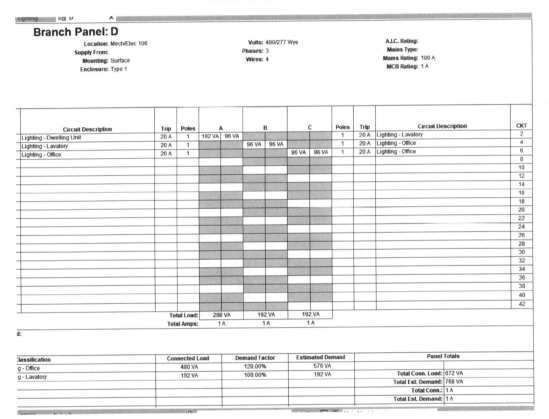

Branch Panel: D

Location: Mech/Elec 106			**Volts:** 480/277 Wye			**A.I.C. Rating:**			
Supply From:			**Phases:** 3			**Mains Type:**			
Mounting: Surface			**Wires:** 4			**Mains Rating:** 100 A			
Enclosure: Type 1						**MCB Rating:** 1 A			

Circuit Description	Trip	Poles	A	B	C	Poles	Trip	Circuit Description	CKT
Lighting - Dwelling Unit	20 A	1	192 VA	96 VA			1	20 A	Lighting - Lavatory	2
Lighting - Lavatory	20 A	1		96 VA	96 VA		1	20 A	Lighting - Office	4
Lighting - Office	20 A	1			96 VA	96 VA	1	20 A	Lighting - Office	6
									8
									10
									12
									14
									16
									18
									20
									22
									24
									26
									28
									30
									32
									34
									36
									38
									40
									42
Total Load:			288 VA	192 VA	192 VA				
Total Amps:			1 A	1 A	1 A				

Classification	Connected Load	Demand Factor	Estimated Demand	Panel Totals	
g - Office	480 VA	120.00%	576 VA		
g - Lavatory	192 VA	100.00%	192 VA	**Total Conn. Load:**	672 VA
				Total Est. Demand:	768 VA
				Total Conn.:	1 A
				Total Est. Demand:	1 A

7. The schedule opens. You can see which circuits are on each panel and the loads for each panel.

   In Lesson 8, I show you how to create load classifications and assign them to families so they appear properly in schedules.

8. Save as *ex6-6.rvt*.

Lab Exercises

Open *lab_6.rvt*.

Create a lighting fixture schedule as shown.

<Lighting Fixture Schedule>			
A	**B**	**C**	**D**
Family and Type	Panel	Circuit Number	Description
Plain Recessed Lighting Fixture: 2x2 - 120	LIGHTING PANEL 'B'	2	
Plain Recessed Lighting Fixture: 2x2 - 120	LIGHTING PANEL 'B'	2	
Plain Recessed Lighting Fixture: 2x2 - 120	LIGHTING PANEL 'B'	2	
Plain Recessed Lighting Fixture: 2x2 - 120	LIGHTING PANEL 'B'	2	
Plain Recessed Lighting Fixture: 2x2 - 120	LIGHTING PANEL 'B'	2	
Plain Recessed Lighting Fixture: 2x2 - 120	LIGHTING PANEL 'B'	2	
Plain Recessed Lighting Fixture: 2x2 - 120	LIGHTING PANEL 'B'	3	
Plain Recessed Lighting Fixture: 2x2 - 120	LIGHTING PANEL 'B'	3	
Plain Recessed Lighting Fixture: 2x2 - 120	LIGHTING PANEL 'B'	3	
Plain Recessed Lighting Fixture: 2x2 - 120	LIGHTING PANEL 'B'	3	
Plain Recessed Lighting Fixture: 2x2 - 120	LIGHTING PANEL 'B'	3	
Plain Recessed Lighting Fixture: 2x2 - 120	LIGHTING PANEL 'B'	3	
Plain Recessed Lighting Fixture: 2x2 - 120	LIGHTING PANEL 'B'	3	

If any of the lighting fixtures are not connected to a panel, use the HIGHLIGHT IN MODEL tool to locate the fixture in the model.

Select in the view.
Select POWER from the ribbon.

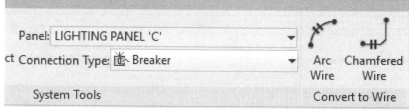

Select the panel to assign.
Add wires.

Return to the schedule and continue assigning light fixtures to panels.

MAIN FLOOR LIGHTING FIXTURES

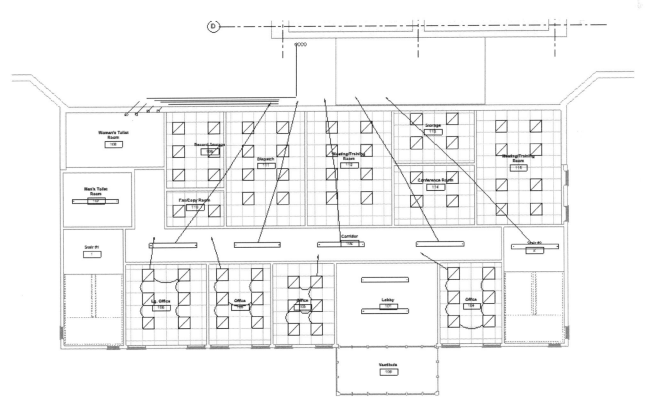

GROUND FLOOR LIGHTING FIXTURES

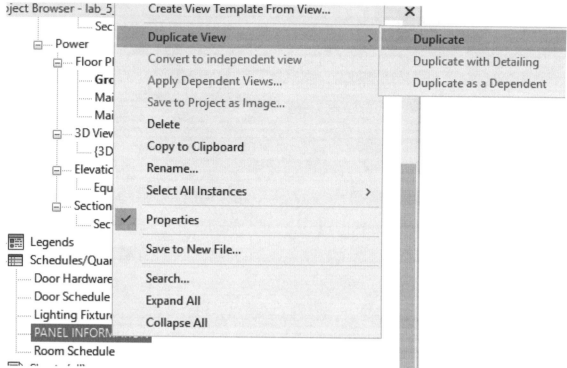

Create a Duplicate view of the PANEL INFORMATION schedule. Repeat to create a second copy.

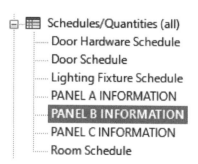

Rename the schedules as shown.

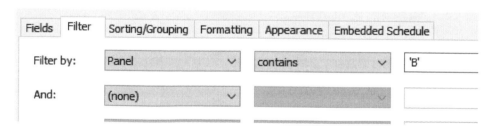

Apply filters so each schedule only lists the required panel.

\<PANEL B INFORMATION\>

A	B	C	D	E	F
Panel	Load Name	Apparent Load	Rating	Receptacle Connected	Voltage Drop
PWR PNL. 'B'	Receptacle	180 VA	20 A	180 VA	0 V
PWR PNL. 'B'	Receptacle	180 VA	20 A	180 VA	0 V
PWR PNL. 'B'	Receptacle	360 VA	20 A	360 VA	0 V
PWR PNL. 'B'	Receptacle	360 VA	20 A	360 VA	0 V
PWR PNL. 'B'	Receptacle	360 VA	20 A	360 VA	0 V
PWR PNL. 'B'	Receptacle	900 VA	20 A	900 VA	2 V
PWR PNL. 'B'	Receptacle	180 VA	20 A	180 VA	0 V
PWR PNL. 'B'	Receptacle	360 VA	20 A	360 VA	0 V
PWR PNL. 'B'	Receptacle	360 VA	20 A	360 VA	0 V
PWR PNL. 'B'	Receptacle	720 VA	20 A	720 VA	1 V
PWR PNL. 'B'	Receptacle	1080 VA	20 A	1080 VA	2 V
LIGHTING PANEL 'B'	Lighting – Dwelling Unit	372 VA	20 A		0 V
LIGHTING PANEL 'B'	Lighting – Dwelling Unit	372 VA	20 A		0 V
LIGHTING PANEL 'B'	Lighting – Dwelling Unit	372 VA	20 A		0 V
LIGHTING PANEL 'B'	Lighting – Dwelling Unit	372 VA	20 A		0 V
LIGHTING PANEL 'B'	Lighting – Dwelling Unit	372 VA	20 A		1 V

\<PANEL C INFORMATION\>

A	B	C	D	E	F
Panel	Load Name	Apparent Load	Rating	Receptacle Connected	Voltage Drop
LIGHTING PANEL 'C'	Lighting – Dwelling Unit	372 VA	20 A		0 V
LIGHTING PANEL 'C'	Lighting – Dwelling Unit	930 VA	20 A		1 V
LIGHTING PANEL 'C'	Lighting – Dwelling Unit	2046 VA	20 A		2 V
LIGHTING PANEL 'C'	Lighting – Dwelling Unit	300 VA	20 A		0 V
LIGHTING PANEL 'C'	Lighting – Dwelling Unit	300 VA	20 A		0 V
LIGHTING PANEL 'C'	Lighting – Dwelling Unit	450 VA	20 A		1 V
LIGHTING PANEL 'C'	Lighting – Dwelling Unit	150 VA	20 A		0 V
LIGHTING PANEL 'C'	Lighting – Dwelling Unit	150 VA	20 A		0 V
LIGHTING PANEL 'C'	Lighting – Dwelling Unit	150 VA	20 A		0 V
LIGHTING PANEL 'C'	Lighting – Dwelling Unit	150 VA	20 A		0 V
LIGHTING PANEL 'C'	Lighting – Dwelling Unit	150 VA	20 A		0 V
LIGHTING PANEL 'C'	Lighting – Dwelling Unit	372 VA	20 A		0 V

Notes:

Lesson

07

Views

You can create different views of the building model, such as plans, sections, elevations, and 3D views.

In the building model, every drawing sheet, 2D view, 3D view, and schedule is a presentation of information from the same underlying building model database.

When you change the building model in one view, Revit propagates those changes throughout the project.

Revit has many different types of views available:

- **Plan Views**
  Two-dimensional views provide a traditional method for viewing a model. These views include floor plans, reflected ceiling plans, and structural plans.

- **Elevation Views**
  View your model from numerous elevation perspectives.

- **Section Views**
  Sections views cut through the model. You can draw them in plan, section, elevation, and detail views. Section views display as section representations in intersecting views.

- **Callout Views**
  A callout shows some portion of another view at a larger scale. In a construction document set, use callouts to provide an orderly progression of labeled views at increasing levels of detail.

- **3D Views**
  Create perspective and orthographic 3D views, and enhance them by adding a background, adjusting the camera position or extents, or changing view properties.

- **Legend Views**
  Create legends to list the building components and annotations used in a project.

- **Drafting Views**
  Drafting Views are details of the building model consisting of 2D elements. The detail views can usually be used across projects to show construction details. They are not associated to the 3D model but can be linked to a callout view. If you have a legacy library of AutoCAD details, they can be imported into Revit drafting views for use in your project.

Exercise 7-1:

Creating a Plan View

Drawing Name: *views.rvt*
Estimated Time: 15 minutes

This exercise reinforces the following skills:
- ❑ Duplicate a View
- ❑ Crop a View
- ❑ Rename a View

1.      Go to **File**.

Select **Open→Project**.

Open *views.rvt*.

2.   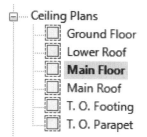   Activate the **Main Floor** ceiling plan.

3.      Highlight the Main Floor view name in the Project Browser.

Right click and select **Duplicate View→Duplicate with Detailing**.

When you duplicate a view with detailing, this means you are copying any annotation elements to the new view in addition to the model elements.

4. 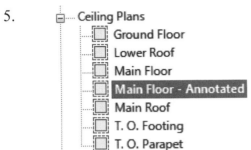 Highlight the copied view.

Right click and select **Rename**.

5.  Rename **Main Floor - Annotated**.

6.  Highlight the **Main Floor- Annotated** view name in the Project Browser.

Right click and select **Duplicate View→Duplicate as a Dependent**.

When you duplicate a view as dependent, it will display the annotations created in the parent view and vice versa.

7. 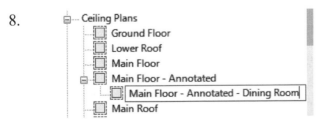 Highlight the dependent view.

Right click and select **Rename**.

8. Rename **Main Floor – Annotated – Dining Room.**

9.

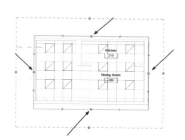

Use the grips on the viewport to adjust the viewport size to only include the Dining Room and Kitchen.

10.

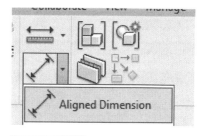

Select the **Aligned Dimension** tool.

11.

On the Options bar:

Enable **Wall faces**.

12.

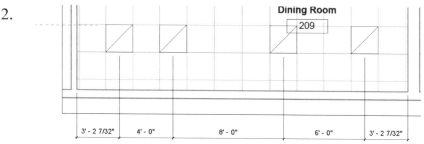

Place dimensions between the lighting fixtures and the walls.

13.

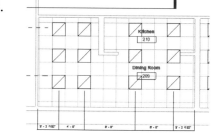

Activate the **Main Floor – Annotated** view.

This is the parent view for the Dining Room view.

Note that the dimensions are visible in this view.

14.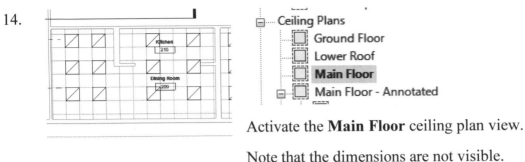

Activate the **Main Floor** ceiling plan view.

Note that the dimensions are not visible.

15.   Save as *ex7-1.rvt*.

Elevation Views

Elevation views are controlled by elevation markers. You can control the visibility of the markers in the Visibility/Graphics dialog on the Annotation tab.

Elevations help clarify relationships between elements as you design and document the building model. In a default template, Revit provides four exterior elevation views. To create additional interior or exterior elevation views, on the View tab, click the Elevation tool. As you place the elevation marker in a plan view, Revit detects model geometry and helps align the elevation marker. To cycle through possible directions for the elevation marker, Click Tab to rotate the elevation marker.

To open the elevation view, double-click the elevation arrow in a drawing or double-click the view name in the Project Browser. As your model changes, Revit automatically updates elevation views.

Electrical designers use elevation views to designate the location of outlets, light fixtures, panels, and other electrical devices.

Exercise 7-2:

Creating an Elevation View

Drawing Name: *elevations.rvt*
Estimated Time: 15 minutes

This exercise reinforces the following skills:
- Views
- Elevation
- Rename View
- Crop View

1.  Go to **File**.

   Select **Open→Project**.

   Open *elevations.rvt*.

2. Open the **Main Floor** floor plan.

3. Zoom into the area to the left of **STORAGE 215**.

4. Activate the **View** ribbon.

5. Select **Elevation**.

6.

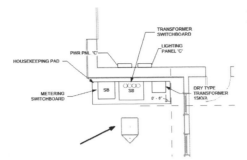

Place the elevation marker below the equipment.

Cancel out of the command.

7.

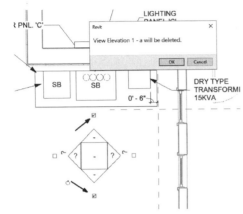

Left click on the square part of the elevation marker.

Place a check mark in the top small square.

This indicates the direction for the view.

Uncheck the small bottom square to delete that view direction.

A dialog will pop up advising the view will be deleted.

Click **OK**.

8.

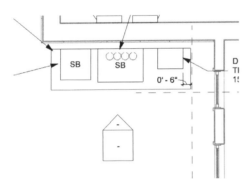

The elevation marker updates with the triangle located at the top of the symbol.

Single left click on the triangle to wake up the crop region.

9.

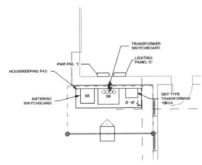

A crop region displays to show the defined view area.

Use the grips to adjust the view area.

Double left click on the triangle to open the elevation view.

10.

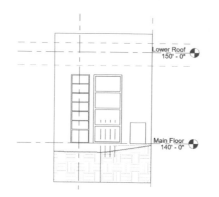

Select the viewport and adjust the outline.

Type **VV** to launch the Visibility/Graphics dialog.

11.

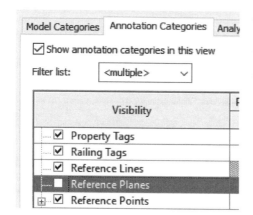

Activate the Annotation Categories tab.

Disable **Reference Planes** to hide them.

Click **OK**.

12.

Toggle **Crop Region** off to hide the viewport.

13.

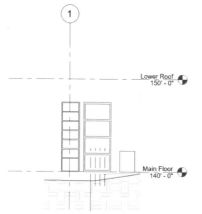

The view should appear similar to this.

If you see the wall behind the equipment, return to the Main Floor view and adjust the elevation view depth.

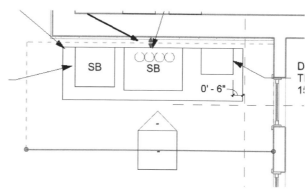

14. 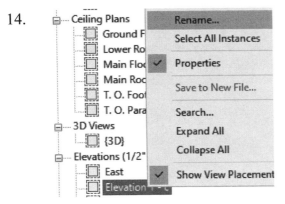 Highlight the Elevation view in the Project Browser.

Right click and select **Rename**.

15. Rename **Equipment**.

16. Save as *ex7-2.rvt*.

Exercise 7-3:

Creating a Section View

Drawing Name: *section_view.rvt*
Estimated Time: 15 minutes

This exercise reinforces the following skills:
- Creating a Section View
- Controlling the view's crop region window
- Rename a view
- Assign a sub-discipline to a view
- Controlling the visibility of linked file elements
- Adding Tags

1.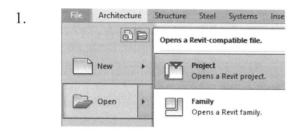
   Go to **File**.

   Select **Open→Project**.

   Open *section_view.rvt*.

2.
   Open the Insert ribbon.

   Use Manage Links to reload the *Floor Plan Imperial – Arch.rvt* file.

3.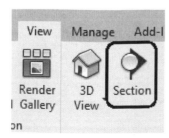
   Open the **3rd Floor Lighting Plan** floor plan.

4.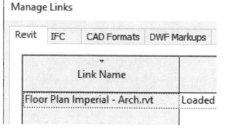
   Activate the View ribbon.

   Select the **Section** tool.

5.

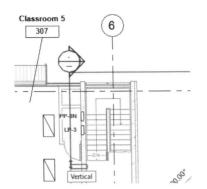

Left click above Classroom 5 to start the section line.

Left click above the door of Classroom 5 to end the section line.

6.

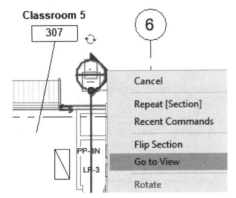

Right click on the section bubble.

Select **Go to View**.

7.

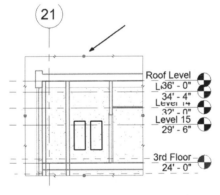

Use the grips on the view crop region to adjust the view so you are only looking at the electrical panels.

8.

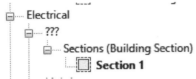

Locate the view in the browser.

Notice that it is missing a category assignment.

9.

Section: Section 1

Color Scheme	
Default Analysis Di...	None
Sub-Discipline	
Sun Path	HVAC
Extents	Lighting
Crop View	Plans
Crop Region Visible	Plumbing
Annotation Crop	Power

Locate the Sub-Discipline field in the Properties palette.

Set it to **Power**.

10.

- Power
  - Floor Plans
    - 1st Floor Power Plan
    - 2nd Floor Power Plan
  - 3D Views
    - 3D Elec
  - Elevations (Building Elevation)
    - East - Elec
    - North - Elec
    - South - Elec
    - West - Elec
  - Sections (Building Section)
    - Classroom 5 Panels

The Section View is reassigned and moves under the Power sub-discipline.

Rename the Section 1 view to **Classroom 5 Panels.**

To rename, right click and select Rename or Click F2 to enter Edit Text mode.

11. Annotate

Activate the **Annotate** ribbon.

12. Tag by Category

Select **Tag by Category.**

13.

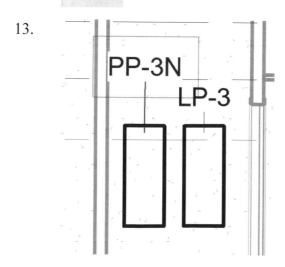

Tag each of the panels with a leader.

14. 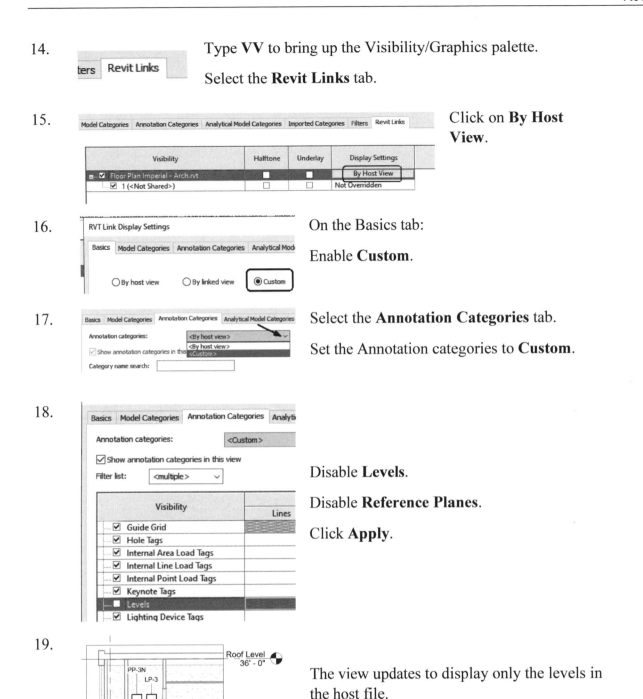 Type **VV** to bring up the Visibility/Graphics palette.

Select the **Revit Links** tab.

15. Click on **By Host View**.

16. On the Basics tab:

Enable **Custom**.

17. Select the **Annotation Categories** tab.

Set the Annotation categories to **Custom**.

18. Disable **Levels**.

Disable **Reference Planes**.

Click **Apply**.

19. The view updates to display only the levels in the host file.

Click **OK** twice to close the dialog.

20. Save as *ex7-3.rvt*.

Exercise 7-4:

Creating a Call-out View

Drawing Name: *callout_views.rvt*
Estimated Time: 10 minutes

This exercise reinforces the following skills:
- ❑ View
- ❑ Callouts
- ❑ Crop Region
- ❑ Rename View
- ❑ Dimensions
- ❑ Units

1.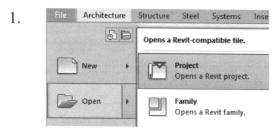

    Go to **File**.

    Select **Open→Project**.

    Open *callout_views.rvt*.

2. | Manage Links | | | |
    |---|---|---|---|
    | Revit | IFC | CAD Formats | DWF Markups |

Link Name	
Floor Plan Imperial - Arch.rvt	Loaded

    Open the Insert ribbon.

    Use Manage Links to reload the *Floor Plan Imperial – Arch.rvt* file.

3. ⊟ Sections (Building Section)
            ▣ **Classroom 5 Panels**

    Verify that the **Classroom 5 Panels** section view is open.

4.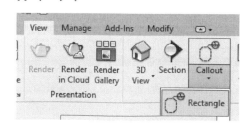

    Activate the View ribbon.

    Select **Callout →Rectangle**.

5.

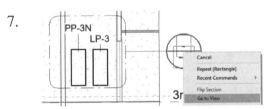

Left click once on the upper left of the panels.

Left click once on the lower right of the panels.

This defines the size of the rectangular callout.

Notice that we see the levels on the linked Revit file. Can you figure out how to turn off their visibility?

6.

You can use the small grip at the bubble's quadrant to position the bubble on the callout.

7.

Right click on the callout bubble.

Select **Go to View**.

8.

Locate the callout view in the Project Browser.

Rename the callout view – **Classroom 5 Panels Detail.**

9.

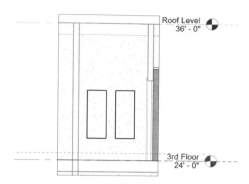

Adjust the crop region so you can see the floor and roof levels.

Notice the panel tags are not visible in the callout view. This is because annotations are view specific. They are only visible in the view in which they are placed.

10.

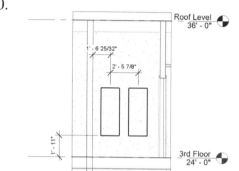

Use **ALIGNED DIMENSION** to add dimensions to indicate the placement of the panels.

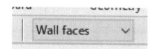

Set **Wall faces** on the Options bar.

11.

Select one of the dimensions.

On the Properties palette, select **Edit Type**.

12.

Text Offset	1/16"
Read Convention	Up, then Left
Text Font	Arial
Text Background	Opaque
Units Format	1' - 5 11/32" (Default)
Alternate Units	None
Alternate Units F	1235 [mm]
Alternate Units Pr	

Left click on the **Units** format field.

13.

Disable **Use project settings**.

Set the Rounding **To the nearest 1/8"**.

Click **OK**.

Use project settings

Units: Feet and fractional inches

Rounding: To the nearest 1/8"

Rounding increment:

14.

Type Parameters

Parameter	
Color	■ Black
Dimension Line Snap	1/4"
Text	
Width Factor	1.000000
Underline	☐
Italic	☐
Bold	☐
Text Size	3/32"
Text Offset	1/16"
Read Convention	Horizontal
Text Font	Arial
Text Background	Opaque

Set the Read Convention to **Horizontal**.

Click **OK** to close the Type Properties dialog.

15. Save as *ex7-4.rvt*.

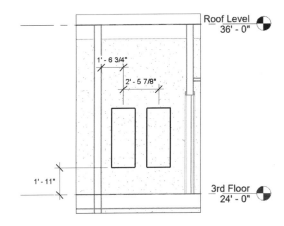

Detail Views

A detail view is a view of the model that appears as a callout or section in other views. This type of view represents the model at finer scales of detail than in the parent view. Typically, a detail view uses both model and annotation elements.

A detail view can be created as a section or a callout. When created as a callout, the detail can have both section and callout annotations assigned to it. That is, a detail view made as a callout can also show up as a section in views that intersect the callout view extents. For example, you may use a callout to create a detail view of a wall intersection. This same callout can appear as a section view in the overall building section view.

For a detail callout to display in the overall building section view, you must select the Intersecting Views option for the Show In instance parameter. You set this parameter on the Properties palette.

Note: The Show In property is not available for a detail created as a section. By default, the section tag will be visible in all intersecting views.

Visibility of a detail view tag depends on the scale of the parent view and whether the crop boundary of the detail view intersects or is entirely within that of the parent view. The detail view parameter Hide at Scales Coarser Than establishes a scale at which details are either shown or hidden in other views. For example, if a detail tag is set to hide at scales coarser than 1/4"=1'0", then a view with a scale set to 1/8" = 1'-0" would not show the detail tag.

All detail views, regardless of whether you draw them as a callout or section, show up in the Project Browser as a detail view.

Exercise 7-5:

Creating a Detail View

Drawing Name: *views_detail.rvt*
Estimated Time: 30 minutes

This exercise reinforces the following skills:
- ❏ Section View
- ❏ Add Tags
- ❏ Detail Lines
- ❏ Add Text Notes and callouts

1. 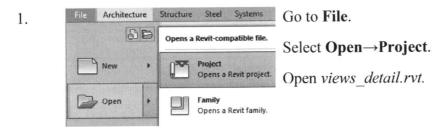 Go to **File**.

   Select **Open→Project**.

   Open *views_detail.rvt*.

2. Open the **Level 1 Power Plan** view.

3.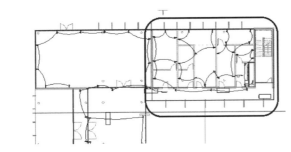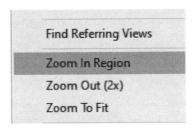

   Use **Zoom In Region** to zoom to the upper right of the model.

4. Activate the View ribbon.

   Select the **Section** tool.

5.

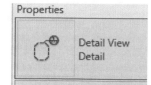

On the Properties palette:

Verify that **Detail View: Detail** is selected.

6.

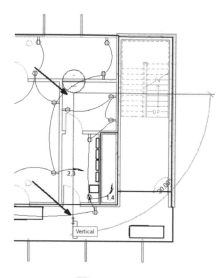

Place a vertical section line looking at the panels.

7.

Locate the Detail View in the Project Browser.

Rename to **Electrical Riser Diagram**.

Double left click on the detail view to open.

8.

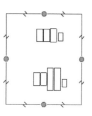

Adjust the crop region to show the electrical panels.

9.

Type **VV** to open the Visibility/Graphics dialog.

Select the Annotation Categories tab.

Enable **Generic Annotations**.

Click **OK**.

10. Activate the Annotate ribbon.

Select the **Tag by Category** tool.

11. 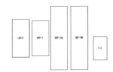 Disable **Leader** on the Options bar.

12. Left click inside of each panel/rectangle to tag.

Click **ESC** to exit the command.

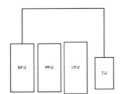

13. Select the **Detail Line** tool from the Annotate ribbon.

14. Draw a detail line from EP-2 to T-2 as shown.

You can use the midpoint snap to locate the end points of the line.

*Hint: Enable **chain** on the option bar to draw the lines interconnected.*

15.   Add the detail lines as shown.

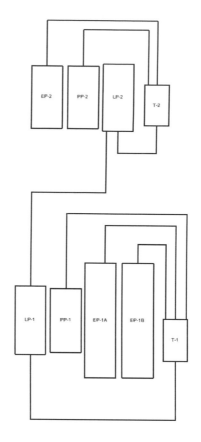

16.   Select the **Detail Line** tool from the Annotate ribbon.

17.   Select the **Rectangle** tool from the Draw panel.

18.   Draw a 4' 3" wide by 3' 6" high rectangle to the left of the LP-1 panel.

Do you remember how to modify temporary dimensions?

Hint: *Select the vertical line to wake up the horizontal dimension and select the horizontal line to wake up the vertical dimension.*

19. 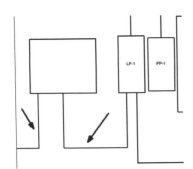 Add detail lines to connect the rectangle to the LP-1 panel and two lines from the rectangle to the left side of the crop region.

    Hint: *Use temporary dimensions to adjust the line locations.*

20. Select the **Symbol** tool on the Annotate ribbon.

21. No Generic Annotations family is loaded in the project. Would you like to load one now?

    Yes | No

    Click **Yes**.

22. File name: | ground symbol.rfa

    Files of type: | All Supported Files (*.rfa, *.adsk)

    Locate the *ground symbol* in the downloaded exercise files.

    Click **Open.**

23. Clipboard | Geometry

    Number of Leaders: 1

    On the Options bar:

    Set the Number of Leaders to **1**.

24. 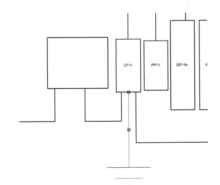 Adjust the leader position to the bottom midpoint of the LP-1 panel.

    Position the symbol below LP-1.

25. A Text Select the **Text** tool from the Annotate ribbon.

26. Left click inside the rectangle you placed to locate the test.

Type UTILITY TRANSOCKET.

Left click outside of the text box to exit the edit box.

Click ESC or right click and select CANCEL twice to exit the TEXT command.

27. Click on the text and use the MOVE icon to position the text in the center of the rectangle.

28. 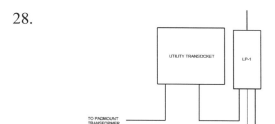 Add text to the wire coming from the left of the Utility Transocket.

The text should read **TO PADMOUNT TRANSFORMER.**

29. Select the **Text** tool from the Annotate ribbon.

30. On the ribbon:

Enable leader.

31.

Left click on the vertical line connecting to the ground to start the leader.

Left click to place the start point of the shoulder.

Left click to start the text.

Type **GROUND ROD**.

Left click outside the text box to exit the text.

Click **ESC** to exit the command.

32.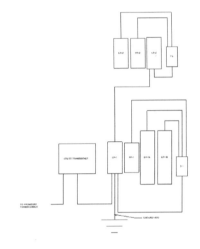

Turn off the crop rectangle.

33.   Save as *ex7-5.rvt*.

Exercise 7-6:

Creating a 3D View

Drawing Name: *3d_view.rvt*
Estimated Time: 20 minutes

This exercise reinforces the following skills:
- ❑ View
- ❑ Camera
- ❑ Properties palette

1.   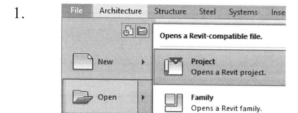 Go to **File**.

        Select **Open→Project**.

        Select 3d_view.rvt.

        Click **Open**.

2.   Activate the **Main Floor** floor plan.

- Views (all)
  - Floor Plans
    - Ground Floor
    - Lower Roof
    - **Main Floor**
    - Main Roof

3.   Activate the **View** ribbon.

4.    Select **3D View→Camera**.

5.

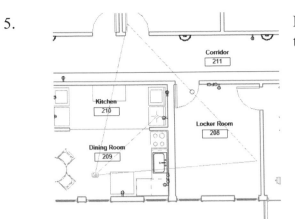

Place a camera in the Dining Room looking towards the kitchen.

6.

Using grips, adjust the size of the viewport so you can see the kitchen.

7.

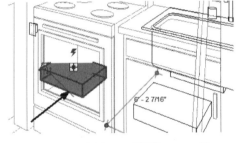

Select the block indicated.

On the Properties panel, it shows that it is a lighting fixture located on the ground floor.

8.

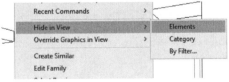

Right click and select **Hide in View→Elements.**

Repeat for the other elements which should not be visible in the view.

9.

You can use the Navigation Wheel located on the right side of the display window to adjust the view further.

Select the Navigation Wheel.

10.

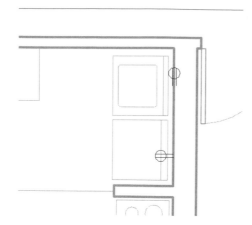

Return to the Main Floor view.

The duplex receptacles need to be adjusted.

11.

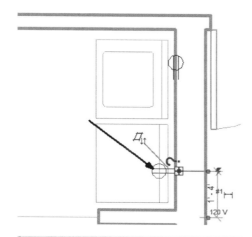

Select the lower receptacle.

On the Properties palette:

12.

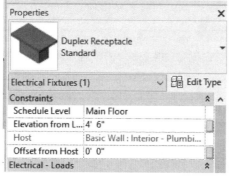

Change the Elevation from Level to **4' 6"**.

Click **ESC** to release the selection.

13.

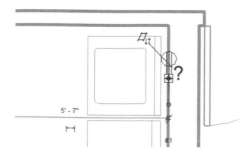

Select the upper receptacle.

14. Select **Pick New** from the ribbon.

15. 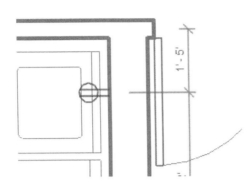 Place it so it is perpendicular to the wall.

16. 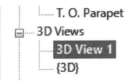 Change the Elevation from Level to **4' 6"**.

Click **ESC** to release the selection.

Electrical Fixtures (1)	⌄
Constraints	
Schedule Level	Main Floor
Elevation from Level	4' 6"
Host	Basic Wall : Inte
Offset from Host	0' 0"

17.
```
          T. O. Parapet
   ⊟  3D Views
        3D View 1
          {3D}
```
Open the 3D view.

18. 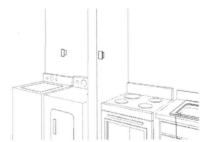 The receptacles now look like they are located properly.

Save as *ex7-6.rvt*.

Legends

Legends appear as tables and typically have two columns: one column for graphic symbols and one column with an explanatory description of the symbol. Typical legends include annotations, model elements, line styles, filled regions, and detail lines.

Legends can be placed on multiple sheets in a project. They can also be shared across projects.

Guidelines for Creating Legends

- Set up legends based on the standard documentation of your organization and load them in the project templates. Pre-loaded legends save time and ensure consistency across documentation packages.
- Import CAD files with legends to re-use legacy data.

Exercise 7-7:

Creating a Legend

Drawing Name: *legends.rvt*
Estimated Time: 10 minutes

This exercise reinforces the following skills:
- Legends
- Views
- Families
- Project Browser

1.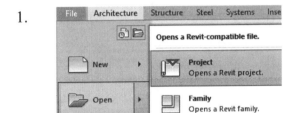

   Go to **File**.

   Select **Open→Project**.

   Open *legends.rvt*.

2.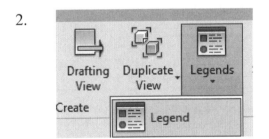

   Activate the View ribbon.

   Select the **Legend** tool.

3.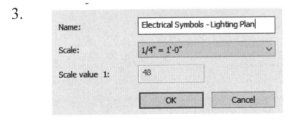

   Type **Electrical Symbols – Lighting Plan**.

   Click **OK**.

4.

   Note in the Project Browser there is now a category called Legends with the new view listed.

5.

- Electrical Fixtures
  - Duplex Receptacle
    - GFCI
    - Standard
  - Light_Switch_-_Parametric_14926
    - 1 - Single
    - 2 - Double
    - 3 - Triple

In the Project Browser:
Scroll down to the Families category.
Locate the Light_Switch_-_Parametric.
Place the left mouse button over 1- Single.
Drag and drop it into the Legend Window.

6.

The switch symbol is placed.

7.

Repeat to place the two types of Duplex
Receptacles: GFCI and Standard.

8.

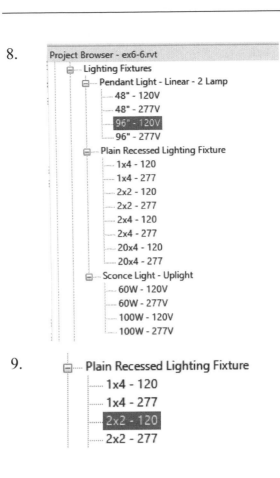

Expand the Lighting Fixtures category.

Drag and drop the **96" – 120V Pendant Light** into the legend.

9.

Drag and drop the **Plain Recessed Lighting Fixture: 2x2 - 120** into the legend.

This is what the legend should look like now.

10.

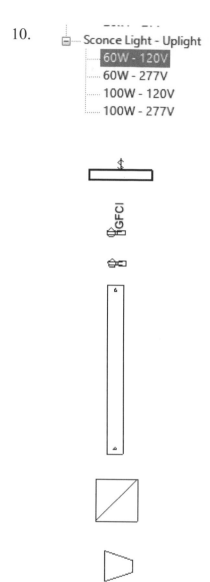

Drag and drop the **Sconce Light – Upright: 60W – 120V** into the legend.

This is what the legend should look like now.

Unfortunately, there is no way to rotate legend components at this time.

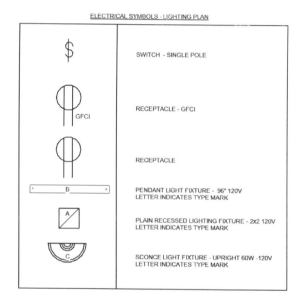

Because Revit does not have good electrical symbology yet for electrical components, many users create their own detail symbols to use in legends. In Lesson 2, I walked you through how to create the symbol for the receptacle.

You can use the detail component families created in Lesson 2 to create a nice-looking legend as shown. I walk you through how to do that in the next exercise.

11.   Save as *ex7-7.rvt*.

Exercise 7-8:

Creating a Legend Using Detail Components

Drawing Name: *legends.rvt*
Estimated Time: 15 minutes

This exercise reinforces the following skills:
- ❑ Views
- ❑ Legend
- ❑ Detail Components
- ❑ Load Families
- ❑ Detail Lines
- ❑ Text

1.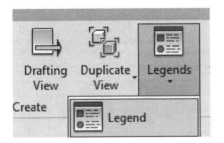

    Go to **File**.

    Select **Open→Project**.

    Open *legends.rvt*.

2.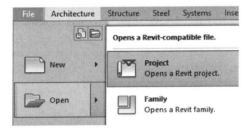

    Activate the View ribbon.

    Select the **Legend** tool.

3.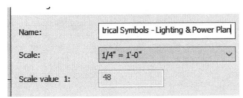

    Type **Electrical Symbols – Lighting & Power Plan**.

    Click **OK**.

4.

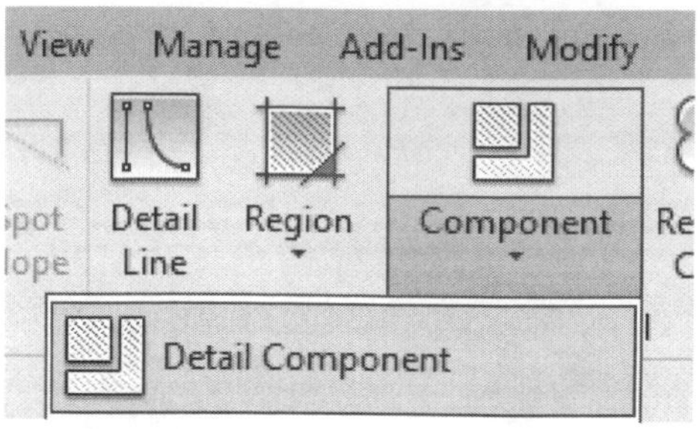

Activate the Annotate ribbon.

Select **Detail Component** from the ribbon.

5.

Select **Load Family**.

6.

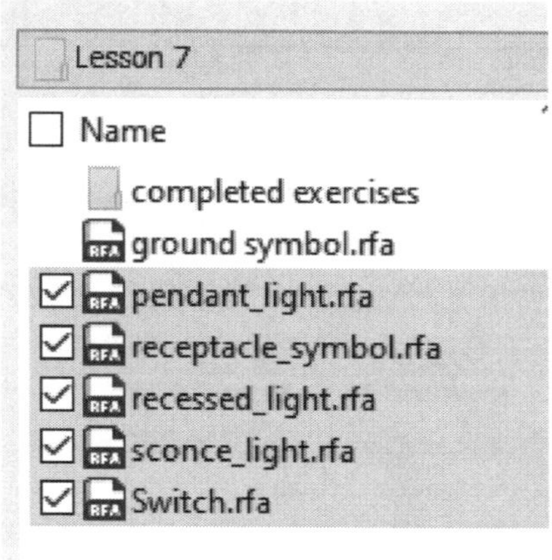

Hold down the **CTL** key.

Select the following files to load:

- pendant_light
- recessed_light
- sconce_light
- switch
- receptacle_symbol

Click **Open**.

7.

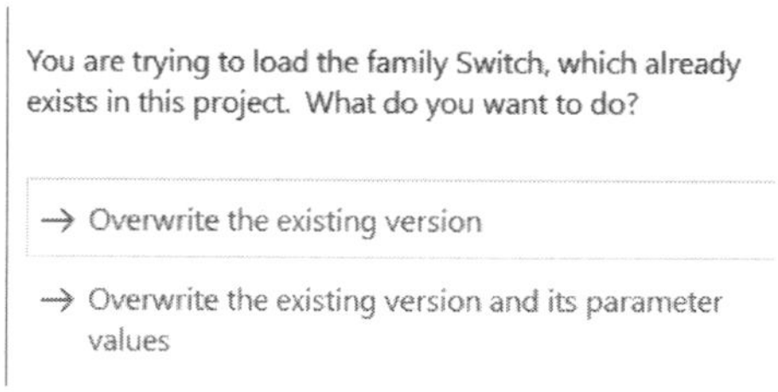

You are trying to load the family Switch, which already exists in this project. What do you want to do?

→ Overwrite the existing version

→ Overwrite the existing version and its parameter values

Click **Overwrite the existing version and its parameter values**.

8.

Use the Type Selector to place the symbols as shown in the view.

9.

A
Text

Select the **Text** tool from the ribbon.

10.

Add the text as shown.

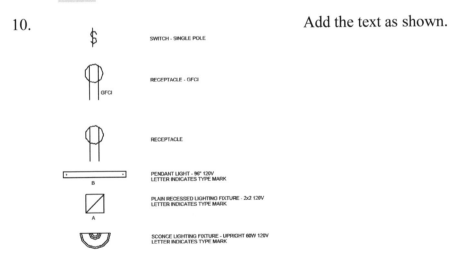

SWITCH - SINGLE POLE

RECEPTACLE - GFCI

RECEPTACLE

PENDANT LIGHT - 96" 120V
LETTER INDICATES TYPE MARK

PLAIN RECESSED LIGHTING FIXTURE - 2x2 120V
LETTER INDICATES TYPE MARK

SCONCE LIGHTING FIXTURE - UPRIGHT 60W 120V
LETTER INDICATES TYPE MARK

11. 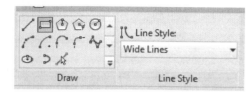 Select the **Detail Line** tool.

12. 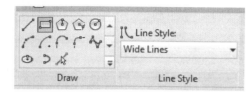 Select the **Rectangle** tool.

Select **Wide Lines**.

13. 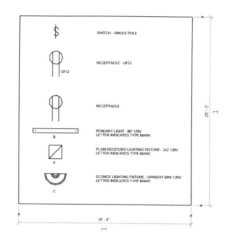 Draw a rectangle around the symbols and text for a border.

14. 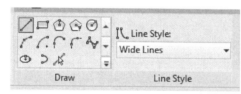 Select the **LINE** tool.

Select **Wide Lines**.

15. 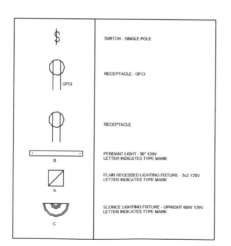 Draw a vertical line to separate the symbols and text.

16. 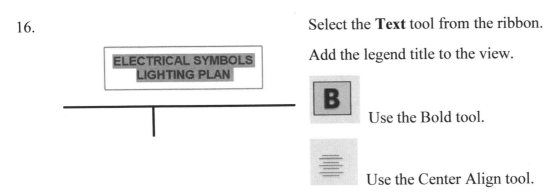 Select the **Text** tool from the ribbon.

Add the legend title to the view.

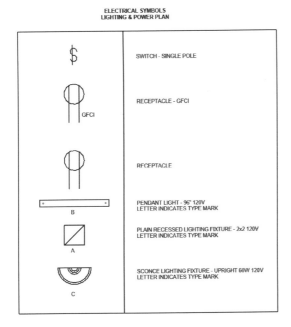 Use the Bold tool.

Use the Center Align tool.

17. Save as *ex7-8.rvt*.

Drafting Views

Use a drafting view to create unassociated, view-specific details that are not part of the modeled design.

Rather than create a callout and add details to it, you may want to create detail conditions where the model is not needed (for example, a view with detailed instructions on how to wire a panel or install a lighting fixture). For this purpose, create a drafting view.

In a drafting view, you create details at differing view scales (coarse, medium, or fine) and use 2D detailing tools: detail lines, detail regions, detail components, insulation, reference planes, dimensions, symbols, and text. These are the exact same tools used in creating a detail view. However, drafting views do not display any model elements. When you create a drafting view in a project, it is saved with the project. Because drafting views are not associated with the model, they are often used across projects. You can transfer drafting views from one project to another using the Insert From File tool on the Insert ribbon.

Exercise 7-9:

Creating a Drafting View

Drawing Name: *drafting_view.rvt*
Estimated Time: 60 minutes

This exercise reinforces the following skills:
- ❑ Drafting View
- ❑ Import AutoCAD Dwg file
- ❑ Line Style
- ❑ Detail Line
- ❑ Filled Region
- ❑ Text
- ❑ Family Types
- ❑ Organizing the Project Browser using Disciplines and Sub-Disciplines

1.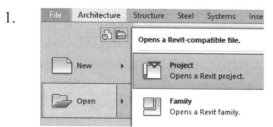
    Go to **File**.

    Select **Open→Project**.

    Open *drafting_view.rvt*.

2. Activate the View ribbon.

    Select the **Drafting View** tool.

3. Type **Ground Bar Detail**.

    Click **OK**.

4. The view is located in the Project Browser under Drafting Views.

5.

Visibility/Graphics Overrides	
Discipline	Electrical
Sub-Discipline	Power
Visual Style	Hidden Line

Change the Discipline to **Electrical**.

Change the Sub-Discipline to **Power**.

6.

- Power
  - Floor Plans
    - Level 1 Power Plan
    - Level 2 Power Plan
    - Level 3 Power Plan
  - Detail Views (Detail)
    - Electrical Riser Diagram
  - Drafting Views (Detail)
    - **Ground Bar Detail**

Notice in the Project Browser that the Drafting View has moved position to be under Power.

7. Activate the **Insert** ribbon.

8. Select **Import CAD.**

9.

Locate the *ground bar.dwg* file.

Set Colors to **Preserve**.

Set Layers to **All**.

Set Import Units to **Inch**.

Set Positioning to **Auto-Origin to Internal Origin**.

Click **Open**.

Note you see a preview of the drawing in the Preview window.

10.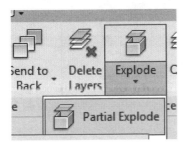

Select the imported drawing.

Select **Partial Explode** from the ribbon.

11.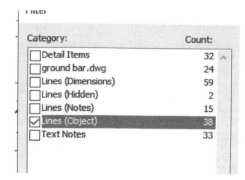

Window around the top view.

Filter

Use FILTER to select **Lines (Object).**

Click **OK.**

Note that the AutoCAD objects are designated by their layer name.

12.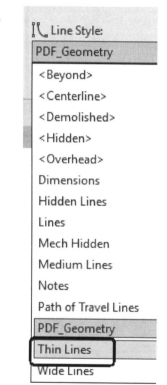

Select **Thin Lines** from the ribbon.

13.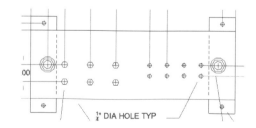

The object looks cleaner since the lines are now using Revit line styles.

14. 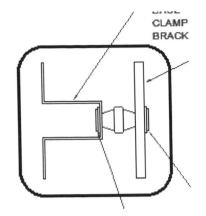 Window around the bottom view.

Use FILTER to select **Lines (Object).**

Click **OK.**

15. Select **Thin Lines** from the ribbon.

16. Activate the Annotate ribbon.

Select **Filled Region**.

17. Select the **Rectangle** tool.

18. Draw a rectangle behind the base clamp/bracket.

19. 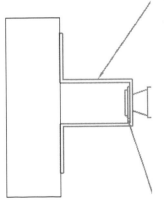 Use ALIGN/MOVE to position the rectangle so it touches the bracket.

20.  Select **Edit Type** to create a new filled region style.

21. 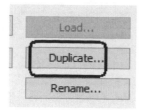 Select **Duplicate**.

22.  Type **Plaster**.

    Click **OK**.

23.  Left click in the **Foreground Fill Pattern** value column.

    Select the **Gypsum/Plaster** pattern.

    Click **OK**.

24.  *The fill pattern is displayed in the dialog.*

    Click **OK** to close the Type Properties dialog.

25. Click the **Green Check** on the ribbon to close the command.

26. 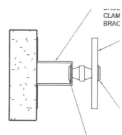 Left click anywhere in the display window to release the selection.

    Review the filled region that was placed.

27. 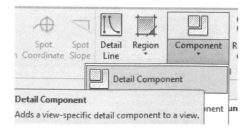 Select **Detail Component** from the Annotate ribbon.

28.  Select **Load Family**.

29. 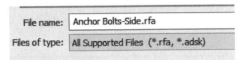 Browse to the *Metal Fastenings* folder under *Detail Items\Div 05-Metals\Common Work Results for Metals*.

30.  Locate the *Anchor Bolts-Side.rfa*.

Click **Open**.

31. On the Type Selector, select the **3/8"** size.

Anchor Bolts-Side
3/8"

32. 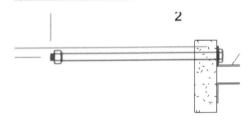 Click the SPACE BAR to rotate the anchor bolt.

Place the bolt at the midpoint at the top side of the bracket.

33. 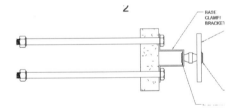 Left click to place a second bolt at the bottom of the bracket.

34. 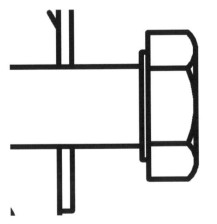 Move the bolt slightly away from the bracket.

35. 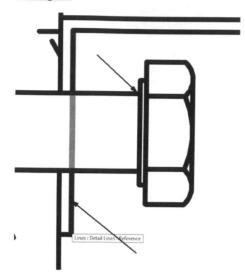 Select the **ALIGN** tool on the Modify ribbon.

36. 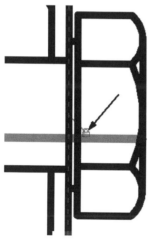 Select the top face of the bracket as the target (the item to be ALIGNED TO).

Select the bottom face of the bolt as the item to shift.

37. 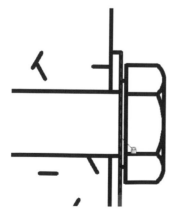 Left click on the LOCK icon to ensure the bolt remains aligned to the bracket.

38. Use the **ALIGN** tool to position the head of the second bolt to the top of the bracket and lock it into position.

By using the LOCK, you ensure the bolt remains in that position.

Click **ESC** to cancel out of the ALIGN command.

39. 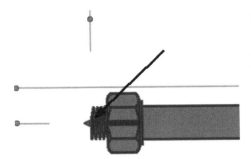 Select the top bolt.

Note there is a grip at the left end that allows you to adjust the length.

40. 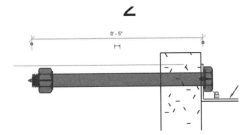 Position the extension lines of the temporary dimension so you can view the length of the bolt.

Then use the left grip to adjust the length to 5".

41. 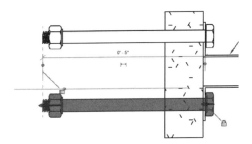 Repeat for the second bolt.

42. 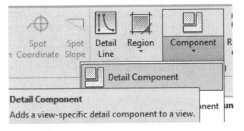 Select **Detail Component** from the Annotate ribbon.

43.  Select **Load Family**.

44. 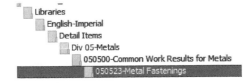 Browse to the *Metal Fastenings* folder under *Detail Items\Div 05-Metals\Common Work Results for Metals*.

45.

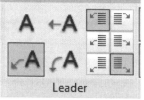

Locate the *A325 Bolts-Head.rfa*.

Click **Open**.

46.

A325 Bolts-Head
5/8"

On the Type Selector, select the **5/8"** size.

47.

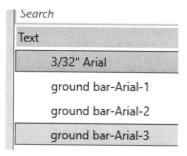

Place a bolt on the middle circle on each side of the ground bar.

48.

A

Text

Select the **Text** tool from the Annotate ribbon.

49.

A ←A

A ⤸A

Leader

Enable **Leader**.

50.

Search

Text

3/32" Arial

ground bar-Arial-1

ground bar-Arial-2

ground bar-Arial-3

Select **ground bar – Arial-3** from the Type Selector.

51.

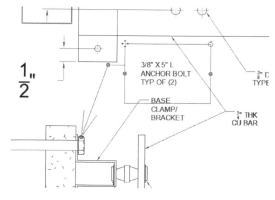

Type **3/8" x 5" L ANCHOR BOLT TYPE OF (2)**.

Exit the text command.

52. Save as *ex7-9.rvt*.

Controlling View Display

One method for managing construction documentation is to create a ceiling view that shows only the ceiling objects, not the ceiling grid. This view can be placed on a sheet in the same location as the lighting floor plan view. Revit will automatically snap the view to the same location as the floor plan view so that the views are aligned.

Many architects don't put in the ceilings until later in the design. This can hold up the electrical design. You can place a reference plane at the same height as the future ceiling and mount your lighting fixtures to the reference plane. This will eliminate coordination issues when the architect finally gets around to placing ceilings.

Early in the project, the architect and the electrical designer should agree on who will model the ceilings and place the light fixtures and whether the light fixtures should reside in the architectural model or in the electrical model.

Exercise 7-10:

Controlling the Display in Views

Drawing Name: *view_display.rvt*
Estimated Time: 15 minutes

This exercise reinforces the following skills:
- ❑ Duplicate View
- ❑ Visibility/Graphics
- ❑ Rename View
- ❑ Tag All

1.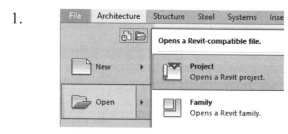

   Go to **File**.

   Select **Open→Project**.

   Open *view_display.rvt*.

2. Open the **Level 2 Lighting Plan** floor plan.

3.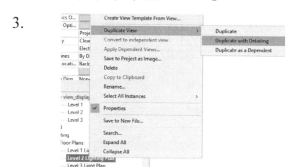

   Highlight the **Level 2 Lighting Plan** floor plan view in the Project Browser.

   Right click and select **Duplicate View→Duplicate.**

   This will copy the view without annotations.

4. Highlight the copied view.

Right click and select **Rename**.

5. Rename the view **Level 2 Lighting Plan - Annotated**.

6. 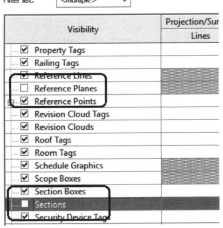 Select the **Visibility/Graphics** tool from the View ribbon.

7. Disable **Reference Planes** and **Sections** on the Annotation Categories tab.

Click **OK**.

8. Activate the **Annotate** ribbon.

Select **Tag All**.

9.

All objects in current view
Only selected objects in current view
☐ Include elements from linked files

☑ Category	Loaded Tags
☐ Air Terminal Tags	Diffuser Tag
☐ Duct Tags	Duct Size Tag
☑ Electrical Equipment Tags	Panel Name
☑ Lighting Fixture Tags	Lighting Fixture Tag : Standard
☐ Mechanical Equipment Tags	Mechanical Equipment Tag
☑ Room Tags	Room Tag
☐ Space Tags	Space Tag
☐ Wire Tags	Wire Tag
☐ Zone Tags	Zone Tag

Enable

- Electrical Equipment Tags

- Lighting Fixture Tags

- Room Tags

Click **OK**.

10.

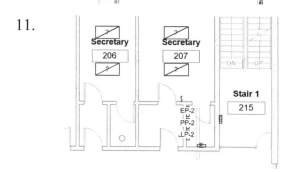

Zoom in to see the lighting fixtures and rooms have been tagged.

11.

? symbols will appear on tags where the device name has not been designated in the Properties palette.

12.

⊟···· Floor Plans
···· Level 1 Lighting Plan
···· Level 2 Lighting Plan
···· Level 2 Lighting Plan Annotated

Open the **Level 2 Lighting Plan** floor plan.

Notice how it is different from the Annotated plan.

13.   Save as *ex7-10.rvt*.

Browser Organization

Many users find it more productive to organize or sort the project browser using view parameters.

Use the Browser Organization tool to group and sort views, sheets, and schedules/quantities in the way that best supports your work. You can specify six levels of grouping. Within groups, items are sorted in ascending or descending order of a selected property.

By default, the Project Browser displays all views (by view type), all sheets (by sheet number and sheet name), and all schedules and quantities (by name).

In addition to grouping and sorting views, you can limit the views that display in the Project Browser by applying a filter. This approach is useful when the project includes many views, sheets, and schedules/quantities, and you want the Project Browser to list a subset. You can specify up to 3 levels of filtering.

The combination of criteria used to filter, group, and sort items in the Project Browser is an organization scheme. You can create separate schemes for views, sheets, and schedules/quantities in the project. Create as many schemes as needed to support different phases of your work.

When creating an organization scheme, you can use any properties of the view, sheet, or schedule/quantity as the criteria for grouping, sorting, and filtering. You can also use project parameters and shared parameters for grouping and filtering.

When working on a project, you can quickly change the organization schemes applied to the Project Browser at any time. Switch between organization schemes whenever needed based on your current work.

Exercise 7-11:

Organize Views in the Project Browser

Drawing Name: *project_browser.rvt*
Estimated Time: 20 minutes

This exercise reinforces the following skills:
- ❏ Project Parameters
- ❏ Project Browser
- ❏ Properties

1.
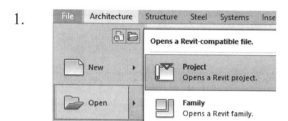
Go to **File**.

Select **Open→Project**.

Open *project_browser.rvt*.

2.
Activate the Manage ribbon.

Select **Project Parameters**.

3.
Select **New**.

4.
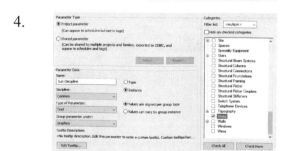
Type **Sub-Discipline** in the Name field.

Enable **Instance**.

Select **Text** as Type of Parameter.

Select **Graphics** for Group parameter under.

Enable **Views** in the Categories pane.

Click **OK**.

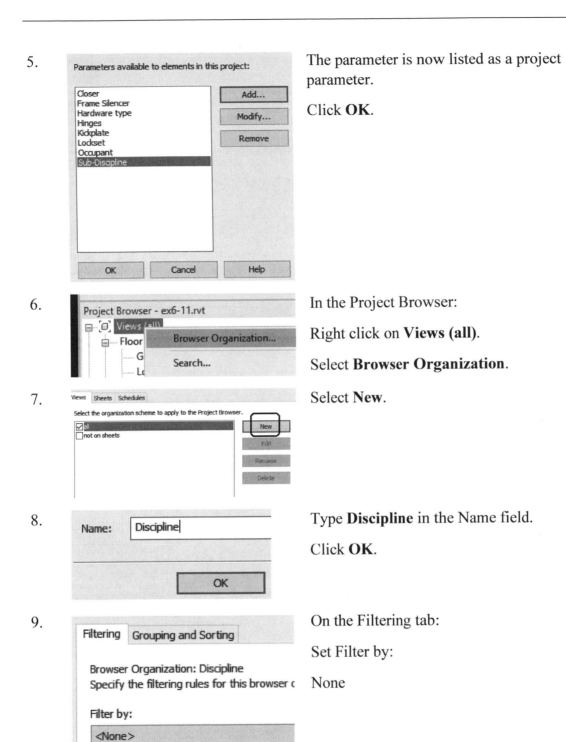

5.   The parameter is now listed as a project parameter.

Click **OK**.

6.   In the Project Browser:

Right click on **Views (all)**.

Select **Browser Organization**.

7.   Select **New**.

8.   Type **Discipline** in the Name field.

Click **OK**.

9.   On the Filtering tab:

Set Filter by:

None

10.

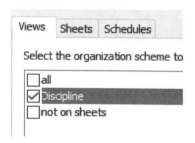

On the Grouping and Sorting tab:

Group by **Discipline**

Then by **Sub-Discipline**

Then by **Family and Type**

Sort by **View Name**

Click **OK**.

11.

Place a check next to **Discipline**.

Click **OK**.

12.

The browser is re-organized.

There is a ??? because no sub-discipline values have been assigned to any of the views.

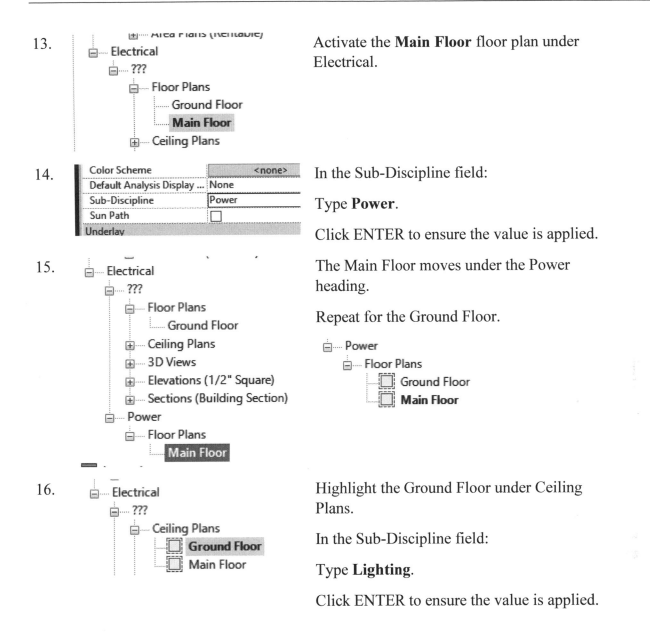

13. Activate the **Main Floor** floor plan under Electrical.

14. In the Sub-Discipline field:

Type **Power**.

Click ENTER to ensure the value is applied.

15. The Main Floor moves under the Power heading.

Repeat for the Ground Floor.

16. Highlight the Ground Floor under Ceiling Plans.

In the Sub-Discipline field:

Type **Lighting**.

Click ENTER to ensure the value is applied.

17.

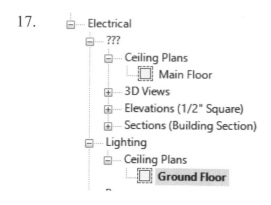

The Ground Floor moves under Lighting.

Repeat for the Main Floor ceiling plan.

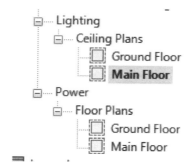

18.   Save as *ex7-11.rvt*.

View Lists

A View List schedule is a type of schedule that helps you keep track of the views within a project.

A View List schedule allows you to quickly view the properties of several views at one time. It allows for quick comparison and editing of multiple views. You can use the View List to verify that the correct views have been placed on each sheet for a submittal package. This means you don't have to inspect each sheet or print out a bunch of paper to check your sheets.

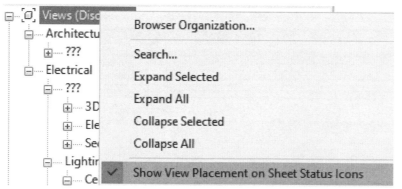

Keep in mind that the Project Browser can be set to show which views have been placed on sheets by enabling Show View Placement on Sheet Status icons.

If enabled, then the square next to the view name will be colored if the view has been placed on a sheet.

Exercise 7-12:

Create a View List

Drawing Name: *view_list.rvt*
Estimated Time: 10 minutes

This exercise reinforces the following skills:
- ❑   Schedules
- ❑   Views
- ❑   Sheets

1.   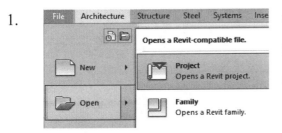   Go to **File**.

Select **Open→Project**.

Open the *view_list.rvt*.

2.   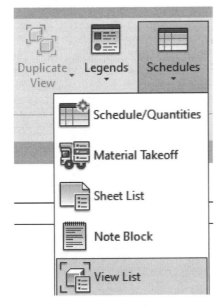   Activate the View ribbon.

Select **View List**.

3.

Select the following fields:

- View Name
- View Template
- Title on Sheet
- Sheet Number
- Sheet Name

4.

On the Sorting/Grouping tab:

Sort by: **Sheet Number**.

Click **OK**.

5.

You can see which views have been placed on sheets and which views are still unassigned.

6.   Save as *ex7-12.rvt*.

View Templates

View templates control the settings of a view. Instead of constantly going into the Visibility/Graphics dialog and turning on and off the visibility of elements, you can save those settings, set up a view template and apply the template to the view.

A view template is a collection of view properties, such as view scale, discipline, detail level, and visibility settings.

Use view templates to apply standard settings to views. View templates can help to ensure adherence to office standards and achieve consistency across construction document sets.

Before creating view templates, first think about how you use views. For each type of view (floor plan, elevation, section, 3D view, and so on), what styles do you use? For example, an electrical designer may use many styles of floor plan views, such as power and signal, lighting fixture, wiring, coordination, and equipment.

You can create a view template for each style to control settings for the visibility/graphics overrides of categories, view scales, detail levels, graphic display options, and more.

Guidelines for View Templates

- Use view templates to ensure project views are consistent.

- Set up view templates in any project templates so new projects have access to those settings.

- Copy view templates from other projects using the Transfer Project Standards tool. This allows you to reuse templates without having to start over from scratch.

- Apply view templates to all views on a sheet by right-clicking on the sheet name in the Project Browser and selecting the Apply View Template to All Views.

- Enable Show Views when applying a view template. This allows you to apply the view settings from another view without saving those settings as a view template.

- Define a default view template for each view type.

Exercise 7-13:

Using a View Template

Drawing Name: *view_template.rvt*
Estimated Time: 30 minutes

This exercise reinforces the following skills:
- ❏ Duplicate View
- ❏ View Range
- ❏ Visibility/Graphics Settings
- ❏ View Template
- ❏ Edit Circuit

1. 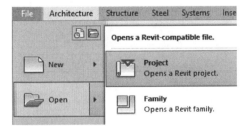   Go to **File**.

   Select **Open→Project**.

   Open *view_template.rvt*.

2.

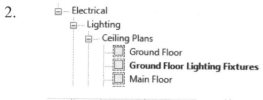

Open the **Ground Floor Lighting Fixtures** view under Ceiling Plans.

3.

Highlight the view.

Right click and select **Duplicate→Duplicate with Detailing**.

This duplicates the view and includes annotations.

4.

Right click on the copied view.

Select **Rename**.

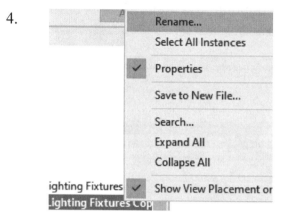

5.

Rename **Ground Floor Lighting Fixtures & Switches.**

6.

In the Properties palette:

Select **Edit** next to Visibility/Graphics overrides.

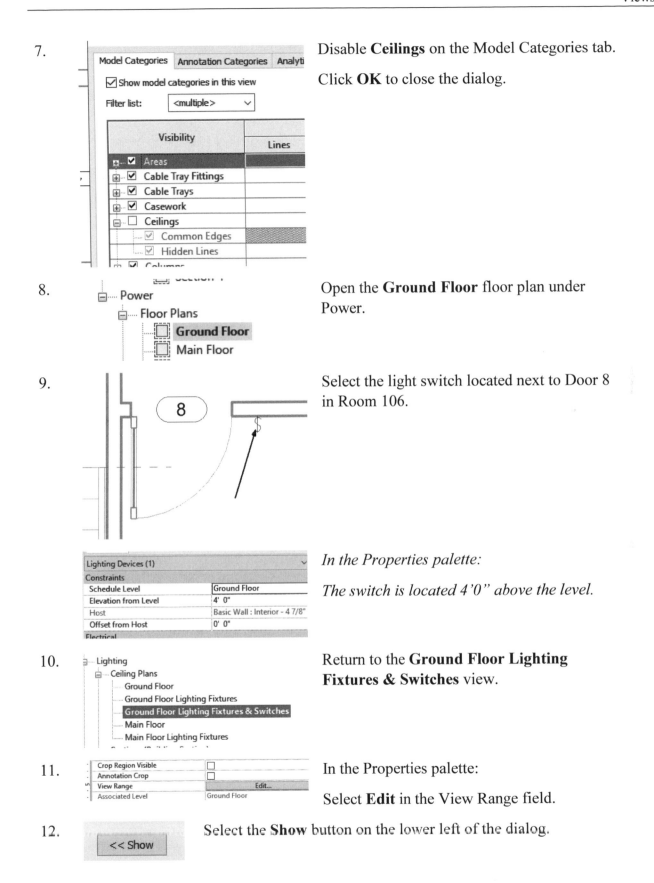

7.

Disable **Ceilings** on the Model Categories tab.

Click **OK** to close the dialog.

8.

Open the **Ground Floor** floor plan under Power.

9.

Select the light switch located next to Door 8 in Room 106.

In the Properties palette:

The switch is located 4'0" above the level.

10.

Return to the **Ground Floor Lighting Fixtures & Switches** view.

11.

In the Properties palette:

Select **Edit** in the View Range field.

12.

Select the **Show** button on the lower left of the dialog.

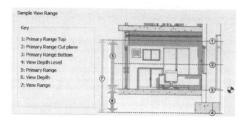

The image on the left explains how view range is defined.

13.

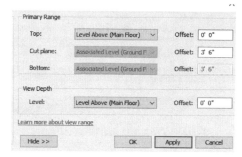

Change the Cut plane to **3' 6"**.

Click **Apply**.

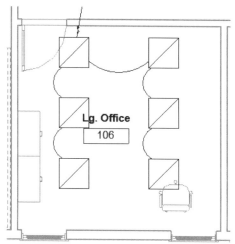

The view updates so that the switches, doors, and furniture are now visible. This is because the cut plane is set lower than the height of the switches.

Click **OK**.

14.

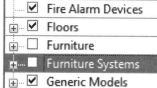

Type **VV** to launch the Visibility/Graphics dialog.

Disable **Furniture and Furniture Systems** on the Model Categories tab.

Click **OK** to close the dialog.

15.

Select the light switch located next to Door 8 in Room 106.

Select the **Power** button on the ribbon.

16.

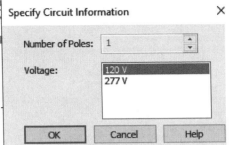

Assign it to the **120V**.

Click **OK**.

Click **ESC** to exit the selection.

17.

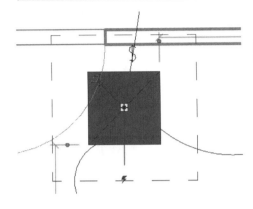

Select one of the light fixtures in Room 106.

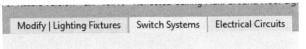

Select the Switch Systems tab on the ribbon.

Select **Select Switch** from the ribbon.

18.

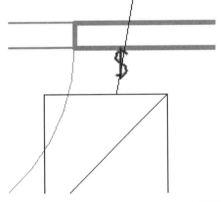

Select the Switch in the room to add it to the circuit.

Left click anywhere in the window to release the selection.

19.

Activate the View ribbon.

Select **View Templates→Create Template from Current View**.

This saves the view settings of the active view.

20.

Type **Ceiling Lights & Switches**.

Click **OK**.

21.

The view template is now listed.

Click **OK**.

22.

On the Properties palette:

Select **None** in the View Template field.

23.

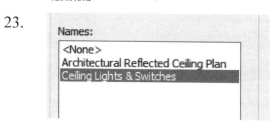

Highlight **Ceiling Lights & Switches**.

Click **OK**.

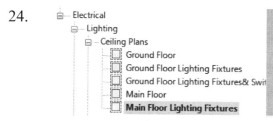

The template is now displayed in the View Template field on the Properties palette.

24.

Open the **Main Floor Lighting Fixtures** ceiling plan view.

25.

Highlight the view.

Right click and select **Duplicate→Duplicate with Detailing**.

26.

| Rename... |
| Select All Instances |
| ✓ Properties |
| Save to New File... |
| Search... |
| Expand All |
| Collapse All |
| ✓ Show View Placement on |

ing Fixtur
ing Fixtur

ɟ Fixtures
ɟ Fixture:

Right click on the copied view.

Select **Rename**.

27.

Ground Floor Lighting Fixtures & Switches
Main Floor
Main Floor Lighting Fixtures
Main Floor Lighting Fixtures & Switches

Rename **Main Floor Lighting Fixtures & Switches.**

28.

Depth Clipping	No clip
Identity Data	
View Template	<None>
View Name	Main Floor Lighting Fixtures & Switc...
Dependency	Independent

On the Properties palette:

Select **None** in the View Template field.

29.

Names:

<None>
Architectural Reflected Ceiling Plan
Ceiling Lights & Switches

Highlight **Ceiling Lights & Switches**.

Click **OK**.

The template is now displayed in the View Template field on the Properties palette.

You can see the switches and doors, but the furniture is not visible.

Kitcl
21

Dining

30.   Save as *ex7-13.rvt.*

Exercise 7-14:

Modifying View Tag Properties

Drawing Name: *view_tags.rvt*
Estimated Time: 20 minutes

This exercise reinforces the following skills:
- ❑ Sheets
- ❑ Views
- ❑ View Tags

1. 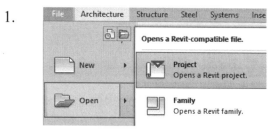 Go to **File**.

   Select **Open→Project**.

   Open *view_tags.rvt*.

2. 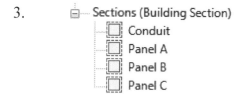 Open the **06 Panels** sheet.

3. 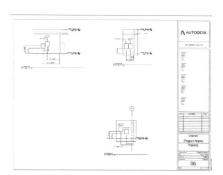 Drag **Panel A, Panel B** and **Panel C** views onto the sheet.

   Did you notice how the icons changed color as the views were placed?

4.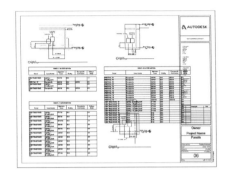

Drag the three Panel Information schedules (A, B and C) onto the sheet.

5.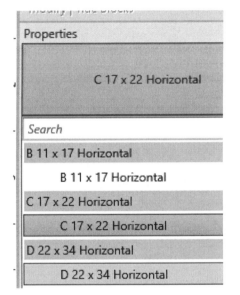

The views don't all fit on the sheet.

Select the titleblock.

6.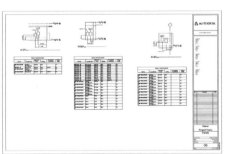

Use the Type Selector on the Properties palette to change to a D size titleblock.

7.

Reposition the views on the sheet.

8.

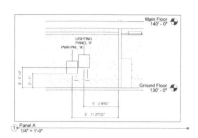

Select the Panel A view.

9.

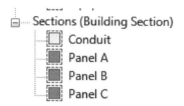

In the Properties palette:

Type **Lighting & Power Panels 'A'** in the Title on Sheet field.

Click **ENTER**.

The View Name remains the same.

The title bar on the view updates with the new name.

10.

Select the titlebar for Panel B.

On the Properties palette: select **Edit Type**

11.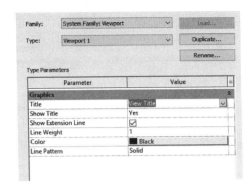

Note that the Title uses an annotation family called View Title.

Click **OK**.

12.

Activate the Insert ribbon.

Select **Load Family**.

13.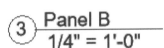

Browse to the Annotations folder.

Open the *View Title – Square w Sheet.rfa*.

14.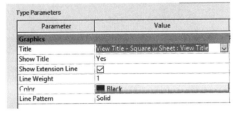

Select the titlebar for Panel B.

On the Properties palette: select **Edit Type**

15.

Select **Duplicate**.

16.

Type **Viewport w Square**.

Click **OK**.

17.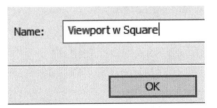

In the Title field:

Select the **View Title – Square w Sheet**.

Click **OK**.

18. |3'-   The titlebar updates.

3	Panel B
06	1/4" = 1'-0"

19.   Click **OK**.

20.   Save as *ex7-14.rvt*.

Exercise 7-15:

Create a View Tag Family

Drawing Name: View_Title.rfa
Estimated Time: 15 minutes

This exercise reinforces the following skills:
- Views
- View Tags

1.   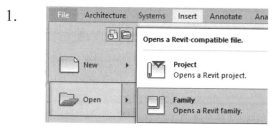   Go to **Open→Family**.

2.   Libraries
       English-Imperial
       Annotations   Browse to the *Annotations* folder under Library.

3.   File name: View Title.rfa
       Files of type: Family Files (*.rfa, *.adsk)   Open *View Title.rfa*.

4.   File name: View Title w Discipline.rfa
       Files of type: Family Files (*.rfa)   Save as *View Title w Discipline.rfa*.

5.

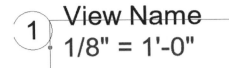

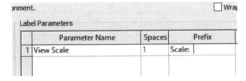

Left click to select the scale label.

On the Properties palette:

Select **Edit** in the Label field.

6.

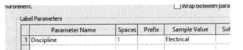

In the Prefix column:

Type **Scale:**

Add two spaces after the colon.

Click **OK**.

7.

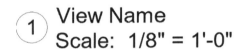

The label updates.

Note the spaces.

Left click anywhere in the display window to release the selection.

8.

Select the **Create** ribbon.

9.

Select the **Label** tool.

Left click in the display window to place the label.

10.

Select **Discipline**.

Type **Electrical** in the Sample Value field.

Click **OK**.

11.

Cancel out of the Label command.

Select the **Discipline** label.

Select **Edit Type** from the Properties palette.

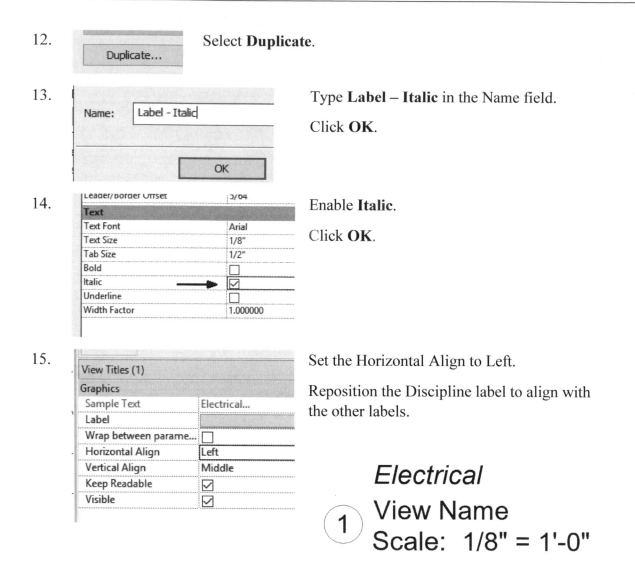

12.   Select **Duplicate**.

13.   Type **Label – Italic** in the Name field.
      Click **OK**.

14.   Enable **Italic**.
      Click **OK**.

15.   Set the Horizontal Align to Left.
      Reposition the Discipline label to align with the other labels.

Electrical

① **View Name**
   Scale: 1/8" = 1'-0"

16.   Save the family as **View Title w Discipline**.

Scope Boxes

Scope boxes are used to control the extents of datum elements, like grids, levels and reference planes. Each of these elements can be assigned to a specific scope box, limiting the 3D extents to the dashed green line limit. Scope boxes are not visible when printing.

To create a scope box, you have to use either a Floor Plan View or a Reflected Ceiling Plan View. However, once a scope box is created, it is going to be visible in the other views: sections, callouts, elevations and 3D views. In elevations and sections, the scope box is only going to be visible if it intersects the cut line. You can adjust the extents of the scope box in all views.

The moment a scope box is assigned to a view, the **Crop Region** is locked and can't be modified. Also, you can't use the **Do Not Crop View** tool. To see the whole project in a view, you'll have to create a different plan or remove the scope box temporarily.

A scope box can also be used to control the angle of a view. If you duplicate the view, the extents are assigned to the new scope box. The crop region is automatically adjusted to fit the angle. Removing the scope box from a view will revert the crop angle back to default.

Check the option bar when creating a scope box: you can assign a name to the scope box and enter a height value. If you need to change the height value, switch to an elevation view and use the grips on the scope box to adjust the height.

Exercise 7-16:

Using Scope Boxes

Drawing Name: scope_box.rvt
Estimated Time: 20 minutes

This exercise reinforces the following skills:
- ❑ Duplicate Views
- ❑ Apply Dependent Views
- ❑ Scope Box

1.

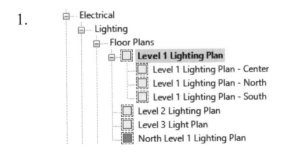

In the Project Browser:

Notice that Level 1 Lighting Plan has three dependent views:

- Center

- North

- South

We want to add three dependent views for Level 2 and Level 3.

2.

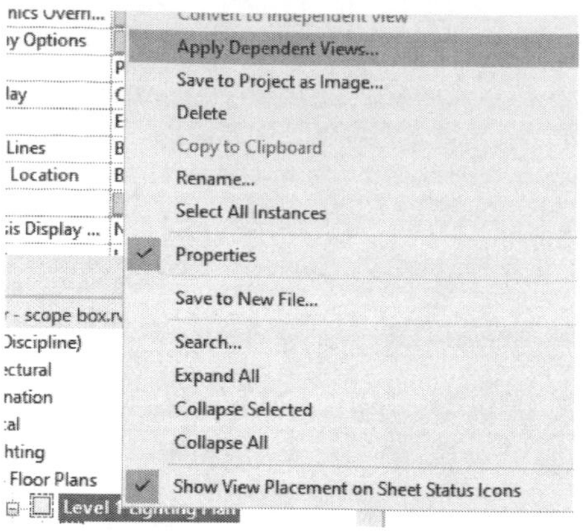

Highlight the **Level 1 Lighting Plan** floor plan.

Right click and select **Apply Dependent Views**.

3.

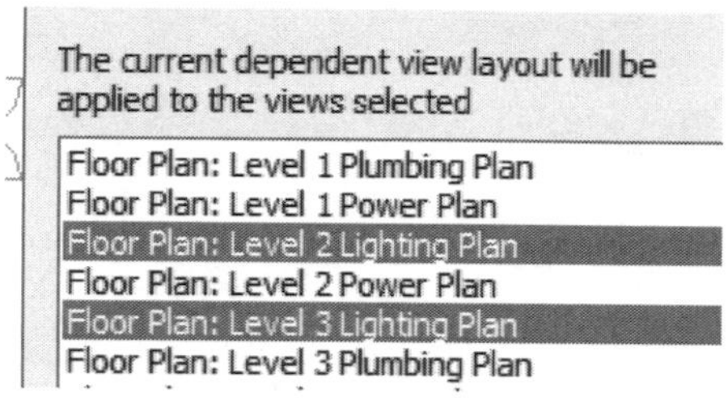

Hold down the **CTL** key.

Select the **Level 2: Lighting Plan** and the **Level 3: Lighting Plan**.

Click **OK**.

4.

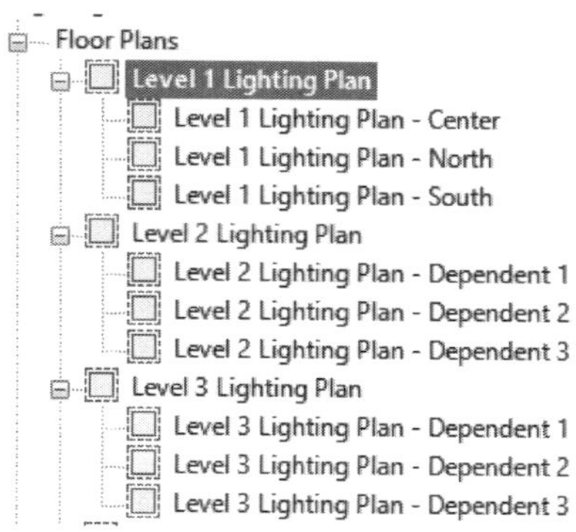

The dependent views are created.

Rename the Dependent views:

- Center
- North
- South

You can use F2 to edit the view name.

If you open the views, you will see they are already cropped and set up to the correct view.

5.

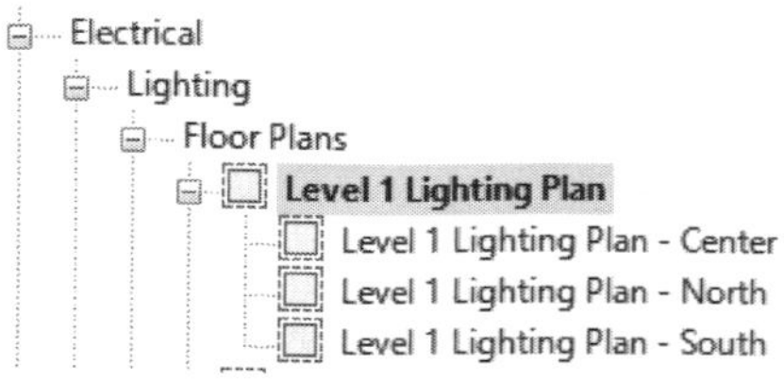

Open the **Level 1 Lighting Plan**.

6. 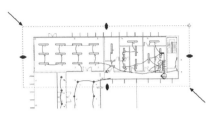 Switch to the **View** ribbon.

Select **Scope Box** on the Create panel on the ribbon.

7. 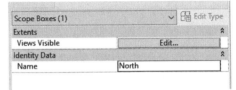 Draw a rectangle around the top section of the floor plan.

8. 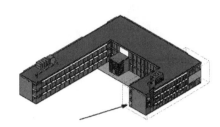 In the Name field, type **North**.

9.

Switch to a 3D view.

You can see the scope box around the model.

10. 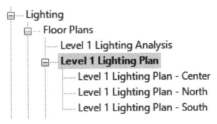 Return to the **Level 1 Lighting Plan** floor plan.

11. 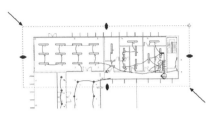 Switch to the **View** ribbon.

Select **Scope Box** on the Create panel on the ribbon.

12.

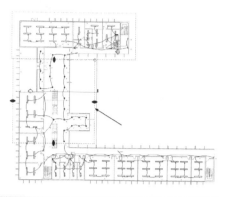

Draw a rectangle around the center section of the floor plan.

13.

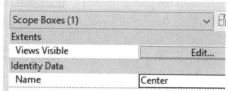

In the Name field, type **Center**.

14.

Switch to the **View** ribbon.

Select **Scope Box** on the Create panel on the ribbon.

15.

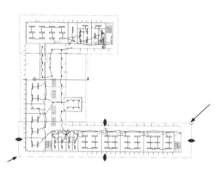

Draw a rectangle around the south section of the floor plan.

16.

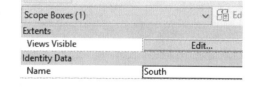

In the Name field, type **South**.

17.

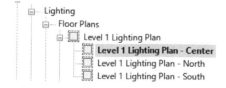

Open the **Level 1 Lighting Plan-Center**.

18.

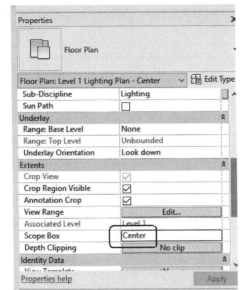

In the Properties palette:

Locate the Scope Box field:

Select **Center** from the drop-down list.

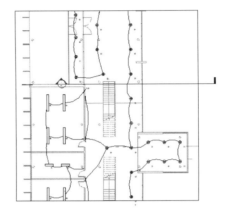

The view displays only the area inside the center scope box.

19.

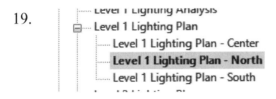

Open the **Level 1 Lighting Plan-North**.

20.

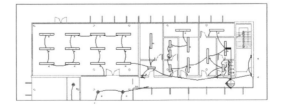

In the Properties palette:

Locate the Scope Box field:

Select **North** from the drop-down list.

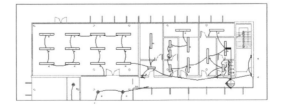

The view displays only the area inside the North scope box.

21.

Level 1 Lighting Plan

    Level 1 Lighting Plan - Center

    Level 1 Lighting Plan - North

    Level 1 Lighting Plan - South

Open the **Level 1 Lighting Plan-South**.

22.

Extents	
Crop View	☑
Crop Region Visible	☑
Annotation Crop	☑
View Range	Edit...
Associated Level	Level 1
Scope Box	South
Depth Clipping	No clip

In the Properties palette:

Locate the Scope Box field:

Select **North** from the drop-down list.

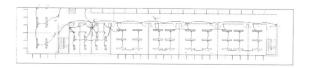

The view displays only the area inside the South scope box.

23.

Floor Plan: Level 2 Lighting Plan - North ⌄   Edit T⟨

Sub-Discipline	Lighting
Sun Path	☐
Underlay	
Range: Base Level	Level 1
Range: Top Level	Level 2
Underlay Orientation	Look down
Extents	
Crop View	☑
Crop Region Visible	☑
Annotation Crop	☑
View Range	Edit...
Associated Level	Level 2
Scope Box	North
Depth Clipping	No clip

Repeat for the Level 2 Lighting Plan dependent views.

24.

Floor Plan: Level 3 Lighting Plan - South ⌄   Edit Ty⟨

Sub-Discipline	Lighting
Sun Path	☐
Underlay	
Range: Base Level	None
Range: Top Level	Unbounded
Underlay Orientation	Look down
Extents	
Crop View	☑
Crop Region Visible	☑
Annotation Crop	☑
View Range	Edit...
Associated Level	Level 3
Scope Box	South
Depth Clipping	No clip

Repeat for the Level 3 Lighting Plan dependent views.

25.   Save as *ex7-16.rvt*.

Exercise 7-17:

Using Scope Boxes to Control Grid Display

Drawing Name: scope_box_2.rvt
Estimated Time: 20 minutes

This exercise reinforces the following skills:
- Grids
- Scope Box
- Visibility

1.

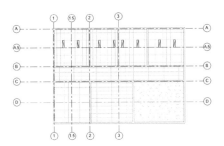

Locate the ceiling plans under Electrical→Lighting.

2.

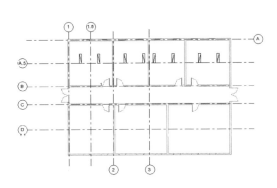

The plans show several grid lines.

We can use a scope box to control which grid lines are displayed in a view.

3.

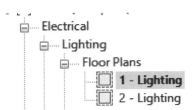

Activate the **Level 1** floor plan under Lighting so you can see how the grids are displayed.

4.

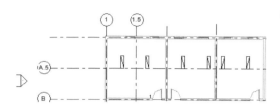

Some of the bubbles on the grid lines are disabled.

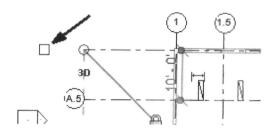

To make the bubbles visible, select the grid line and place a check in the box at the end of the grid line.

5.

Scope
Box

Select **Scope Box** from the View ribbon.

6.

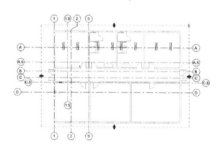

Draw the scope box around the entire building.

7.

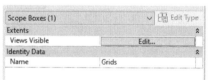

Name the Scope Box **Grids** in the Properties palette.

8.

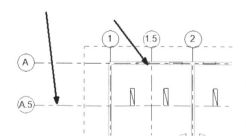

Select Grids **1.5** and **A.5**.

In the Properties palette:

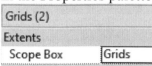

Assign the grids to the Grids **Scope Box**.

9.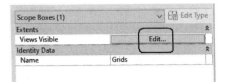

Select the Scope Box.

In the Properties palette:
Select **Edit** next to Views Visible.

10.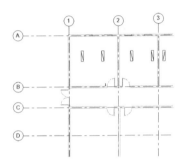

Set the grids assigned to the scope box to be invisible in all the ceiling plans.

11.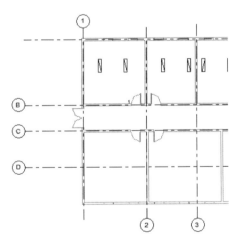

Set the selected grids to be invisible in Lighting and Power floor plans.

Click **OK**.

12.

Notice the view updated to not display the selected grids.

13.

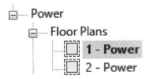

Open the **1-Power** floor plan.

Notice the view updated to not display the selected grids.

14.

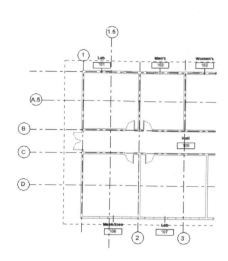

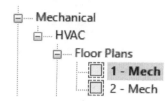

Open the **1-Mech HVAC Mechanical** floor plan.

Notice the grids and scope box are visible in this view.

15.   Save as *ex7-17.rvt*.

Lab Exercises

Open Lesson_7_lab.rvt.

Place the Sheet List schedule on the 00 – Sheet List sheet.

Copy the legend from ex7-8.rvt to the active project.

- Start a new legend in the active project. Name it **Electrical Symbols – Lighting & Power**.
- Open ex7-8.rvt. Open the legend view.
- Window around the legend to select all the elements.
- Click **CTL-C**.
- Switch back to the lab project.
- Left click in the legend view.
- Click **Ctl-V**.

Open the Ground Floor Ceiling Plan.

Create a section view of the plain recessed lighting fixture located in ROOM 106.

Assign LIGHTING as the sub-discipline in the view's property palette.

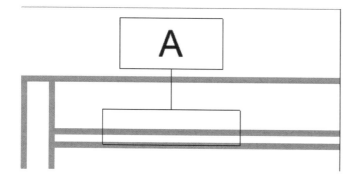

Family: Plain Recessed Lighting Fixture

Type: 2x2 - 120

Type Parameters

Parameter	
Manufacturer	
Type Comments	
URL	
Description	
Cost	
Assembly Description	
Type Mark	A
OmniClass Number	23.80.70.11.11

Edit the Plain Recessed Lighting Fixture Type Property to assign a Type Mark.

Repeat for the remaining light fixtures used in the project.

An easy way to locate the lighting fixtures is to go to the Families category in the Project Browser.
Select the lighting fixture, right click and select Type Properties.

Lighting Fixtures
  Pendant Light - Linear - 2 Lamp
  Plain Recessed Lighting Fixture
  Sconce Light - Uplight

Another method is to create a lighting fixture schedule using the fields for Family & Type and Type Mark.

Use type marks that match up with what appears on the legend that was copied over.
Tag the light fixture in the new section view.

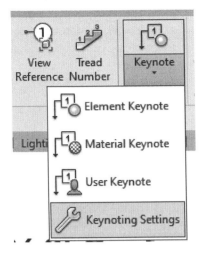

You may need to load the keynote reference database.

Access Keynoting Settings on the Annotate ribbon. Browse to the library and load the correct file.

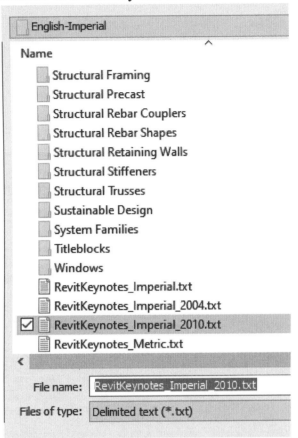

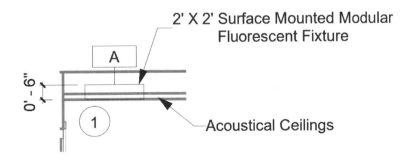

2' X 2' Surface Mounted Modular Fluorescent Fixture

Acoustical Ceilings

Add annotations including an element keynote, note block, and a dimension.

The Note Block.rfa family is included in the exercise folder.

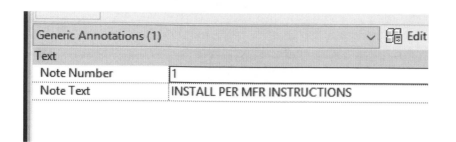

Modify the properties of the note block as shown.

Repeat process to create section views of the other lighting fixtures.

To locate lighting fixtures, you can use the lighting fixture schedule.
Highlight the lighting fixture, then select SHOW IN MODEL from the ribbon.

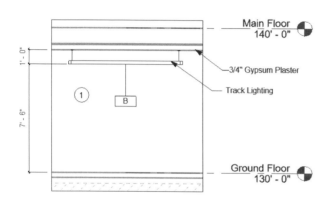

Add annotations including an element keynote, tag, note block, and a dimension.

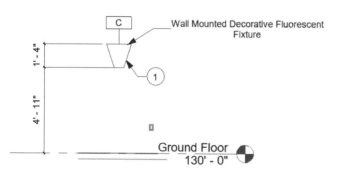

Add annotations including an element keynote, tag, note block, and a dimension.

Projects

Revit files are called "projects". A project is a building model, but also includes metadata such as the size and location of systems, materials used, and annotations. The display settings in a project define the appearance of the model in different views. You have control over how elements are displayed in each view.

Every project starts with a template. The template provides the initial settings, such as display defaults, units, and families which are pre-loaded into the file.

A project is the entire building design as well as the associated documentation (sheets).

In most cases, you will be working with an architectural building model provided by an architectural or AEC firm and then linking to that file or modifying the file to add the necessary electrical elements.

You add the parametric building components, such as lighting fixtures and switches, to the project – either working in a separate host file or in the file provided. You then create plan, section, elevation, and 3D views. Any changes made in one view propagate throughout the project and all associated views automatically update reflecting any changes. You then create sheets and place views on the sheets.

Discipline Settings are used to define the appearance and behavior of the system components in electrical systems. After standard settings have been established in your company, they can be configured in a template file, so you don't have to reset them for every project.

Electrical Settings determine the voltage, power distribution systems, wiring, and demand factors for the electrical systems. The Electrical Settings dialog box is used to specify these settings. Electrical settings can be created in template files or copied to other projects using the Transfer Project Settings dialog.

Hidden Line

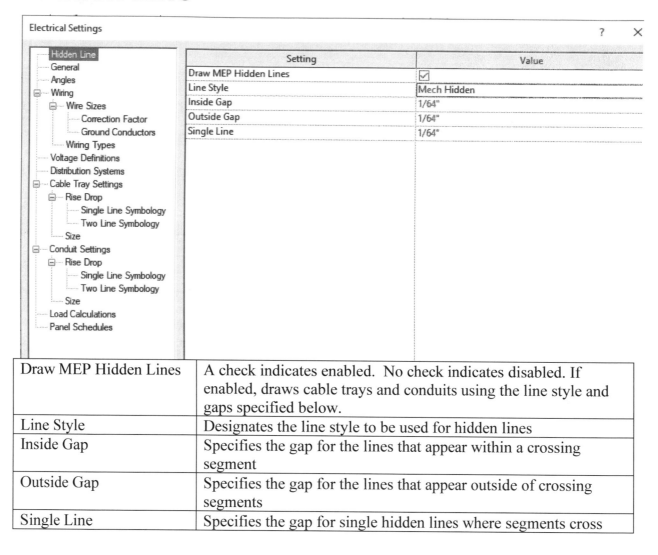

Draw MEP Hidden Lines	A check indicates enabled. No check indicates disabled. If enabled, draws cable trays and conduits using the line style and gaps specified below.
Line Style	Designates the line style to be used for hidden lines
Inside Gap	Specifies the gap for the lines that appear within a crossing segment
Outside Gap	Specifies the gap for the lines that appear outside of crossing segments
Single Line	Specifies the gap for single hidden lines where segments cross

General

Setting	Value
Electrical Connector Separator	-
Electrical Data Style	Connector Description Voltage / Number of Poles –...
Circuit Description	480V-3P/30A
Circuit Naming by Phase - Phase A Label	A
Circuit Naming by Phase - Phase B Label	B
Circuit Naming by Phase - Phase C Label	C
Capitalization for Load Names:	From Source Parameters
Circuit Sequence:	Numerical (1,2,3,4,5,6,7,8,9,10,11,12)
Circuit Rating	20 A
Circuit Path Offset	9' - 0"

Tree navigation:
- Hidden Line
- General
- Angles
- Wiring
  - Wire Sizes
    - Correction Factor
    - Ground Conductors
  - Wiring Types
- Voltage Definitions
- Distribution Systems
- Cable Tray Settings
  - Rise Drop
    - Single Line Symbology
    - Two Line Symbology
  - Size
- Conduit Settings
  - Rise Drop
    - Single Line Symbology
    - Two Line Symbology
  - Size
- Load Calculations
- Panel Schedules

Electrical Connector Separator	Specifies the symbol used to separate rating values
Electrical Data Style	Specifies the style for the electrical data parameter in the Element Properties dialog box
Circuit Description	Specifies the format for circuit descriptions that appear as the Load Name in the panel schedules.
Circuit Naming by Phase – Phase Label (A,B,C)	These values are only used if you specify circuit naming by phase for the panel using the Properties palette.
Capitalization for Load Names	Specifies the format for the Load Name parameter in the instance properties for circuits
Circuit Sequence	Specifies the sequence in which power circuits are created, enabling creation of circuits grouped by phase
Circuit Rating	Specifies the default rating when creating circuits
Circuit Path Offset	Specifies the default offset when placing a circuit path

Voltage Definitions

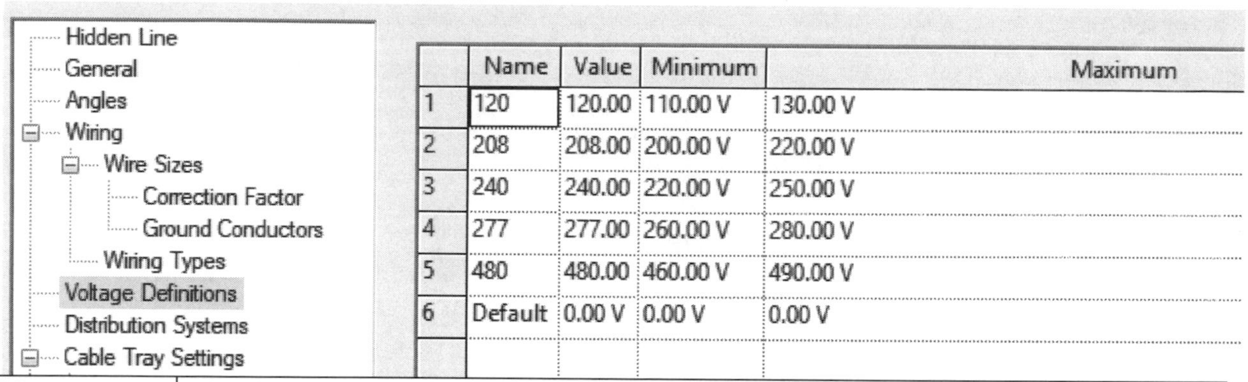

	Name	Value	Minimum	Maximum
1	120	120.00	110.00 V	130.00 V
2	208	208.00	200.00 V	220.00 V
3	240	240.00	220.00 V	250.00 V
4	277	277.00	260.00 V	280.00 V
5	480	480.00	460.00 V	490.00 V
6	Default	0.00 V	0.00 V	0.00 V

Tree navigation panel (left side):
- Hidden Line
- General
- Angles
- Wiring
  - Wire Sizes
    - Correction Factor
    - Ground Conductors
  - Wiring Types
  - Voltage Definitions
  - Distribution Systems
- Cable Tray Settings

Name	Identifies a voltage definition
Value	The actual voltage for the voltage definition
Minimum	The lowest voltage rating for electrical devices and equipment that can be used with the voltage definition.
Maximum	The highest voltage rating for electrical devices and equipment that can be used with the voltage definition.

The Voltage Definitions table defines the ranges of voltages that can be assigned to the Distribution Systems available in your project.

Each voltage definition is specified as a range of voltages to allow for differing voltage ratings on devices from various manufacturers. For example, devices used on a 120V distribution system may carry ratings of anywhere from 110V to 130V.

You can create Voltage Definitions and you can delete definitions that are not currently in use with any distribution system.

Revit does not prevent you from specifying unfeasible voltage values. For example, you could configure a distribution system with a L-L Voltage value of 120 and an L-G Voltage value of 480, even though this is physically impossible.

Distribution Systems

```
···· Hidden Line
···· General
···· Angles
⊟··· Wiring
   ⊟··· Wire Sizes
       ···· Correction Factor
       ···· Ground Conductors
   ···· Wiring Types
···· Voltage Definitions
···· Distribution Systems
⊟··· Cable Tray Settings
```

	Name	Phase	Configuration	Wires	L-L Voltage	L-G Voltage
1	120/208 Wye	Three	Wye	4	208	120
2	120/240 Single	Single	None	3	240	120
3	480/277 Wye	Three	Wye	4	480	277
4	Default	Single	None	2	None	None

Name	Identifies a distribution system
Phase	Either Three or Single, selected from the drop-down list
Configuration	After you click the value, you can select Wye or Delta, from the drop-down list (three-phase systems only).
Wires	Specifies to the number of conductors (3 or 4 for three-phase, 2 or 3 for single-phase).
L-L Voltage	After you click the value, you can select a Voltage Definition that represents the voltage measured between any two phases. The specification of this parameter depends on the Phase and Wire selections. For example, L-L Voltage is not applicable for a single-phase, 2-wire system.
L-G Voltage	After you click the value, you can select a Voltage Definition that represents the voltage measured between a phase and ground. L-G is always available.

Cable Tray Settings

Hidden Line
General
Angles
⊟ Wiring
⊟ Wire Sizes
Correction Factor
Ground Conductors
Wiring Types
Voltage Definitions
Distribution Systems
⊟ Cable Tray Settings

Setting	Value
Use Annot. Scale for Single Line Fittings	☑
Cable Tray Fitting Annotation Size	0' 0 1/8"
Cable Tray Size Separator	x
Cable Tray Size Suffix	ø
Cable Tray Connector Separator	-

Use Annot. Scale for Single Line Fittings	Specifies whether cable tray fittings are drawn at the size specified by the Cable Tray Fitting Annotation Size parameter. Changing this setting does not change the plotted size of components already placed in a project.
Cable Tray Fitting Annotation Size	Specifies the plotted size of fittings drawn in single-line views. This size is maintained regardless of the drawing scale.
Cable Tray Size Separator	Specifies the symbol to be used in showing cable tray sizes. For example, when an x is used, a cable tray that is 12 inches high and 4 inches deep would be shown as 12" x 4".
Cable Tray Size Suffix	Specifies the symbol appended to the cable tray size.
Cable Tray Connector Separator	Specifies the symbol used to separate information between 2 different connectors.

Rise Drop

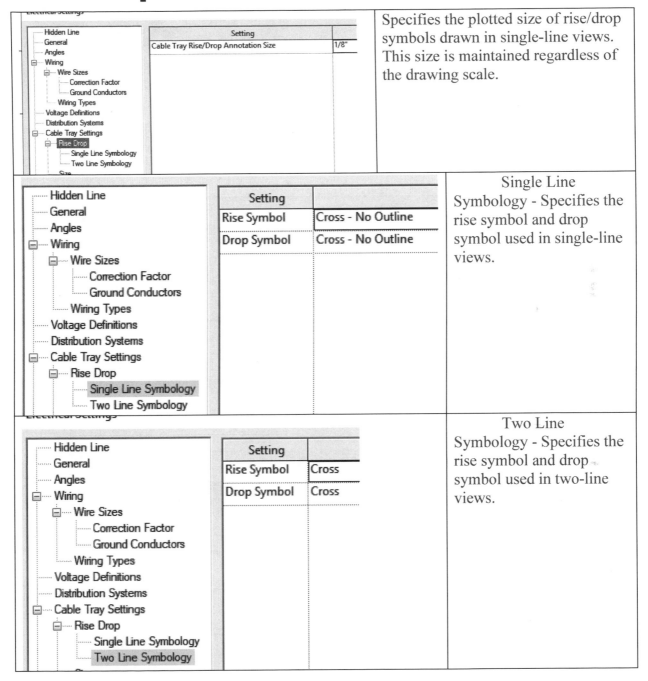

	Specifies the plotted size of rise/drop symbols drawn in single-line views. This size is maintained regardless of the drawing scale.
	Single Line Symbology - Specifies the rise symbol and drop symbol used in single-line views.
	Two Line Symbology - Specifies the rise symbol and drop symbol used in two-line views.

Cable Tray Settings – Size

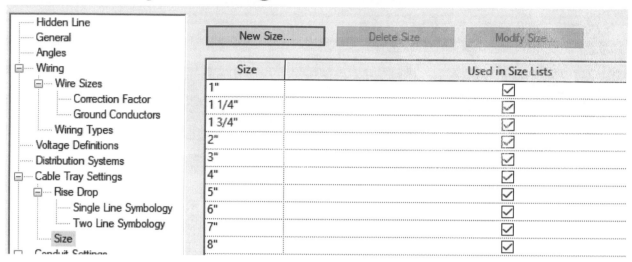

You use the Size table to specify the cable tray sizes that can be used in your project.

You can add, modify, or delete sizes as needed. For each cable tray size, the Used in Size Lists parameter specifies that the size is displayed in lists throughout Revit, including the cable tray layout editor and cable tray modify editor.

Load Calculations

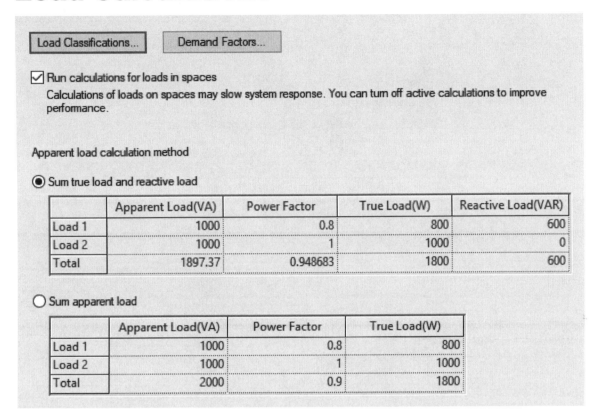

Use the check box to specify whether to enable load calculations for loads in spaces.

Running calculations may slow system response.

- Load Classifications - Click this button to open the Load Classifications dialog.

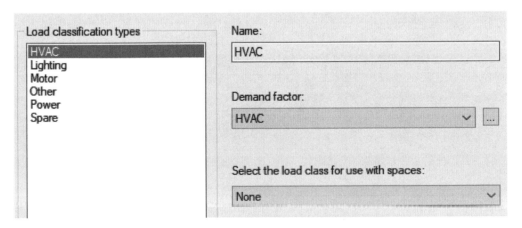

- Demand Factors - Click this button to open the Demand Factors dialog.

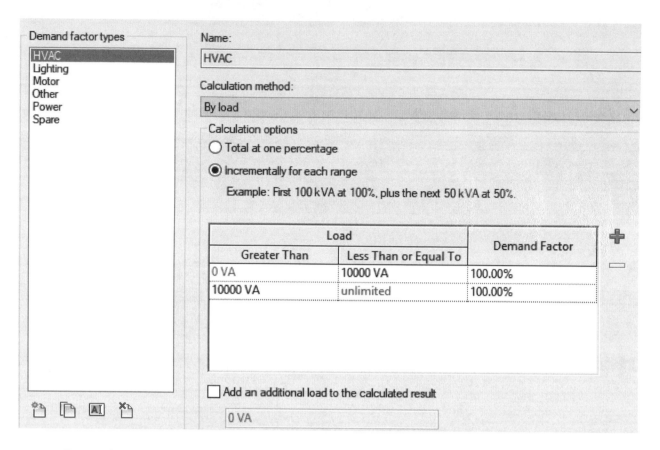

-     Run calculations for loads in spaces - Since calculations of loads on space may slow system response, you can disable calculations to improve performance.

-     Apparent load calculation method - Specifies how Revit sums electrical loads: Sum true load and reactive load or Sum apparent load. The selected method applies to all loads, circuits, and panels in the model.

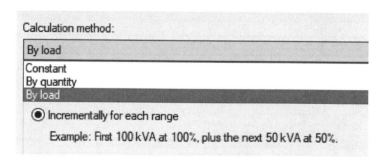

Electrical circuit load capacity is the total amount of power that the building actually will use. In order to decide what size electrical service is needed, one has to do a little math homework. Older buildings often only had a 60-amp electrical service, connected to a fuse panel. Newer buildings have 100- or 200-amp electrical services.

Understanding Electrical Loads

Calculating how much power both you and any electrical appliances use is necessary to calculate this number. As technology continues to advance, more and more electrical loads are added.

Circuits should only be loaded up to 80% of the total circuit capabilities. Having said that, it doesn't mean that you should keep adding additional loads until you get to *% capacity. Instead, aim for a more reasonable amount, say %0- 60% load if at all possible, allowing for future additional loads. It's better to have too many circuits than too few.

Calculating Capacity

To help you understand the concept, if you have a 15-amp circuit, the safe operating amperage would be no greater than 12 amps. The total wattage would be 1,800 watts, meaning the safe wattage usage would be 1,440 watts. If you had a 1,100-watt hairdryer plugged into this circuit, just one device uses almost the entire desired load capabilities.

If you have a 20-amp circuit, the safe operating amperage would be no greater than 16 amps. The total wattage would be 2,400 watts, meaning the safe wattage usage would be 1,920 watts. In this instance, you could have a hairdryer, radio, and electric razor running on the same circuit, but not much else. That's why there should be additional bathroom circuits to cover lighting, exhaust fans, and heat lamps for drying.

On a 30-amp circuit, the safe operating amperage would be no greater than 24 amps. The total wattage would be 3,600 watts, meaning the safe wattage usage would be 2,880 watts. This information comes in handy with central air conditioners, electric dryers, electric ranges, and electric ovens.

To determine the wattage, you take the voltage times the amperage. Check the tags on all installed appliances for the required amperage rating. Add all of the lighting load by adding the total wattage of the light fixtures.

Most buildings will likely also have 240-volt appliances like water heaters, air conditioners, electric dryers, and electric ranges. These will have an amperage rating label and the wattage can be calculated. The voltage, 240 volts, times the amperage, say 30 amps, will equal the wattage requirements.

Main Circuit Breaker Panel and Sub-Panels

Once you've determined the total load for the building, you'll know what size electrical service you need. Most homes have either a 100-amp or a 200-amp circuit breaker. There may also be additional sub-panels feeding off of the main circuit breaker panel. People often put sub-panels on each floor for easy access to breakers in the event of trouble.

With a large enough main circuit breaker panel, you can add sub-panels to any outbuildings as well. A good rule of thumb is to always have more outlets and service space available for adding circuits and electrical loads.

Panel Schedules

Setting	
Spare Label:	Spare
Space Label:	Space
Include Spares in Panel Totals	☑
Merge multi-poled circuits into a single cell	☐

Spare Label - Specifies default label text to apply to the Load Name parameter for any spare in a panel schedule.

Space Label - Specifies default label text to apply to the Load Name parameter for any space in a panel schedule.

Include Spares in Panel Totals - Specifies whether to include spares in the panel totals when you add load values to spares in a panel schedule.

Merge multi-poled circuits into a single cell - Specifies whether to merge 2 or 3 pole circuits into a single cell in a panel schedule.

Circuit Naming

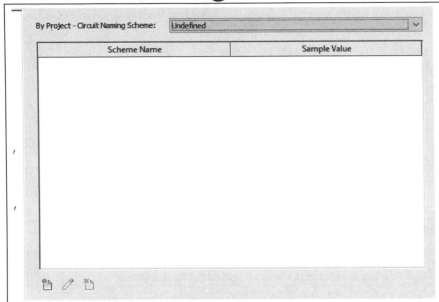

You can assign a default circuit naming scheme for your model by selecting a scheme in the By Project - Circuit Naming Scheme drop down. Use the Circuit Naming instance parameter to assign a circuit naming scheme. The By Project - Circuit Naming Scheme is used when the panel's instance parameter is set to By Project.

Linking Revit Architectural Projects

In most cases, you will work in your own project overlaid over the floor and building plans provided by an architect or an outside party. Close coordination between the two files is essential because if the supplied file changes it may directly affect the placement of electrical components.

When linking files, it is important to consider the hierarchy. Normally, you have a main project file which acts as the host file and then other files are loaded into the host file. Positioning the files is important to ensure that the components line up and are placed correctly. Most users use shared positioning to ensure that the linked files are aligned properly. Another method is to select a specific point or grid intersection and use that to ensure proper alignment.

When you link a Revit project file into a host Revit project file, Revit opens the linked model and retains it in memory. The more linked files, the more resource memory will be used. When you link a file into your host project, the path is saved in the host file. After linking, if the path is changed, you can reload it in the host project file to view the updated linked project file. You can update the path using the Manage Links dialog box.

The Copy tool is used to copy elements from the linked file to your project. Because many electrical components require a host (for example, an electrical panel needs a wall to mount on), you need to copy any necessary elements into your project. Any copied elements will be monitored by Revit to ensure that if they move or are modified, you are alerted.

Exercise 8-1:

Linking Files

Drawing Name: *linking_revit_models.rvt*
Estimated Time: 25 minutes

This exercise reinforces the following skills:

- ❑ Linking Revit files
- ❑ Changing the display of linked files
- ❑ Monitoring linked files

1.  Open *linking_revit_models.rvt*.
   This file is downloaded from the publisher's website.

2. Activate the **Insert** ribbon.

3. Select **Manage Links**.

4. Verify that the Revit tab is active.
   Select the **Add** button at the bottom of the dialog.

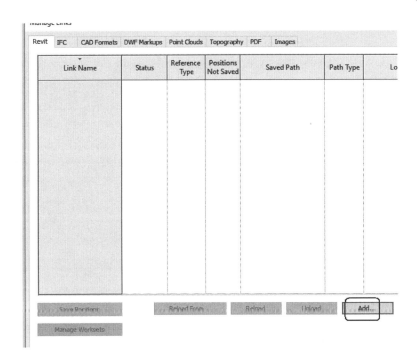

5. Select the *link_revit_building.rvt* file. Set the Positioning to **Auto – Internal Origin to Internal Origin**. Click **Open**.

File name:	link_revit_building.rvt
Files of type:	RVT Files (*.rvt)
Positioning:	Auto - Internal Origin to Internal Origin

6.

Revit	IFC	CAD Formats	DWF Markups	Point Clouds	Topography	PDF	Images

Link Name	Status	Reference Type	Positions Not Saved	Saved Path	Path Type
link_revit_building.rvt	Loaded	Overlay	☐	link_revit_building.rvt	Relative

Click **OK** to close the dialog.

7.
```
Elevations (Building Elevation)
    East - Mech
    North - Mech
    South - Mech
    West - Mech
```
Open the **South – Mech** elevation.

8.

```
Level 3
24' - 0"


Level 2                  Level 2
12' - 0"                 10' - 0"


Level 1       Level 1
0"            0' - 0"
```

Note that the elevation values for Level 2 are different between the two files.

9.

```
Level 2
12' - 0"
    RVT Links : Linked Revit Model : i_link_revit_building.rvt :
    2 : location <Not Shared>
```

Hover your cursor over the Level 2 with the 12'-0" value.
Note that the 12'-0" value belongs to the linked file.

10.

```
Level 2                      3D
12' - 0"
                     Level 2         ☑
                     12' 0"
```

Double left click on the 10'-0" value for the host file's Level 2 and change it to 12'-0".
Click **ENTER** to accept the change.

11. Collaborate

Activate the **Collaborate** ribbon.

12.

```
Manage    Publish    Copy/
ud Models Settings   Monitor

Mod
        Use Current Project

        Select Link
```

Select **Copy/Monitor→Select Link**.

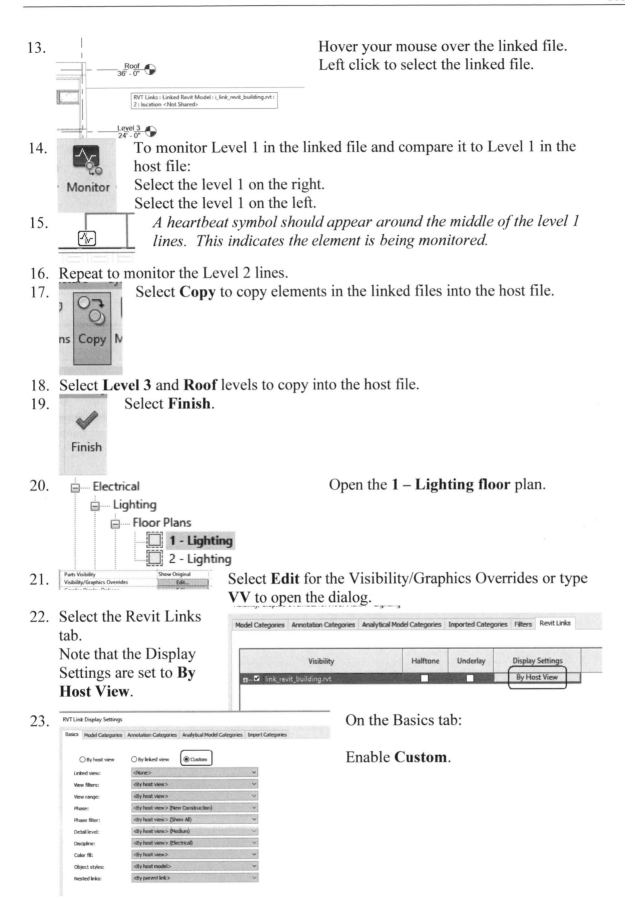

13. Hover your mouse over the linked file. Left click to select the linked file.

RVT Links : Linked Revit Model : i_link_revit_building.rvt : 2 : location <Not Shared>

14. To monitor Level 1 in the linked file and compare it to Level 1 in the host file:
Select the level 1 on the right.
Select the level 1 on the left.

15. *A heartbeat symbol should appear around the middle of the level 1 lines. This indicates the element is being monitored.*

16. Repeat to monitor the Level 2 lines.

17. Select **Copy** to copy elements in the linked files into the host file.

18. Select **Level 3** and **Roof** levels to copy into the host file.

19. Select **Finish**.

20. Open the **1 – Lighting floor** plan.

21. Select **Edit** for the Visibility/Graphics Overrides or type **VV** to open the dialog.

22. Select the Revit Links tab.
Note that the Display Settings are set to **By Host View**.

23. On the Basics tab:

Enable **Custom**.

24.

| Basics | Model Categories | Annotation Categories | Analytical Model Categories | In |

Annotation categories: <Custom>
  <By host view>
  <Custom>

☑ Show annotation categories in this

Filter list: <multiple>

Select the **Annotation Categories** tab.

Select Custom for **Annotation Categories.**

25.

Visibility/Graphic Overrides for Floor Plan: 1 - Lighting

| Model Categories | Annotation Categories | Analytical Model Categories | Imported Categori |

☑ Show annotation categories in this view

Filter list: <multiple>

Visibility	Projection/Surface Lines	Halftone
☑ Electrical Fixture Tags		☐
☑ Elevations		☐
☑ Fire Alarm Device Tags		☐
☑ Floor Tags		☐
☑ Furniture System Tags		☐
☑ Furniture Tags		☐
☐ Generic Annotations		☐
☑ Generic Model Tags		☐
☐ Grids		■
☑ Guide Grid		
☑ Keynote Tags		☐
☑ Levels		☐

Disable **Grids**.
Click **OK**.
Note that the grids are turned off on the linked file.

26. Close without saving.

Exercise 8-2:

Working In a Host File

Drawing Name: *wiring.rvt*
Estimated Time: 30 minutes

This exercise reinforces the following skills:

- ❑ Place Components
- ❑ Create Circuit
- ❑ Add Wires

1. 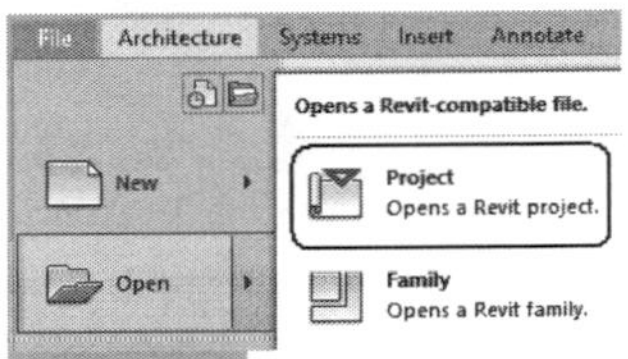 Open *wiring.rvt*.

2. 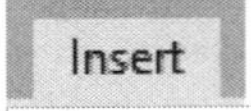 Activate the **Insert** ribbon.

3. 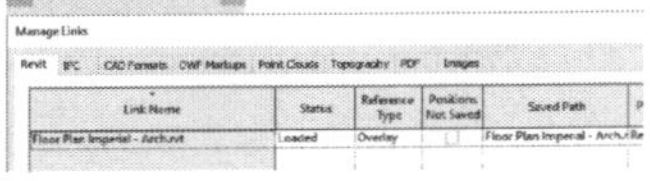 Select the **Manage Links** tool.
    Note that there is a floor plan file linked to the project.
    Click **OK**.

    You may need to reload the file to point it to the correct link.

4. 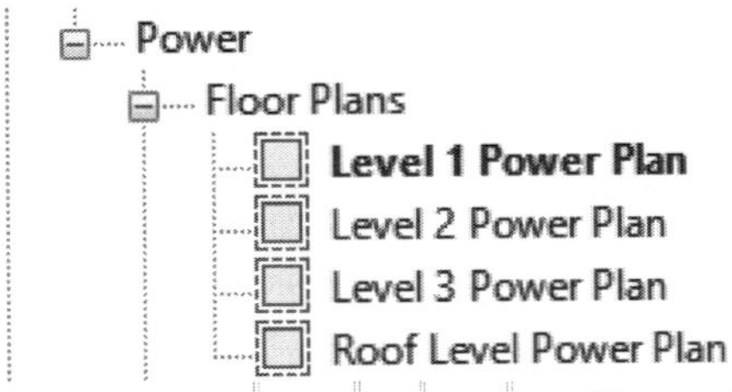 Verify that the **Level 1 Power Plan** floor plan is the active view.

5. 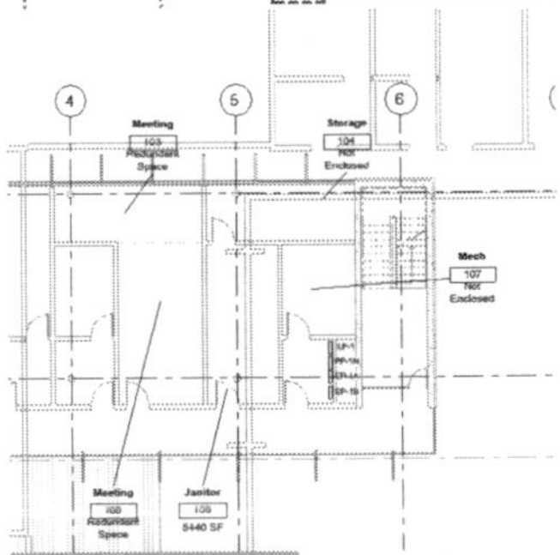 Zoom into the area between Grids 4-6 and 19-21.
    There should be several rooms labeled in this area.

6. 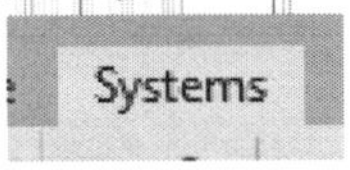 Activate the **Systems** ribbon.

7. Select the **Place a Component** tool.

8. 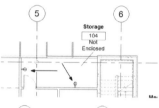 Using the Type Selector, verify that the **Duplex Receptacle – Standard** is active.

9. Place two receptacles in Rooms 104.

10. 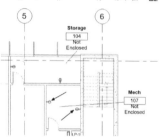 Place two receptacles in Rooms 107.

11.

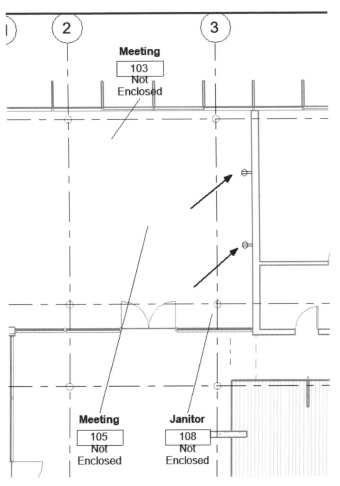

Place two receptacles in Rooms 108.

Click **ESC** to exit the command.

12.

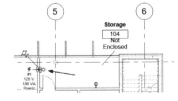

Select one of the receptacles placed in Room 104.

13.

Select the **Power** tool on the Create Systems panel on the ribbon.
This creates an electrical circuit.

14.

Select the **Edit Circuit** tool on the ribbon.

15. Verify that **Add to Circuit** is enabled on the ribbon.

16. Select the remaining receptacles to be added.
On the Options bar verify that 6 elements have been selected.

17. Assign the circuit to Panel: **PP-1N**.

18. Select **Finish Editing Circuit** from the ribbon.

19. Hover the cursor over one of the power receptacles in Room 104.
Click the **TAB** key.
Dashed lines will appear to indicate the intended wiring for the new circuit.
Left click on the receptacle to accept the default wiring.

20. *Wiring is displayed between the receptacles and the panel.*
Note the home run symbol between the panel and one of the receptacles.

21. Locate the arced wiring symbol.
Left click to place the wiring.

22.

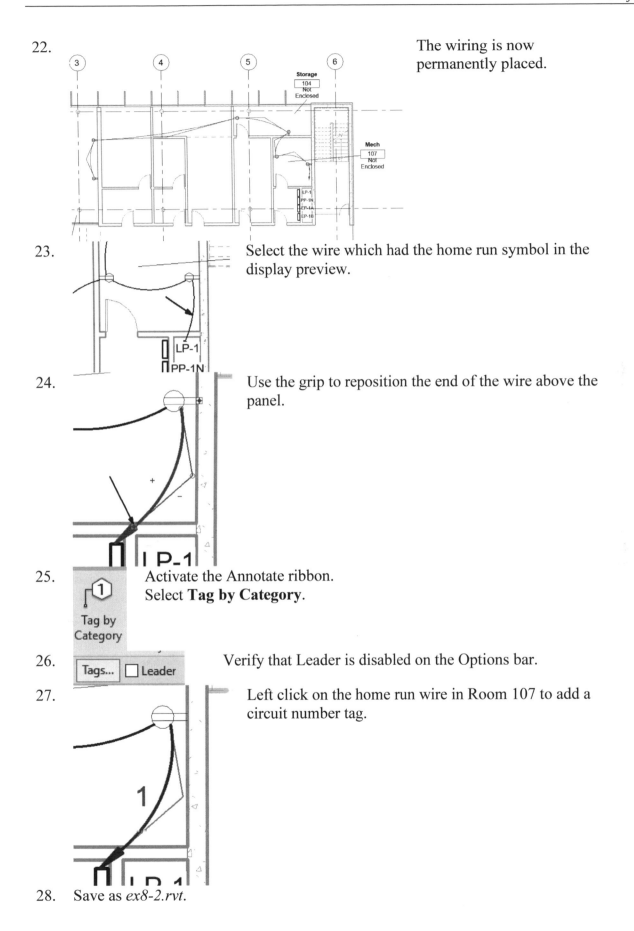

The wiring is now permanently placed.

23. Select the wire which had the home run symbol in the display preview.

24. Use the grip to reposition the end of the wire above the panel.

25. Activate the Annotate ribbon.
Select **Tag by Category**.

26. Verify that Leader is disabled on the Options bar.

27. Left click on the home run wire in Room 107 to add a circuit number tag.

28. Save as *ex8-2.rvt*.

Coordination Review Tool

The Coordination Review tool is used when monitored/copied elements in a linked file have been modified. When you re-load a linked file that has been modified, warnings will appear to notify you of any violations. Warnings can also appear when the original element in the linked file is deleted or the copied element in the host file is deleted or modified.

The Coordination Review tool is available on the Collaborate ribbon.

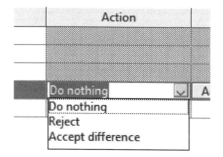

In the Coordination Review dialog box, you can select any of the changes made and select an action to address the change.

Action	Description
Do nothing	Takes no action on the element. This allows you time to message another team member to determine the best course of action regarding the change.
Reject	This command is used when the change made to the host file is correct and the linked file needs to be updated. This option is only available in the Host file.
Accept Difference	This option accepts the changes made in the linked file and updates the affected element in the host file. This option is only available in the Host file.

Guidelines for Monitoring Changes in Linked Files

- Use Copy and Monitor to monitor level lines, walls, doors, ceilings, and windows. This ensures that floor to floor heights remain coordinated as well as elements which are used to position electrical components.
- Use the Coordination Review tool to perform coordination reviews and generate reports of any changes made to the host or linked file(s). The reports ensure a paper trail of any changes made by team members and the decisions made regarding those changes.
- Use the Add Comment option in the Review dialog to add comments regarding any changes. These comments are visible to other team members and enhances communication.
- Reload the linked file periodically to check for any changes. Verify that you are using the latest version of the model prior to doing any major work on your host file.

Exercise 8-3:

Coordination Review

Drawing Name: *multiple_disciplines.rvt*
Estimated Time: 30 minutes

This exercise reinforces the following skills:

- ❑ Linked Files
- ❑ Manage Links
- ❑ Add a Revit project to a host file
- ❑ Reload linked file
- ❑ Coordination Review

1.   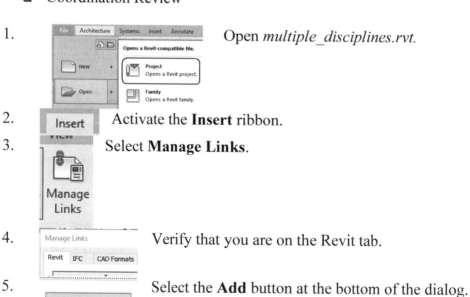   Open *multiple_disciplines.rvt.*

2.   Activate the **Insert** ribbon.

3.   Select **Manage Links**.

4.   Verify that you are on the Revit tab.

5.   Select the **Add** button at the bottom of the dialog.

6.   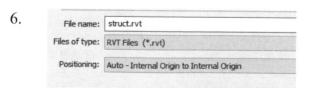   Locate the *struct.rvt file.*
Set the Positioning to **Auto – Internal Origin to Internal Origin.**
Click **Open**.

7. Verify the file is loaded.

    Click **OK** to close the dialog.

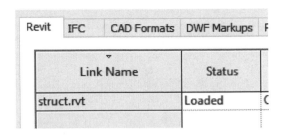

8.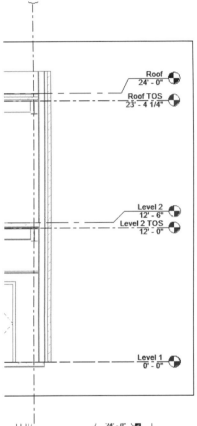

    Verify that you are in the **Section 1** view.

    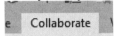

    Activate the **Collaborate** ribbon.

    Select the **Copy/Monitor→Select Link** tool from the ribbon.

9.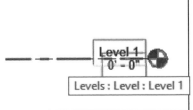

    Left click to select the linked Revit file.

10. Select the **Monitor** tool.

11. The Level 1 lines overlap.
    Left click on each level 1 line.

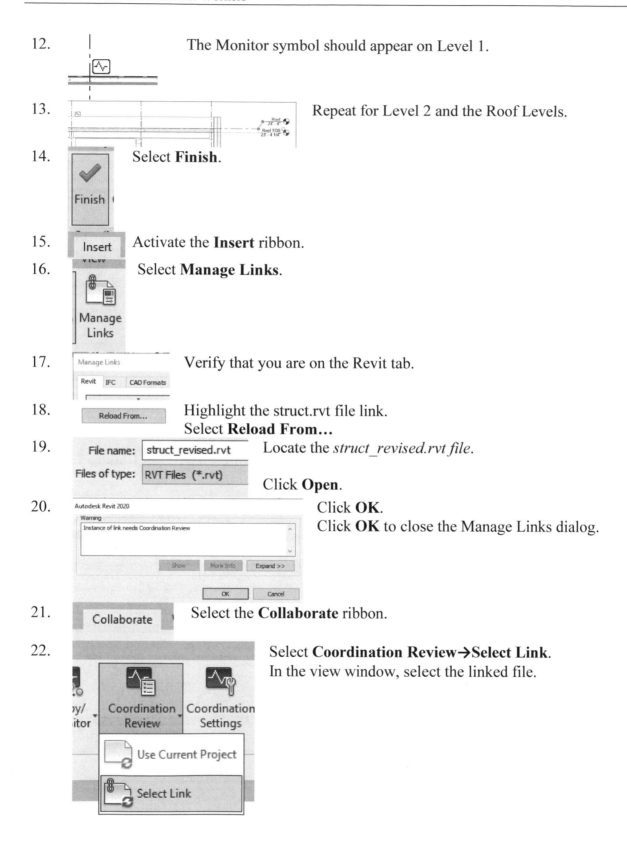

12. The Monitor symbol should appear on Level 1.

13. Repeat for Level 2 and the Roof Levels.

14. Select **Finish**.

15. Activate the **Insert** ribbon.

16. Select **Manage Links**.

17. Verify that you are on the Revit tab.

18. Highlight the struct.rvt file link.
    Select **Reload From...**

19. Locate the *struct_revised.rvt file*.

    Click **Open**.

20. Click **OK**.
    Click **OK** to close the Manage Links dialog.

21. Select the **Collaborate** ribbon.

22. Select **Coordination Review→Select Link**.
    In the view window, select the linked file.

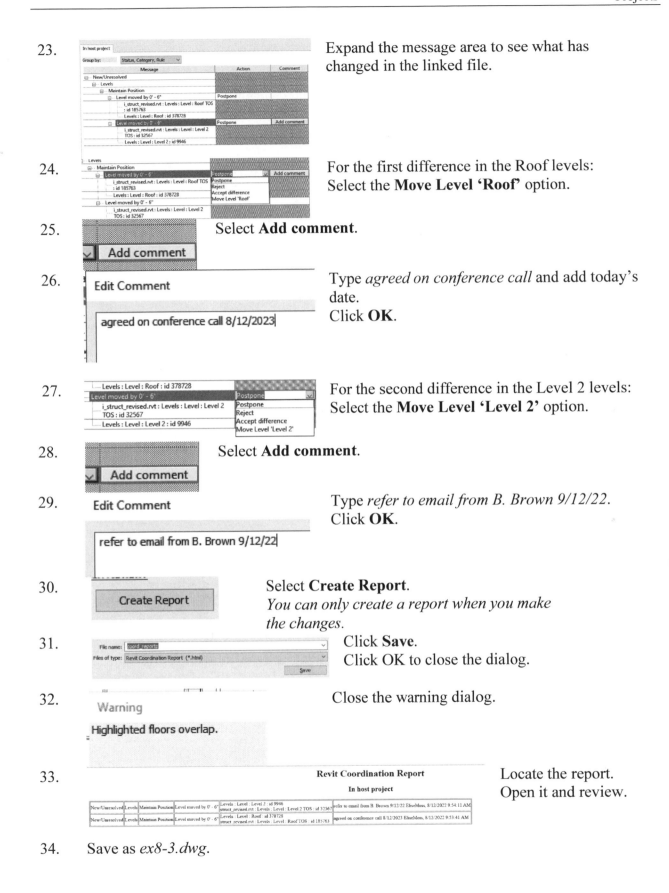

23. Expand the message area to see what has changed in the linked file.

24. For the first difference in the Roof levels:
Select the **Move Level 'Roof'** option.

25. Select **Add comment**.

26. Type *agreed on conference call* and add today's date.
Click **OK**.

27. For the second difference in the Level 2 levels:
Select the **Move Level 'Level 2'** option.

28. Select **Add comment**.

29. Type *refer to email from B. Brown 9/12/22.*
Click **OK**.

30. Select **Create Report**.
You can only create a report when you make the changes.

31. Click **Save**.
Click OK to close the dialog.

32. Close the warning dialog.

33. Locate the report.
Open it and review.

34. Save as *ex8-3.dwg*.

Interference Checks

Interference checks detect overlapping geometry between elements of selected categories. Based on the selection made, the check scans the building to identify pairs of elements that conflict or interfere with each other. You can perform a check within a project or between linked projects.

Interference checks only apply to model elements, so they apply to any view where the model elements are located.

The check generates an interference report that alerts the user to the number of interferences in the model, which may result in construction errors. You can then review the conflicts and determine the best resolution for each conflict.

Interferences are not always visible in existing views. You may need to create new sections or levels to reveal the interference issue.

Guidelines for Checking and Fixing Interference Conditions

- Check for interference early in the design stage. Correcting conflicts is easier when there are fewer elements to evaluate.

- Select a limited number of elements or categories to check to reduce the processing time.

- Refresh an interference report after correcting any interference conditions and run the report a second time to ensure that all conflicts have been resolved.

- Generate an interference report to gather additional input from other team members when you need guidance resolving conflicts.

Exercise 8-4:

Interference Checking

Drawing Name: *interference_checking.rvt*
Estimated Time: 30 minutes

This exercise reinforces the following skills:

- ❑ Interference Check
- ❑ Manage Links
- ❑ Resolving Conflicts

1.   Open *interference_checking.rvt*.

2.  Select the **Collaborate** ribbon.

3.   Select **Run Interference Check** from the ribbon.

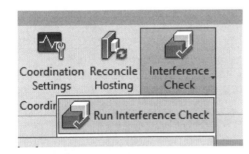

4.   Select **Air Terminals, Ceilings, Floors**, and **Walls** from the left pane.

Categories from
Current Project

- ☑ Air Terminals
- ☑ Ceilings
- ☐ Curtain Panels
- ☐ Curtain Wall Mullions
- ☐ Doors
- ☐ Duct Fittings
- ☐ Ducts
- ☐ Electrical Equipment
- ☐ Electrical Fixtures
- ☐ Flex Ducts
- ☑ Floors

5.

Select **Electrical Equipment, Lighting Devices** and **Lighting Fixtures** from the right pane.
We are looking to see if any of the electrical systems have any conflicts.

Click **OK**.

6.

There are several conflicts being reported.

7.

Expand the first conflict under Air Terminals.

8.

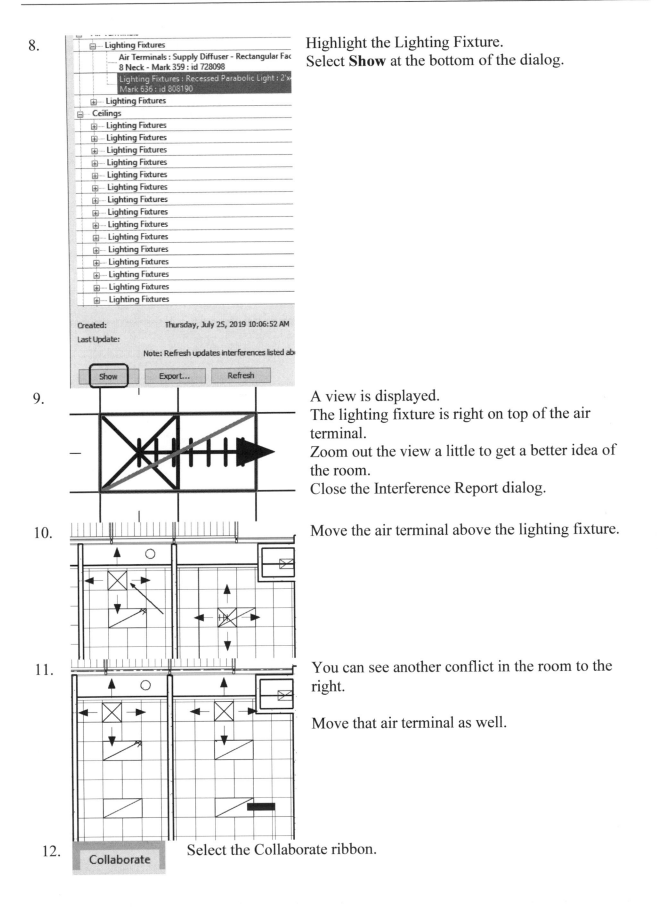

Highlight the Lighting Fixture.
Select **Show** at the bottom of the dialog.

9.

A view is displayed.
The lighting fixture is right on top of the air terminal.
Zoom out the view a little to get a better idea of the room.
Close the Interference Report dialog.

10.

Move the air terminal above the lighting fixture.

11.

You can see another conflict in the room to the right.

Move that air terminal as well.

12.

Select the Collaborate ribbon.

13.

Select **Show Last Report**.

14.

The report still shows there is a conflict between the air terminals and the lighting fixtures.

Click **Refresh**.

15.

The report clears the conflicts with the air terminals, but several conflicts remain.

16.

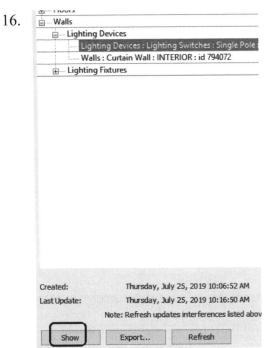

Highlight the first Lighting Device conflict under Walls.
Select **Show**.

17.

There is no open view that shows any of the highlighted elements. Searching through the closed views to find a good view could take a long time. Continue?

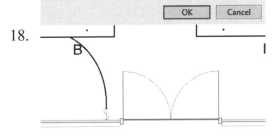

Click **OK** until you find a good view where you can see the conflict.

18.

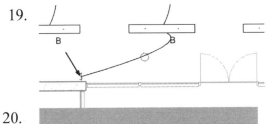

The Level 1 Lighting Plan floor plan opens to show the light switch mounted on the curtain wall is not a good location.
Close the Interference Check dialog.

19.

Move the switch to the wall adjacent to the curtain wall.

20.

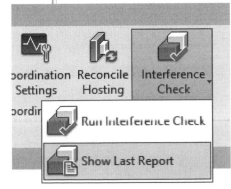

Select **Show Last Report**.

21.

Ceilings
Floors
Walls
    Lighting Devices
    Lighting Fixtures

The report still shows there is a conflict between the air terminals and the lighting fixtures.
Click **Refresh**.

22.

Walls
  Lighting Fixtures
    Lighting Fixtures : Recessed Parabolic Light : 2'x2
    359 : id 620269
    Walls : Basic Wall : Generic - 6" : id 794143

The report clears the first conflict with a wall.
Highlight the remaining Lighting Fixture.
Click **Show**.

23.

Coordination
  All
    Ceiling Plans
      Level 1
      Level 2
      Level 3

The Level 2 Ceiling plan opens.

24.

It looks like the lighting fixture doesn't fit properly on the ceiling.
Close the Interference Check dialog.

25.

Use the MOVE tool to shift the air terminal and duct down to the center of the ceiling.
Use the MOVE tool to shift the lighting fixture down.

26. 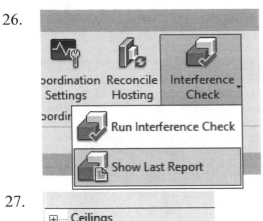 On the Collaborate ribbon:
Select **Show Last Report**.

27. *The report still shows there is a conflict between the walls and the lighting fixtures.*
Click **Refresh**.

⊞ Ceilings
⊞ Floors
⊟ Walls
⠀⠀⊞ Lighting Fixtures

28. The report clears the wall interference.
Remaining are all the conflicts with the ceilings and floors. These can be fixed during a lab period.

⊞ Ceilings
⊞ Floors

29. Save as *ex8-4.rvt*.

Load Classifications

Load classification in Revit segments your loads into different classifications. These classifications will be put on your panel schedules and are useful when filling out utility letters at the end of a project. The utility company uses this information to size the transformer at the service entrance. Typically, this information takes a long time to collect; you need to add up all the receptacle loads, lighting loads, largest motor, as well as heating and cooling loads. When done throughout the course of the project, this information is simple to provide.

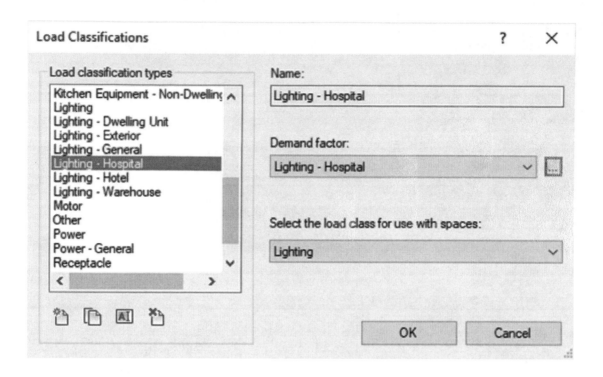

When placing equipment, it is important to make sure that each load is connected to the right classification. Not doing this important step could result in over-sized panels, transformers, and switchboards, which could add tens if not hundreds of thousands of dollars to the electrical costs of the project.

Load classifications also have specific demand factors. Receptacles have a demand factor of 100% for the first 10 kVA then 50% for every VA after. Hospital Lighting has a demand factor of 40% for the first 50 kVA and 20% for every VA after. Make sure you enter the correct demand factor!

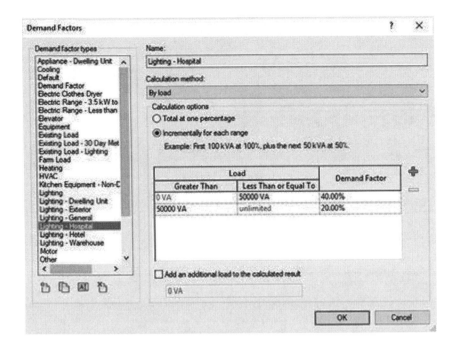

To set your load classifications, you need to:

1. Click on the family.
2. Go to edit type.
3. Click on load classification.
4. There will be a list on the right of the different load classifications in the model.
5. These classifications will be what appears on your panel schedules.
6. Make sure that you have the right demand factor for each classification type; this sets connected vs. demand.
7. Repeat for each family type – lights, receptacles, motor connections, heating and cooling connections.

Exercise 8-5:

Creating Load Classifications

Drawing Name: *load_classifications.rvt*
Estimated Time: 30 minutes

This exercise reinforces the following skills:

- ❑ Load Classifications
- ❑ Demand Factors
- ❑ Families

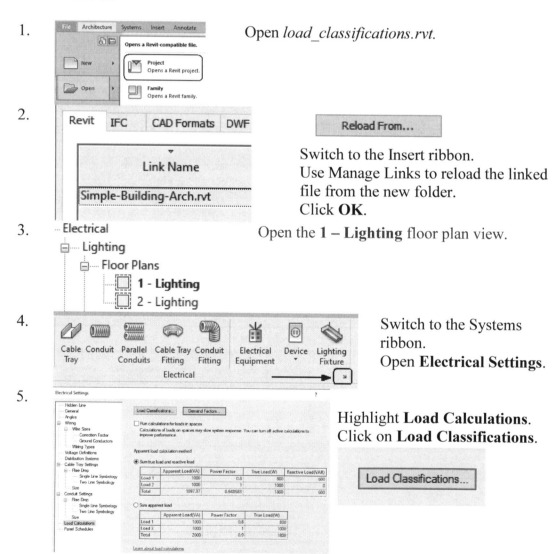

1. Open *load_classifications.rvt*.

2. Switch to the Insert ribbon.
   Use Manage Links to reload the linked file from the new folder.
   Click **OK**.

3. Open the **1 – Lighting** floor plan view.

4. Switch to the Systems ribbon.
   Open **Electrical Settings**.

5. Highlight **Load Calculations**.
   Click on **Load Classifications**.

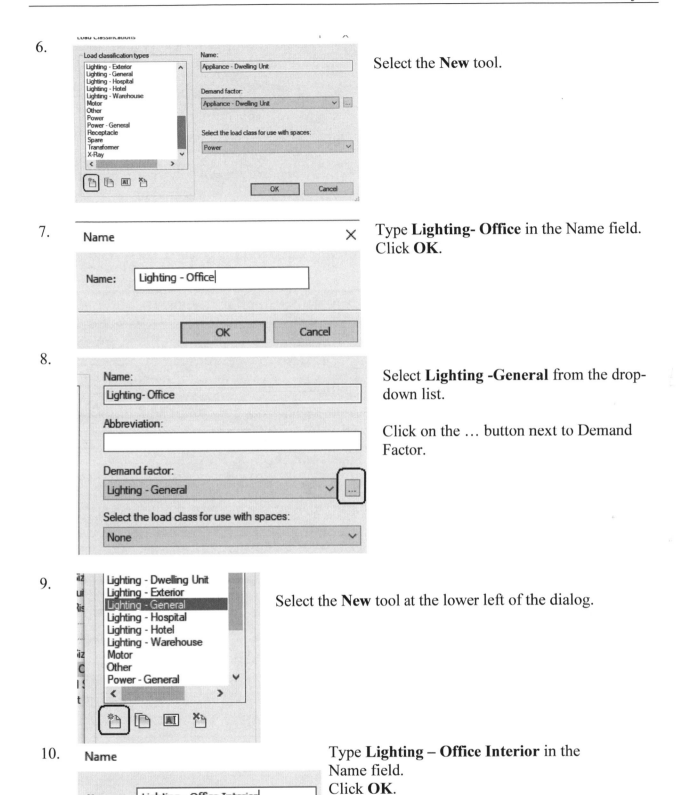

6.   Select the **New** tool.

7.   Type **Lighting- Office** in the Name field. Click **OK**.

8.   Select **Lighting -General** from the drop-down list.

Click on the ... button next to Demand Factor.

9.   Select the **New** tool at the lower left of the dialog.

10.   Type **Lighting – Office Interior** in the Name field.
Click **OK**.

11.

Name:

Lighting - Office Interior

Calculation method:

Constant

Calculation options

Apply a constant demand factor to the load in the load classification.

Demand factor:

120.00%

☐ Add an additional load to the calculated result

10 VA

Set the Calculation method to **Constant**. *This means the lights will stay on during business hours. They won't be switched on and off throughout the day.*
Set the Demand factor to **120.00%**. *This ensures that the panels will not be overloaded if all the lights remain on throughout the day.*
Click **OK**.

12.

Load Classifications   ?   ✕

Load classification types

Existing Load - Lighting
Farm Load
Heating
HVAC
Kitchen Equipment - Non-Dwelling
Lighting
Lighting - Dwelling Unit
Lighting - Exterior
Lighting - General
Lighting - Hospital
Lighting - Hotel
Lighting - Warehouse
Lighting - Office

Name:
Lighting - Office

Abbreviation:

Demand factor:
Lighting – Office Interior

Select the load class for use with spaces:
Lighting

Assign the **Lighting – Office** load classification to use the **Lighting – Office Interior** Demand factor.
Set the load class for use with spaces to **Lighting**.
Click **OK**.

13.

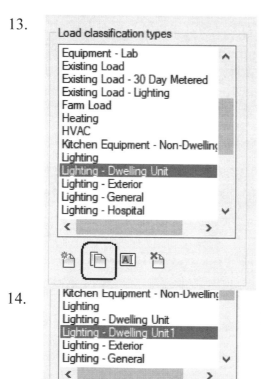

Highlight the **Lighting – Dwelling Unit**.
Select the **Duplicate** tool.

14.

Kitchen Equipment - Non-Dwelling
Lighting
Lighting - Dwelling Unit
Lighting - Dwelling Unit 1
Lighting - Exterior
Lighting - General

Highlight the copied type.
Select **Rename**.

15.

Rename	×
Previous:	Lighting - Dwelling Unit1
New:	Lighting - Lavatory
	OK Cancel

Type **Lighting - Lavatory** in the New Name field.
Click **OK**.

16.

Demand factor:

Lighting - Dwelling Unit   ⌄ | ... |

Click on the … button next to Demand Factor.

17.

Name:

Lighting - Dwelling Unit

Calculation method:

By load

Calculation options

○ Total at one percentage

◉ Incrementally for each range

Example: First 100 kVA at 100%, plus the next 50 kVA at 50%.

Load		Demand Factor
Greater Than	Less Than or Equal To	
0 VA	3000 VA	100.00%
3000 VA	120000 VA	35.00%
120000 VA	unlimited	25.00%

☐ Add an additional load to the calculated result

Because lights in homes are turned on and off, the calculation method used is different.
Click **OK**.

18.

Name:

Lighting -Lavatory

Abbreviation:

Demand factor:

Lighting - Dwelling Unit   ⌄ | |

Select the load class for use with spaces:

Lighting   ⌄

Select the load class for use with spacing.
Set this to **Lighting**.
Click **OK**.

19. Click **OK** to close the dialog.

20.

Select one of the lighting fixtures in **Lab-101**.

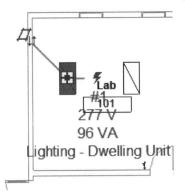

21. Select **Edit Type** on the Properties palette.

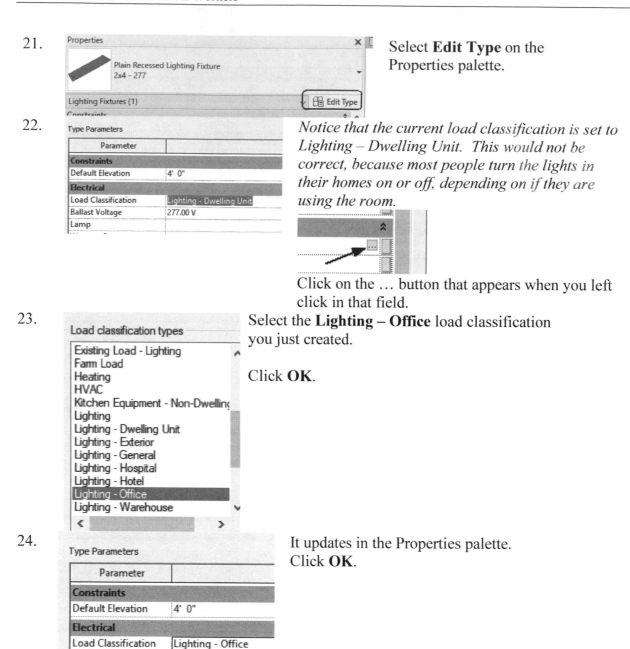

22. *Notice that the current load classification is set to Lighting – Dwelling Unit. This would not be correct, because most people turn the lights in their homes on or off, depending on if they are using the room.*

Click on the … button that appears when you left click in that field.

23. Select the **Lighting – Office** load classification you just created.

Click **OK**.

24. It updates in the Properties palette.
Click **OK**.

25.

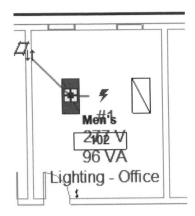

Because this was a type parameter that was changed, it affects all lights that are the same type.
Select a light fixture in the Men's lavatory.

26.

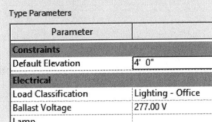

Select **Edit Type** on the Properties palette.

27.

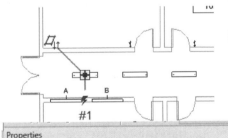

Notice it is using the correct load classification because it is the same type of fixture as it used in Lab - 101.
Some offices have the lights in the bathrooms turn off when not in use. This means you would have to place a different light fixture family using a different load classification.

Click **OK** to close the Type Properties dialog.

28.

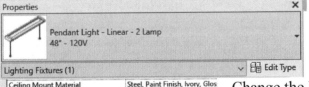

Select one of the pendant lights in the corridor.

29.

Select **Edit Type** on the Properties palette.

30. Change the Load Classification to **Lighting – Office.**

31.   Save as *ex8-5.rvt*.

Exercise 8-6:

Assigning Load Classifications to a Family

Drawing Name: *loads_families.rvt*
Estimated Time: 10 minutes

This exercise reinforces the following skills:

- ❏ Load Classifications
- ❏ Families

1. 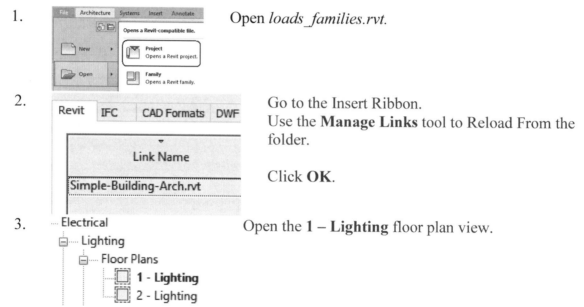 Open *loads_families.rvt*.

2. Go to the Insert Ribbon.
    Use the **Manage Links** tool to Reload From the folder.

    Click **OK**.

3. 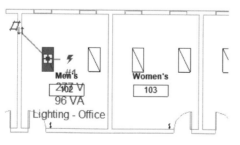 Open the **1 – Lighting** floor plan view.

4. 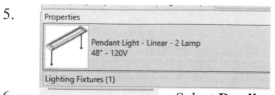 Select one of the lighting fixtures in the Men's lavatory.

5. Select **Edit Type** on the Properties palette.

6. Select **Duplicate**.

7.

Name: | 2x4 - 278 - Lavatory|

Type **2x4 – 278 – Lavatory** in the Name field.
Click **OK**.

8.

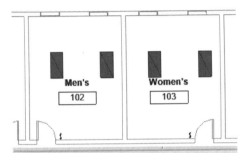

Set the Load Classification to **Lighting –
Lavatory.**

Click **OK.**

9.

Select the lighting fixtures in the two lavatories.

Men's
102

Women's
103

10.

Properties

Plain Recessed Lighting Fixture
2x4 - 278 - Lavatory

Use the Type Selector on the Properties palette to
assign the **Lavatory** lighting fixture to the
selected lighting fixtures.

11.   Save as *ex8-6.rvt.*

Exercise 8-7:

Assigning Load Names to a Circuit

Drawing Name: *load_names.rvt*
Estimated Time: 10 minutes

This exercise reinforces the following skills:

- ❑ Load Classifications
- ❑ Load names
- ❑ Circuits

1. 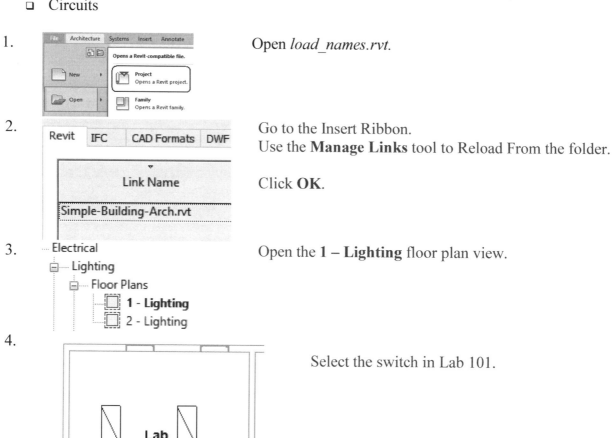 Open *load_names.rvt*.

2. Go to the Insert Ribbon.
    Use the **Manage Links** tool to Reload From the folder.

    Click **OK**.

3. Open the **1 – Lighting** floor plan view.

4. Select the switch in Lab 101.

5. Select the **Electrical Circuits** tab on the ribbon.

6.

Circuit: 2	
Electrical Engineering	
Schedule Circuit Notes	
Electrical - Loads	
Circuit Number	2
Connection Type	Breaker
Load Name	Lighting - Office
Panel	A
System Type	Power

*Note the Load Name is set to **Other**.*
Type the Load Name **Lighting – Office**.
Click ENTER and Apply to ensure the new name is applied.
Click ESC to release the selection.

7.

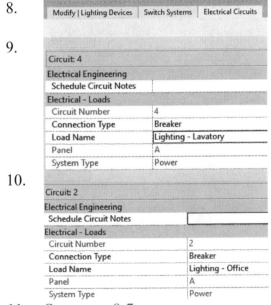

#1

Select the switch in Men's 102.

8.

| Modify | Lighting Devices | Switch Systems | Electrical Circuits |

Select the **Electrical Circuits** tab on the ribbon.

9.

Circuit: 4	
Electrical Engineering	
Schedule Circuit Notes	
Electrical - Loads	
Circuit Number	4
Connection Type	Breaker
Load Name	Lighting - Lavatory
Panel	A
System Type	Power

*Note the Load Name is set to **Other**.*
Type **Lighting – Lavatory** to change the Load Name.
Click ESC to release the selection.

10.

Circuit: 2	
Electrical Engineering	
Schedule Circuit Notes	
Electrical - Loads	
Circuit Number	2
Connection Type	Breaker
Load Name	Lighting - Office
Panel	A
System Type	Power

Select each switch in the building and assign it to the correct load name.
All the circuits should be assigned to Lighting – Office except for the switches in the lavatories.
Each circuit should be assigned to a panel and a load name.

11.   Save as *ex8-7.rvt*.

Shared Parameters

Shared parameters are parameter definitions that can be used in multiple families or projects.

Shared parameters are definitions of parameters that you can add to families or projects. Shared parameter definitions are stored in a text file independent of any family file or Revit project; this allows you to access the file from different families or projects. The shared parameter is a *definition* of a container for information that can be used in multiple families or projects. The *information* defined in one family or project using the shared parameter is not automatically applied to another family or project using the same shared parameter.

In order for information in a parameter to be used in a tag, it must be a shared parameter. Shared parameters are also useful when you want to create a schedule that displays various family categories; without a shared parameter, you cannot do this. If you create a shared parameter and add it to the desired family categories, you can then create a schedule with these categories. This is called creating a multi-category schedule in Revit.

Most companies will create a single text file which contains all the shared parameters used in projects. This file is then placed on the server and distributed to the different team members so they can use this file and not have to create their own individual shared parameters file. The "shared" in shared parameters means the parameters are shared across families and projects, not that the parameters are shared between team members.

The shared parameters file cannot be easily edited, and it is not recommended for editing. Revit automatically adds formatting and spacing to the txt file when it is created. If you open the file in Word, or even in Notepad, you can corrupt the file. Once you define a parameter, it is not easily modified. If you need to modify a parameter, you will most likely have to delete it and create a new one.

I recommend that companies review the column headers they use in schedules and any columns which are not covered by an existing parameter is a candidate to be used as a shared parameter.

Exercise 8-8:

Creating a Shared Parameter

Drawing Name: *shared_parameters.rvt*
Estimated Time: 15 minutes

This exercise reinforces the following skills:

❑   Shared Parameters

1.   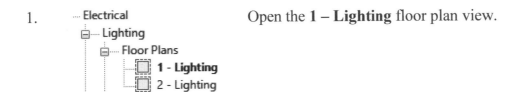   Open the **1 – Lighting** floor plan view.

2.      Go to the Insert Ribbon.
Use the **Manage Links** tool to Reload
From the folder.

Click **OK**.

3.   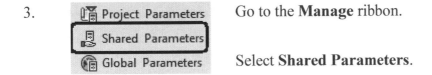   Go to the **Manage** ribbon.

Select **Shared Parameters**.

4.   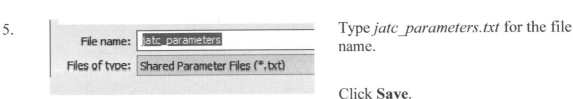   Select **Create.**

5.      Type *jatc_parameters.txt* for the file
name.

Click **Save**.

6.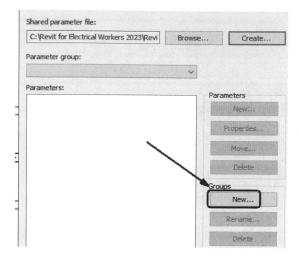

Select **New** under Groups.

Shared parameters are organized under Groups to make it easier for users to locate the desired parameters.

7.

New Parameter Group

Name: Lighting

Type **Lighting** in the Name field.

Click **OK**.

8.

Click **New** under Parameters.

The new parameter will be placed in the Lighting parameter group.

9.

Name:

Lighting Zone

Discipline:

Common

Type of Parameter:

Text

Tooltip Description:
<No tooltip description. Edit this parameter to write a custom...

Edit Tooltip...

Type **Lighting Zone** in the Name field.

Set the Discipline to **Common**.

Set the Type of Parameter to **Text**.

Click **OK**.

Save as *ex8-8.rvt*.

Exercise 8-9:

Add a Shared Parameter to a Family

Drawing Name: *shared_parameters.rvt*
Estimated Time: 15 minutes

This exercise reinforces the following skills:

- ❑ Shared Parameters
- ❑ Families

1. 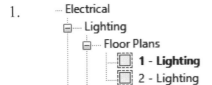 Open the **1 – Lighting** floor plan view.

2. 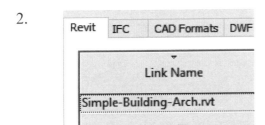 Go to the Insert Ribbon.
Use the **Manage Links** tool to Reload From the folder.

Click **OK**.

3. 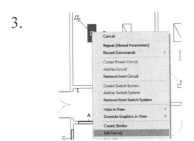 Select the light fixture in Lab 101.

Right click and select **Edit Family**.

4. 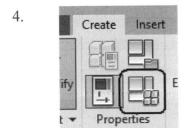 Select **Family Types** on the ribbon.

5.

Type name: 2x4 - 277

| 1x4 - 120 |
| 1x4 - 277 |
| 20x4 - 120 |
| 20x4 - 277 |
| 2x2 - 120 |
| 2x2 - 277 |
| 2x4 - 120 |
| 2x4 - 277 |
| 2x4 - 278 - Lavatory |

Search param

Constraints

Default Eleva

Electrical

Load Classification   Light

This lighting fixture has several types already defined.

We want to add the shared parameter to the 2x4 -277 and the 2x4-277 Lavatory types.

Select the **2x4 – 277** type name from the drop-down list.

6.

Select **New** parameter.

7.

Parameter Type

○ Family parameter

  (Cannot appear in schedules or tags)

◉ Shared parameter

  (Can be shared by multiple projects and families, exported to ODBC, and appear in schedules and tags)

    Select...   Export...

Enable **Shared parameter**.
Click on **Select**.

8.

Parameter group:

Lighting

Parameters:

Lighting Zone

Switch to the **Lighting Parameter** group.

Highlight **Lighting Zone**.

Click **OK**.

9.

Parameter Data

Name:

Lighting Zone

Discipline:

Common

Type of parameter:

Text

Group parameter under:

Electrical - Lighting

Enable **Instance**.

Group the parameter under **Electrical-Lighting**.

The Discipline and Type of parameter are filled in based on how the shared parameter was defined.

Click **OK**.

10.

Wattage Comments		
Electrical - Lighting		
Calculate Coefficient of Utiliz	☑	
Coefficient of Utilization (def		
Lighting Zone (default)		

The Lighting Zone parameter appears under the Electrical category. Because it is an instance parameter, users may wish to fill in a default value.

The parameter is now available to be assigned and used in schedules.

Click **OK**.

11.

pipe-strap-atkore-sing... 11/29/2019 4:37 PM

File name: Plain Recessed Lighting Fixture.rfa

Files of type: Family Files (*.rfa)

Save the family to your exercise folder.

12.

Select **Load into Project and Close**.

13.

You are trying to load the family Plain Recessed Lighting Fixture, which already exists in this project. What do you want to do?

→ Overwrite the existing version

→ Overwrite the existing version and its parameter values

Select **Overwrite the existing version and its parameter values**.

14. Save as *ex8-9.rvt*.

Switch Legs & Lighting Zones

Adding a switch leg to light fixtures allows you to easily create and change lighting zones within a project. Once a lighting control zone has been established and light fixtures are tied to a specific switch, changing the name of the control zone or adding and removing fixtures becomes as simple as editing a circuit. This allows for easy manipulation of zones, labeling and adding of zones. Here's how you add switch legs:

1. Select the fixtures that you want to be switched together.
2. Add the switch to the switch system.
3. You can now edit the switch system, either adding or removing lights from the system.
4. Select the switch you want to control those lights.
5. In that switch under the instance property of 'Switch ID' enter the lowercase letter that you want to call that switch group.
6. Tag the switch and fixtures with their switch ID.

To change the display of switch legs, use the same workflow that was used to create conduit types.

1. Create a wire type for switch legs.
2. Assign a description for the switch leg wires.
3. Create a filter for the switch leg description.
4. Assign the filter to a view.
5. Use overrides to change the display of the switch leg wires.

Exercise 8-10:

Assigning Lighting Zones to Light Fixtures

Drawing Name: *lighting zones.rvt*
Estimated Time: 10 minutes

This exercise reinforces the following skills:

- ❑ Shared Parameters
- ❑ Families

1.

    ···· Electrical
      ⊟···· Lighting
         ⊟···· Floor Plans
            ··· ☐ **1 - Lighting**
            ··· ☐ **2 - Lighting**

Open the **1 – Lighting** floor plan view.

2.

Revit	IFC	CAD Formats	DWF

Link Name

Simple-Building-Arch.rvt

Go to the Insert Ribbon.
Use the **Manage Links** tool to Reload From the folder.

Click **OK**.

3.

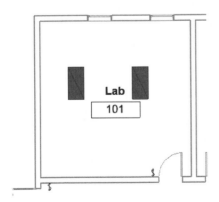

Select the Lighting fixtures in Lab -101.

4.

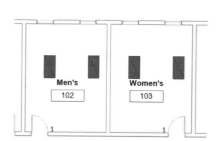

In the Properties palette:

Type **Zone 1** for the lighting zone.

5.

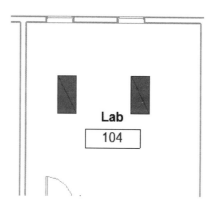

Select the lighting fixtures in the two lavatories.

6.

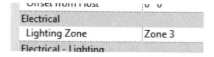

In the Properties palette:

Type **Zone 2** for the lighting zone.

7.

Select the Lighting fixtures in Lab -104.

8.

In the Properties palette:

Type **Zone 3** for the lighting zone.

9. Save as *ex8-10.rvt*.

Exercise 8-11:

Creating a Custom Lighting Fixture Tag

Drawing Name: *lighting_fixture tag.rvt*
Estimated Time: 25 minutes

This exercise reinforces the following skills:

❑ Lighting Fixture Tags
❑ Shared parameters

1. 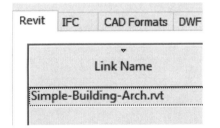 Open the **1 – Lighting** floor plan view.

2. 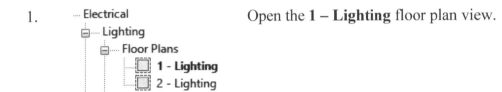 Go to the Insert Ribbon.
Use the **Manage Links** tool to Reload From the folder.

Click **OK**.

3. Switch to the **Annotate** ribbon.
Select **Tag by Category**.

4. 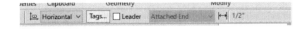 Disable **Leader** on the Options bar.

5. Left click to place on each light fixture in Rooms 101, 102, 103, and 104.

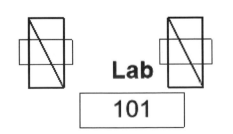

Select **Modify** on the ribbon to exit the command.

6.

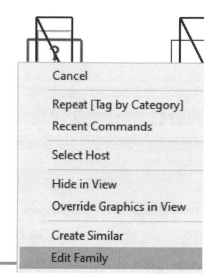

Select one of the tags.
Right click and select **Edit Family**.

7.

Select the text.

Select **Edit** next to Label in the Properties palette.

8.

Highlight the **Type Mark** parameter name.

Select the **Remove** button to remove the parameter.

9.

Select **New parameter**.

10.

Parameter Type

○ Shared parameter

(Can be shared by multiple projects and families, exported to ODBC, and appear in schedules and tags)

Select...   Export...

Parameter Data

Click **Select.**

11.

Choose a parameter group, and a parameter.

Parameter group:

Lighting

Parameters:

Lighting Zone

Highlight the **Lighting Zone** parameter.

Click **OK.**

12.

Parameter Data

Name:

Lighting Zone

Discipline:

Common

Type of Parameter:

Text

Click **OK.**

13.

Label Parameters

	Parameter Name	Spaces	Prefix	Sample Value
1	Lighting Zone	1		Zone 1

Highlight **Lighting Zone** and add it to the Label Parameters.

Type **Zone 1** in the Sample Value column.

Click **OK.**

14.

Zone 1

Expand the text box, so it is all one line.

15.

Label

Open the Create ribbon.

Select the **Label** tool.

Left click below the Zone 1 label to place the new label.

16.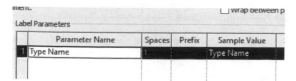

Add **Type Name** to the Label Parameters.
Click **OK**.

17.

Position the Type Name so it is centered under the Zone Name.

18.

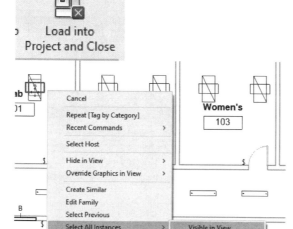

Adjust the box size.

Select a line and then drag it to the desired position.

Save the family as *Lighting Fixture Tag – Zone.rfa.*

19.

Select **Load into Project and Close** from the ribbon.

20.

Select one of the lighting fixture tags.
Right click and select **Select All instances→Visible in View**.
This will select all the lighting fixture tags of that type in this view.

21.

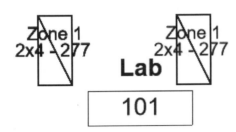

Using the Type Selector, change the tags to **Lighting Fixture Tag – Zone Standard.**

22.

The tags update.
Save the file as *ex8-11.rvt*.

Transfer Project Standards

Transfer Project Standards is used to copy system families from one project to another. System families are elements that are defined inside a project and are not loaded from an external file. System families include view templates, walls, conduits, and wires.

You need to have the project open that you want to import into as well as the project you want to transfer from. If the project you are transferring from has a linked file, you can also import families from the linked file without opening that file.

Items which can be copied from one project to another include:

- Family types (including system families, but not loaded families)
- Line weights, materials, view templates, and object styles
- Mechanical settings, piping, and electrical settings
- Annotation styles, color fill schemes, and fill patterns
- Print settings

You may get a prompt during the transfer process alerting that an item being transferred already exists in the new project. You may opt to overwrite, ignore or cancel for the existing elements. You do not get to pick and choose which elements you overwrite if more than one item has been selected to be transferred.

Exercise 8-12:

Transfer Project Standards

Drawing Name: *export standards.rvt*
Estimated Time: 10 minutes

This exercise reinforces the following skills:

- ❑ Project Standards
- ❑ Conduits
- ❑ View Templates

1.

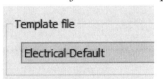

Open the *export standards.rvt file.*
This is the file we will import standards from.

Start a New Project using the Electrical-Default template.

2.
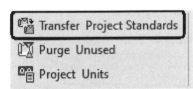

Select the Manage ribbon.

Select **Transfer Project Standards** from the Settings panel on the ribbon.

3.
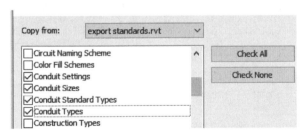

Select *export standards.rvt* to copy from.

This file must be open to be selected.

Click **Check None**.

Place a check next to:
    Conduit Settings
    Conduit Sizes
    Conduit Standard Types
    Conduit Types

4.

Scroll down the list.

Place a check next to:
      Distribution System
      Electrical Demand Factor Definitions
      Electrical Load Classifications
      Electrical Settings
      Filters

5.

Scroll down the list.

Place a check next to:
      View Templates

Click **OK**.

6.

Click **Overwrite**.

This replaces any existing element definitions with the definitions used in the export standards file.

7.

Switch to the View ribbon.

Select **Manage View Templates**.

8.

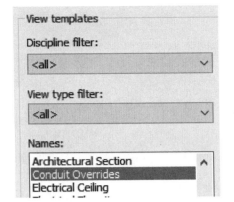

You imported a view template called Conduit Overrides.

Click **OK** to close the dialog box.

9.

Switch to the Systems ribbon.
Select **Conduit**.

10.

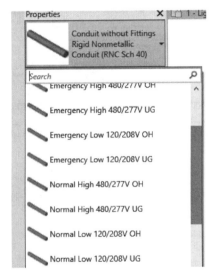

Use the Type Selector to see all the conduit types which were imported.

11.

Save as *ex8-12.rvt*.

Shared Coordinates

Coordinates in a file will only be shared, or the files will have shared coordinates after a process of transferring the coordinate system used in a file into another. It does not matter if in two files the location is exactly the same. They will be not sharing coordinates unless the sharing coordinates process has been carried out.

Normally there is one file, and only one, that is the source for sharing coordinates. From this file the location of different models is transferred to them, and after that all files will be sharing coordinates, and be able to be linked with the "Shared Coordinates option".

Revit uses three coordinate systems. When a new project is opened, the three systems overlap. They are:

- The Survey Point
- The Project Base Point
- The Internal Origin

The Survey Point is represented by a blue triangle with a small plus symbol in the center. This is the point that stores the universal coordinate system, or a defined global system of the project to which all the project structures will be referred. The Survey Point is a real-world relation to the Revit model. It represents a specific point on the Earth, such as a geodetic survey marker or a point of reference based on project property lines. The survey point is used to correctly orient the project with other coordinate systems, such as civil engineering software.

- It is the origin that Revit will use in case of share coordinates between models.
- The origin point used when inserting linked files with the Shared Coordinates option.
- The origin point used when exporting with Shared Coordinates option.
- The origin point to which spot coordinates and spot elevations are referenced, if the Survey Point is the coordinate origin in the type properties.
- The origin point to which level elevation is referred, if the SP is the Elevation Base in the level type properties.
- When the view shows the True North, it is oriented according to the Survey Point settings.

The Project Base Point is represented by a blue circle with an x. The position of this point is unique for each model, and this information is not shared between different models. The Project Base point could be placed in the same location as the Survey Point, but it is not usual to work in that way. This point is used to create a reference for positioning elements in relation to the model itself. By default, the Project Base Point is the origin (0,0,0) of the project. The Project Base Point should be used as a reference point for measurements and references across the site. The location of this point does not affect the shared site coordinates, so it should be located in the model where it makes sense to the model, usually at the intersection of two gridlines or at the corner of a building.

- The origin point to which spot coordinates and spot elevations are referenced, if the Project Base Point is the coordinate origin in the type properties.
- The origin point to which level elevation is referred, if the Project Base Point is the Elevation Base in the level type properties.
- When the view shows the Project North, it is oriented according to the Project Base Point.
- If it is not necessary, it is better not to move this point, so that it always is coincident with the third point: the Internal Origin.
- If we have moved the Project Base Point to a different location, it can be placed back to the initial position by right clicking on it when selected, and use **Move to Startup Location.**

The Internal Origin is now visible starting with the 2021 release of Revit. The Internal Origin is coincident with the Survey Point and the Project Base Point when a new project is started. It is displayed with an XY icon. Prior to this release, users sometimes resorted to placing symbols at the internal origin location to keep track of the origin.

- The project should be modeled in a restricted area around the internal origin point. The model should be inside a 20 miles radius circle around the internal origin, so that Revit can compute accurately.

- This is the origin point that is used when inserting external files using the "Internal Origin to Internal Origin" option.

- This is the origin point that is used when copy/pasting model objects from one file into another using the "aligned" option.

- The origin point to which spot coordinates and spot elevations are referenced, if Relative is the coordinate origin in the type properties

- The origin point used when exporting with Internal Coordinates option.

- Revit API and Dynamo use this point as the coordinates origin point for internal computational calculations.

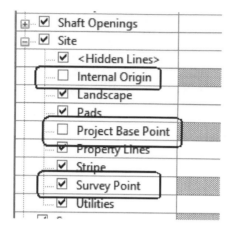

The visibility of these three systems is controlled under the Site category on the Model tab of the Visibility Graphics dialog.

Moving the project base point is like changing the global coordinates of the entire project. Another way of looking at it is if you modify the location of the project base point, you are moving the model around the earth. If you modify the location of the survey point, you are moving the earth around the model.

The benefits of correctly adjusting shared coordinates are not limited to aligning models for coordination but can also position the project in the real world if using proper geodetic data from a surveyor landmark or a provided file with the real-world information present.

To fully describe where an object (a building) sits in 3D space (its location on the planet), we need four dimensions:

- East/West position (the X coordinate)
- North/South position (the Y coordinate)
- Elevation (the Z coordinate)
- Rotation angle (East/West/True North)

These four dimensions uniquely position the building on the site and orient the building relative to a known benchmark as well as other landmarks. Revit allows you to assign a unique name to these four coordinates. Revit designates this collection as a **Shared Site**. You can designate as many Shared Sites as you like for a project. This is useful if you are planning a collection of buildings on a campus. By default, every project starts out with a single Shared Site which is named Internal. The name indicates that the Shared Site is using the Internal Coordinate System. Most projects will only require a single Shared Site, but if you are managing a project with more than one building or construction on the same site, it is useful to define a Shared Site for each instance of a linked file.

For example, in a resort, there may be different instances of the same cabin located across the site. You can use the same Revit file with the cabin model and copy the link multiple times. Each cabin location will be assigned a unique Shared Site, allowing each instance of the linked file to have its own unique location in the larger Shared Coordinate system.

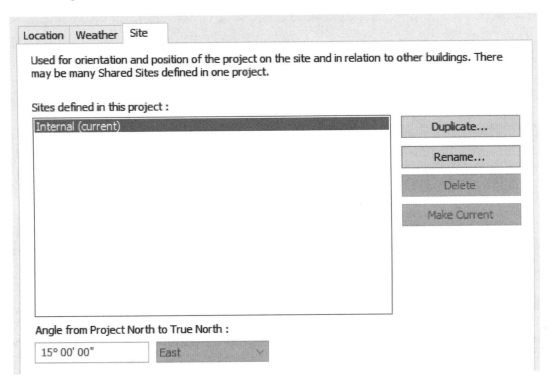

Shared Sites are named using the Location Weather and Site dialog.

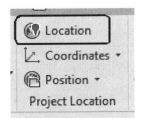

The dialog is opened using the Location tool on the Manage ribbon.

Why setting up shared coordinates is important:

- To properly coordinate the project models to work across platforms, including BIM Track.
- To display the project in its real-world location (including BIM and GIS data overlay).
- To link files together that have different base points (arbitrary coordinate systems).

Shared Coordinates allow you to adjust the reference point from project origin to a shared site point, to make it appear in a different position or with a different angle based on the internal base point. It is important to note that nothing in the Revit model moves when using this system, even if it looks like it moves and/or rotates on the screen. It is just applying an adjustment to the point of origin, so in fact, all elements keep their relation to the internal base point when using this option.

Revit has two different North directions: Project North and True North. True North doesn't change. Project North is used to position the project for sheets and views. This allows the project to be parallel and perpendicular for presentation and modeling purposes. The Project North rotation is referenced to the True North. The model's True North settings are saved and set when shared coordinates are used.

If multiple models are being linked together and need to be positioned appropriately in relation to one another, shared coordinates are required. Most electrical projects use a linked model which contains the architectural model (the floor plan) and sometimes linked models for the plumbing and the structural designs as well.

There are two ways to establish shared coordinates: the push method and the pull method. The push method has the host file determine which shared coordinates should be used and requires all linked files to align with the host file. The pull method takes the shared coordinates from a linked file and adjusts the position of the model elements in the host file. Most BIM projects either have the architect who has designed the floor plan control which coordinates are used when linking to files or the civil engineer who has designed the site survey designate the shared coordinates. Most electrical workers need to "pull" the shared coordinates from the project file provided by either the architect or the civil engineer.

The steps to pull shared coordinates are as follows:

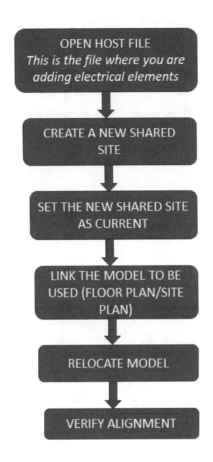

Exercise 8-13:

Understanding Shared Coordinates

Drawing Name: new
Estimated Time: 20 minutes

This exercise reinforces the following skills:

- ❑ Shared Coordinate System
- ❑ Site Point
- ❑ Base Point
- ❑ Internal Origin
- ❑ Views
- ❑ Visibility/Graphics
- ❑ Spot Coordinate
- ❑ Specify Coordinates at a Point
- ❑ Report Shared Coordinates

1.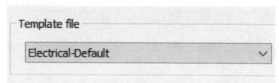

Start a new project using the *Electrical-default* template.

2.  Highlight the **1- Lighting** floor plan.

Right click and select **Duplicate View→Duplicate**.

3.  Set the Discipline to **Coordination**.
Set the Sub-Discipline to **Site**.

You will have to type Site in the Sub-Discipline field.

4.

The view will be placed under Coordination/Site.

Rename the view **Site**.

5.

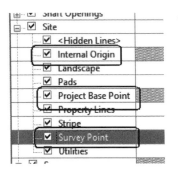

Type **VV** to open the Visibility/Graphics dialog.

On the Model Categories tab:

Enable:
- Internal Origin
- Project Base Point
- Survey Point

These are located under the Site category.

Click **OK**.

The Internal Origin, Project Base Point and Survey Point are all visible.

They are overlaid on top of each other.

6.

Select the **Project Base Point**.
That's the circle symbol.

The coordinates for the base point are displayed. Note that by default the base point is located at 0,0.

It is also centered on the screen relative to the elevation markers.

We can add a spot coordinate symbol to the project base point to keep track of its location.

Spot coordinates report the North/South and East/West coordinates of points in a project. In drawings, you can add spot coordinates on floors, walls, toposurfaces, and boundary lines. Coordinates are reported with respect to the survey point or the project base point, depending on the value used for the Coordinate Origin type parameter of the spot coordinate family.

7.

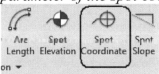

Switch to the Annotate ribbon.

Select the **Spot Coordinate** tool on the Dimension panel.

8.

Select the Project Base Point.

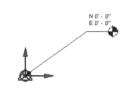

Left click to place the spot coordinate.

9.

Switch to the Manage ribbon.

Select **Report Shared Coordinates** on the Project Location panel.

Select the Project Base Point.

10.

The coordinates are displayed on the Options bar.

Cancel out of the command.

11.

Select the Project Base Point.

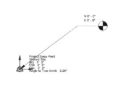

Notice that the coordinates that are displayed are blue. This means they are temporary dimensions which can be edited.

12.

Click on the **E/W** dimension.
Change it to **20' 6"**.
Press **ENTER**.

13.

The location of the project base point shifted to be 20' 6" east of the internal origin/survey point.

Notice that the spot coordinate value updated.

14.

Left click on the Survey Point.
That's the triangle symbol.
Notice that dimensions are black, which means they cannot be edited.

15.

Drag the survey point away from the internal origin (the XY icon).

Notice that the Project Base Point value updates.

Remember the Survey Point represents a real and known benchmark location in the project (usually provided by the project survey). The Project Base Point is simply a known point on the building (usually chosen by the project team).

Think of the Survey Point as the coordinates in the *World* and the Project Base Point as the local building coordinates.

The Survey Point is always located at 0,0, while the Project Base Point is relative to the Survey Point.

It is called a shared coordinate system because it represents the coordinate system of the world around us and is shared by all the buildings on the site.

16.

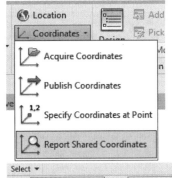

Select **Report Shared Coordinates** on the Project Location panel.

Select the Internal Origin (the XY icon).

17.

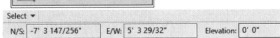

The coordinates are displayed on the Options bar.

Notice that the Internal Origin is located relative to the Survey Point. The Internal Coordinate System represents the building's "local" coordinates.

Cancel out of the command.

18.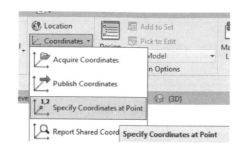

Usually, you will be provided the coordinate information for a project from the civil engineer.

Select **Specify Coordinates at Point**.

Select the Survey Point (the triangle symbol.).

19.

Fill in the Shared Coordinates:

For North/South: **39' 4"**.
For East/West: **16' 0"**.
For Elevation: **20' 0"**.
For Angle from Project North to True North: **15° East**.

Click **OK**.

20.

Zoom out and you will see that the Survey Point's position has shifted.

Note the coordinates for the Project Base Point updated.

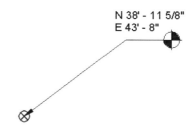

21. Save as *ex8-13.rvt*.

Exercise 8-14:

Understanding Location

Drawing Name: Simple-Building-Coordinates.rvt
Estimated Time: 20 minutes

This exercise reinforces the following skills:

- ❏ Linking Files
- ❏ Shared Coordinate System
- ❏ Survey Point
- ❏ Base Point
- ❏ Internal Origin
- ❏ Visibility/Graphics

1. Open the **Site** floor plan.

2. *The Survey Point and the Project Base Point are visible, but not the Internal Origin.*

    Open the Visibility/Graphics dialog by typing **VV**.

3. 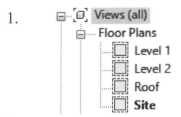 On the Model Categories tab:

    Enable:
    - Internal Origin

    This is located under the Site category.

    Click **OK** and close the dialog.

4. Activate the **Insert** ribbon.

    Select **Link Revit**.

5.

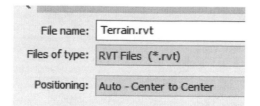

Locate the *Terrain.rvt* file.

Set the Positioning to **Auto – Center to Center**.

Click **Open**.

Many users will be tempted to align the files using Internal Origin to Internal Origin. This is not the best option when we plan to reposition the building on the site plan. Origin to Origin works well when linking MEP files to the Architectural file.

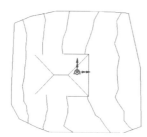

Next, we will reposition the linked file into the correct relative position. If we do it this way, we are not moving the building, we are moving the terrain. We will still use the terrain to establish the Shared Coordinate system, which is the "pull" method. The "pull" method pulls the information from the linked file.

6.

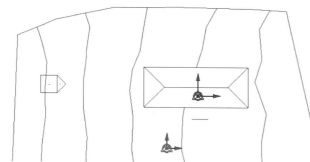

Select the linked file.

Select the **Move** tool on the ribbon.

Select the bottom midpoint of the building as the start point.

Move the cursor down and type 30' to move the terrain down 30'.

Select the Move tool on the ribbon.
Select the bottom midpoint of the building as the start point.
Move the cursor to the left and type 15' to move the terrain left 15'.

7.

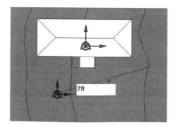

Select the linked file.

Select the Rotate tool on the ribbon.

Start the angle at the horizontal 0° level.
Rotate up and type **20°**.

8.

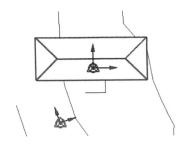

We have repositioned the terrain on the XY plane, but we have not changed the elevation.

Click on the Project Base Point on the building.

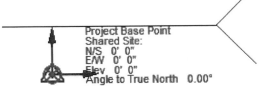

Notice that it has not changed.

9.

Click on the Project Base Point for the terrain file.

It shows as 0,0 still as well.

10.

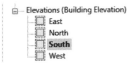

Open the **South** elevation.

11.

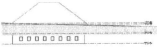

Notice that the levels are not aligned between the files.

12.

*Use the **ALIGN** tool to move the Level 1 in the terrain file to align with Level 1 in the building file.*

Select the **ALIGN** tool.

Select the Level 1 that is on the building file as the target.
Select the Level 1 on the terrain file to be moved.

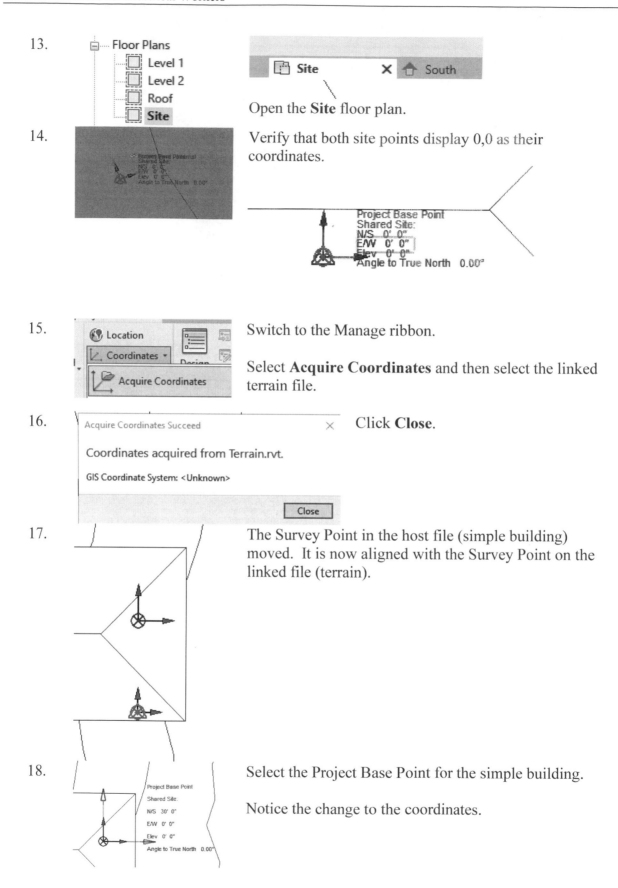

13. Open the **Site** floor plan.

14. Verify that both site points display 0,0 as their coordinates.

15. Switch to the Manage ribbon.

    Select **Acquire Coordinates** and then select the linked terrain file.

16. Click **Close**.

    Acquire Coordinates Succeed

    Coordinates acquired from Terrain.rvt.

    GIS Coordinate System: <Unknown>

    Close

17. The Survey Point in the host file (simple building) moved. It is now aligned with the Survey Point on the linked file (terrain).

18. Select the Project Base Point for the simple building.

    Notice the change to the coordinates.

    Project Base Point
    Shared Site:
    N/S   30' 0"
    E/W   0' 0"
    Elev   0' 0"
    Angle to True North 0.00°

19.

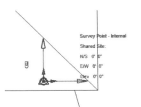

Select the Survey Point.

Notice it is set to 0,0.

20.   Save the file as *ex8-14.rvt*.

Exercise 8-15:

Linking Files using Shared Coordinates

Drawing Name: terrain.rvt
Estimated Time: 10 minutes

This exercise reinforces the following skills:

- Linking Files
- Shared Coordinate System
- Survey Point
- Base Point
- Internal Origin
- Visibility/Graphics

1.   Open the Site floor plan.

2.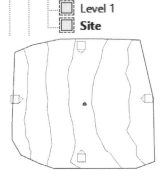

The Survey Point and the Project Base Point are visible, but not the Internal Origin.

Open the Visibility/Graphics dialog by typing **VV**.

3.

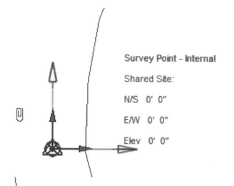

On the Model Categories tab:

Enable:

- Internal Origin

This is located under the Site category.

Click **OK** and close the dialog.

4.

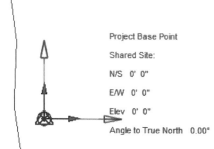

Use **FILTER** to select just the Survey Point.

Notice that it is still located at 0,0.

Press ESC to release the selection.

5.

Select the Project Base Point.

It is also located at 0,0.

The file wasn't changed because we acquired the coordinates from the linked file and applied them to the host file.

6.

Switch to the **Insert** ribbon.

Select **Link Revit**.

7.

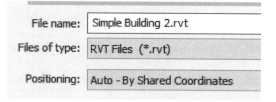

Select *Simple Building 2.rvt*.
Set the Positioning to **Auto -By Shared Coordinates.**

Simple Building 2 has the coordinates assigned that we defined in the previous exercise.

Click **Open**.

8.

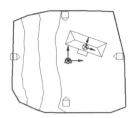

The linked file is positioned in the correct location.

9.

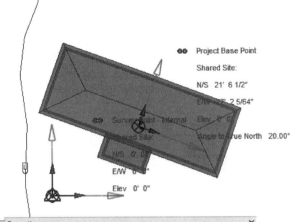

Click on the building.
Notice that Project Base Point
is using the Shared Site
coordinates.

10.

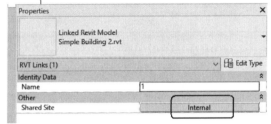

With the linked file selected,
notice in the Properties palette, it is
using the Internal Shared Site definition.

11. Save as *ex8-15.rvt*.

Exercise 8-16:
Defining a Shared Site

Drawing Name: terrain 2.rvt
Estimated Time: 30 minutes

This exercise reinforces the following skills:

- Linking Files
- Shared Coordinate System
- Survey Point
- Base Point
- Shared Site

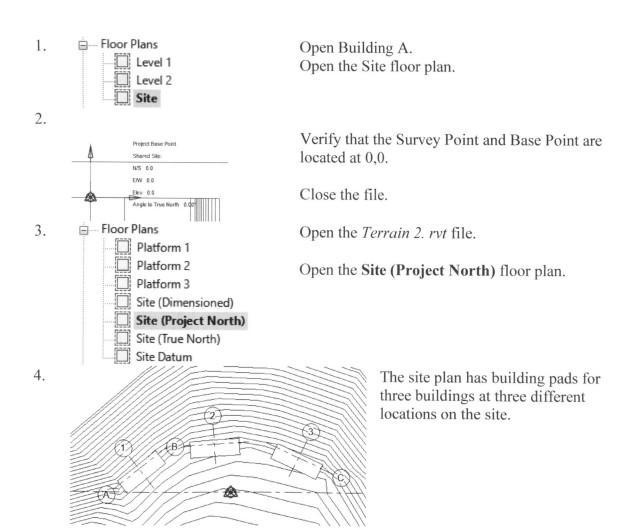

1. Open Building A.
   Open the Site floor plan.

2. Verify that the Survey Point and Base Point are located at 0,0.

   Close the file.

3. Open the *Terrain 2. rvt* file.

   Open the **Site (Project North)** floor plan.

4. The site plan has building pads for three buildings at three different locations on the site.

5.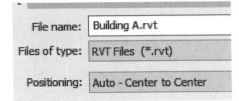

    Switch to the **Insert** ribbon.

    Select **Link Revit**.

6.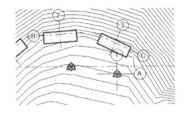

    Select Building A.*rvt*.
    Set the Positioning to **Auto -Center to Center.**

    Click **Open**.

7.

    Note the position of Building A on the site plan.

    Use the **ALIGN** tool to move Building A onto the A platform. Align Grids 1 to each other and Grids A to each other.

    Remember to select the Platform grid first as the target and then the building grid.

8. Open the **South (Full)** elevation view.

    ☐— Elevations (Building Elevation)
    ☐ East
    ☐ North
    ☐ South (Cropped)
    ☐ South (Full)
    ☐ West

9.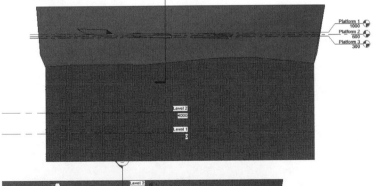

    Notice that the Building A elevation needs to be adjusted.

10.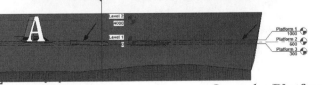

    Use the **ALIGN** tool to align Level 1 for Building A to Platform 1.

11. Open the **Platform 1** view.

    ☐— Floor Plans
    ☐ Platform 1
    ☐ Platform 2
    ☐ Platform 3

12.

Verify that Building A is situated properly on the platform.

13.

Select the linked file (Building A).

Click **<Not Shared>** on the Properties palette.

14.

We can either push coordinates or pull coordinates.

If we select Publish the current shared coordinate system..., we are using the coordinates from the host file and pushing them to the linked file.

If we select Acquire the shared coordinate system...we are using the coordinate system from the linked files and pulling them into the host file.

Regardless of which method we use, after the selection, they will be sharing the same coordinate system.

Since we started with the terrain file, it is standard to let the topo file control which coordinate system is used in a project.

Remember that whichever option you select does not change the linked file or the host file local internal coordinates. We are only defining how we want the files to interact relative to each other.

15.

Enable **Publish**.

Click **Change**.

16.

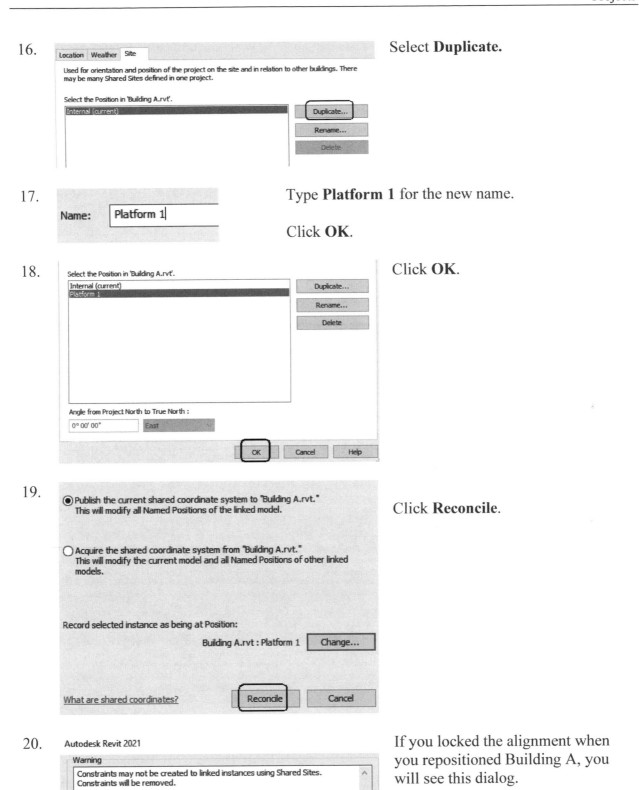

Select **Duplicate.**

17.

Name: Platform 1

Type **Platform 1** for the new name.

Click **OK.**

18.

Click **OK.**

19.

Click **Reconcile.**

20.

If you locked the alignment when you repositioned Building A, you will see this dialog.

Click **OK.**

21.

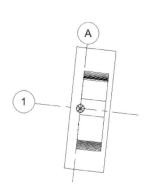

The coordinates on the linked files update.

Save the file as *ex8-16.rvt*.

22.

Location Position Changed

You have changed the "current" Position in Building A.rvt. What do you want to do?

→ Save
Saves the new position back to the link.

→ Do not save
Returns to the previously saved position when the link is reloaded or reopened.

→ Disable shared positioning
Retains the current placement of the link and clears the Shared Position parameter.

We want to save the coordinates to the linked file, so we select **Save**.

The other two choices basically cancel out everything we just did.

23. Close all open files.

Open *Building A.rvt*

24.

Switch to the Site plan view.

Select the Project Base Point. Notice that the value has changed.

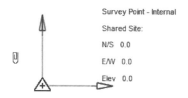

Survey Point - Internal

Shared Site:

N/S 0.0

E/W 0.0

Elev 0.0

Notice that the Survey Point is still at 0,0.

25.

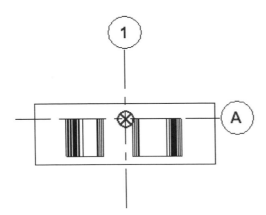

Open the Level 1 view.

The building is oriented horizontally and vertically to the screen to make it easy to work on.

26.

Switch to the **Insert** ribbon.

Select **Link Revit**.

27.

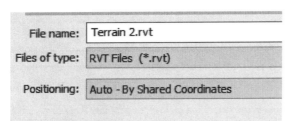

File name:	Terrain 2.rvt
Files of type:	RVT Files (*.rvt)
Positioning:	Auto - By Shared Coordinates

Select *Terrain 2.rvt.*
Set the Positioning to **Auto -By Shared Coordinates.**

Click **Open**.

28. Notice that the terrain is oriented perfectly to the building.

This is because we are using the same shared coordinates between the two files.

Close all files.

eTransmit

eTransmit allows you to copy a Revit project along with any linked/dependent files to a single folder. This is also useful if you want to create snapshots of your project at different stages of construction.

You can elect to:
- Include related dependent files such as linked models and DWF markups.
- Include supporting files such as documents or spreadsheets.
- Upgrade the Revit (.rvt) model and linked models to the current release.
- Disable worksets.
- Remove unused families, materials, and other objects from the Revit models to reduce file size.
- Delete sheets, and specific view types so that the models do not contain unnecessary data.
- Include only the views that are placed on sheets.

Shared parameter files, external font files, and lookup tables are not included in the eTransmit package.

Exercise 8-17:

Transmit a Model

Drawing Name: etransmit.rvt
Estimated Time: 25 minutes

This exercise reinforces the following skills:

❑   eTransmit

1. Verify all files are closed.
    Select the **Add-Ins** tab on the ribbon.
    Select **Transmit a model**.

2. Click on **Browse Model...**

3. Select *etransmit.rvt*.
    Click **Open**.

4. Locate a folder to save the files.

5.

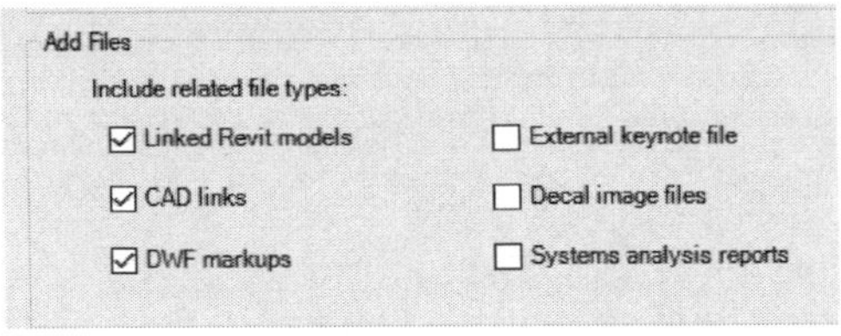

Enable:

- Linked Revit models
- CAD links
- DWF markups

Note that you can also add spreadsheets, reports, etc. to the transmittal package using the Add files... button.

6.

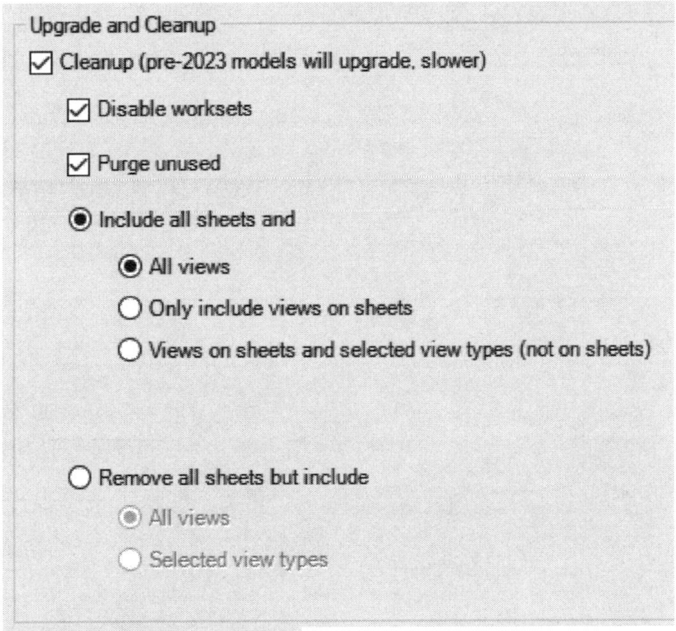

Enable.

Enable **Purge unused**.
This deletes any unused families from the files to reduce file size.
Cleanup

You can also opt to only include the views on sheets.

7.

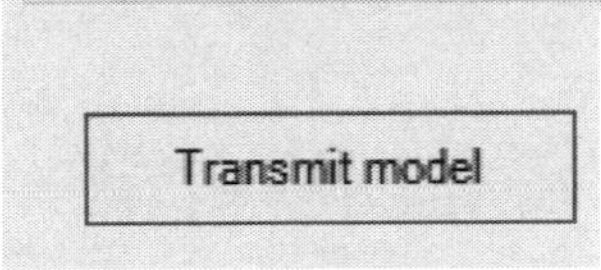

Click **Transmit model**.

8.

Using File Explorer, locate the eTransmit folder that was created.

etransmit_2022-9-4_12.8.25

9.

That folder includes a folder with the files and a txt file.

Open the txt file.

10.

Note the report lists which files were included as well as which files were excluded.

Lesson

09

Annotations, Dimensions, and Symbols

The term 'Annotations' is used in Revit to include several different types of elements:

- Dimensions
- Text Notes
- Keynotes
- Tags
- Symbols

Dimensions

There are three types of dimensions in Revit:

Temporary dimensions are only displayed when elements are selected. They show the position of elements relative to other dimensions. They control the placement of elements. They are not plotted. A dimension symbol appears near the temporary dimension of a selected element. If you left click on the dimension symbol, a permanent dimension is placed.

Permanent dimensions are annotations users place in a view. They are view-specific, and they will be visible when the sheet or view is printed. The value of the permanent dimension is fed from the temporary dimension. In order to change the permanent dimension, the user needs to edit the temporary dimension. If you lock a permanent dimension, the value cannot be changed.

Listening dimensions are the dimensions which appear when placing or modifying an element. Listening dimensions appear in bold and update as elements are created or modified.

Settings can be specified such as snapping points for temporary dimensions.

Guidelines for Working with Dimensions

- Change temporary dimensions to permanent dimensions when you need to refer to the distance frequently while working. You can delete or turn off the visibility of dimensions when they are no longer necessary.
- Add parallel reference planes or model lines at the corners of a room when you want to place dimensions across the corners. You can then turn off the visibility of the planes/model lines.
- Add dimension styles to project templates once you have established a standard.
- Use the Duplicate View tool to create a copy of the view. Have one view with dimensions and annotations and one without. You can also hide individual dimensions by right clicking on a dimension and selecting Hide element.
- Adjust the view scale prior to placing text and dimensions. Text and Dimensions will scale with the view. If you change the view scale, you may need to reposition text and dimensions to maintain clarity.

Exercise 9-1:

Adding Dimensions

Drawing Name: *dimensions_2.rvt*
Estimated Time: 5 minutes

This exercise reinforces the following skills:

❏ Add dimensions to a view

1.  Open *dimensions_2.rvt*.

2.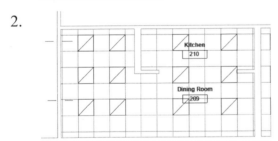

   Main Floor -Annotated
       Main Floor -Annotated - Dining Room
   Main Roof

   Open the **Main Floor – Annotated - Dining Room** ceiling plan.

3. Annotate   Activate the **Annotate** ribbon.

4. Aligned   Select the **ALIGNED** dimension tool on the ribbon.

5. Wall faces   Pick: Individual Reference   Options

   On the Options bar:

   Select **Wall faces.**

6.

Working from left to right, select the inside wall face on the west wall, select the center of the lighting fixture, then left click below the view to place the dimension.

Continue placing dimensions on the lower row of lighting fixtures.

7.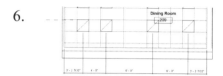

Select the **ALIGNED** dimension tool on the ribbon.

8.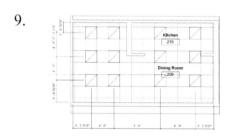

On the Options bar:

Select **Wall faces.**

9.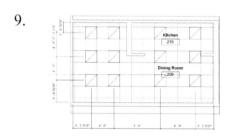

Starting at the south wall, place vertical dimensions from the wall to the lighting fixture.

10. Save as *ex9-1.rvt*.

Exercise 9-2:

Create a Dimension Style

Drawing Name: *dimension_style.rvt*
Estimated Time: 20 minutes

This exercise reinforces the following skills:

❑ Define a new Dimension family type
❑ Activate View
❑ Deactivate View
❑ Type Properties
❑ Type Selector

1. 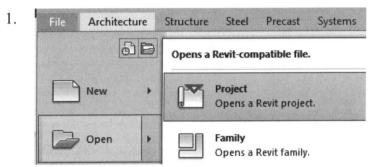 Open *dimension_style.rvt*.

2.  Open the **07 Lighting Fixture Details** sheet in the Project Browser.

3. 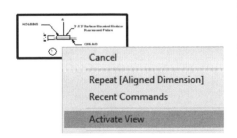 Highlight the middle view with the recessed lighting fixture.

Right click and select **Activate View**.

4.

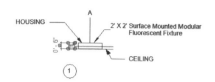

Select the vertical dimension that displays the height of the lighting fixture.

5.

Select **Edit Type** on the ribbon.

6.

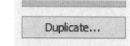

Select **Duplicate**.

7.

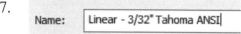

Name: Linear - 3/32" Tahoma ANSI

Rename **Linear – 3/32" Tahoma ANSI**.

Click **OK**.

8.

Type Parameters	
Parameter	Value
Graphics	
Dimension String Type	Continuous
Leader Type	Arc
Leader Tick Mark	None
Show Leader When Text Moves	Beyond Witness Lines
Tick Mark	Arrow Filled 15 Degree

Change the Tick Mark to **Arrow Filled 15 Degree.**

9.

Text Size	3/32"
Text Offset	5/128"
Read Convention	Horizontal
Text Font	Tahoma
Text Background	Opaque

Scroll down the Text Parameters.

Set the Read Convention to **Horizontal**.

Set the Text Font to **Tahoma**.

Click **OK**.

Review the change to the dimension.

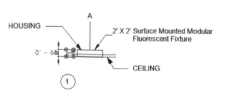

10.

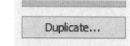

Select the vertical dimension that displays the height of the lighting fixture.

11.  Select **Edit Type** on the ribbon.

12. 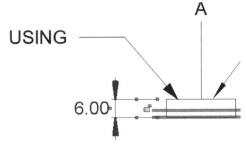 Select **Duplicate**.

13. Rename **Linear – 3/32" Arial ANSI**.

Click **OK**.

14. 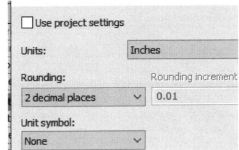 Set the Text Font to **Arial**.

Left click on the **Units Format** button.

15. Disable **Use project settings.**

Set the Units to **Inches**.

Set the Rounding to **2 decimal places**.

Click **OK**.

Click **OK** to close the dialog.

16. The dimension has updated to the new dimension type.

Use the Type Selector to assign the **Linear – 3/32" Tahoma ANSI** dimension type to the dimension.

17. 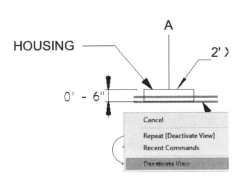 Right click on the view and select **Deactivate View**.

18.

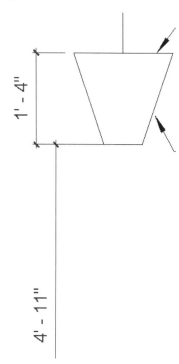

Highlight the right view with the sconce lighting fixture.

Right click and select **Activate View**.

19.

Select the two vertical dimensions.

Linear Dimension Style
Linear - 3/32" Arial ANSI

Use the Type Selector to change the dimensions to use the **Linear – 3/32" Arial ANSI** type.

The dimensions update.

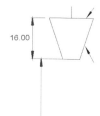

20.

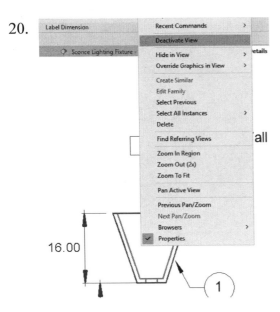

Right click on the view and select **Deactivate View**.

21. Save as *ex9-2.rvt*.

Exercise 9-3:
Modifying Dimensions

Drawing Name: *modify_dimensions.rvt*
Estimated Time: 10 minutes

This exercise reinforces the following skills:

- ❑ Place an ALIGNED dimension
- ❑ Modifying Dimensions
- ❑ Temporary Dimensions

1.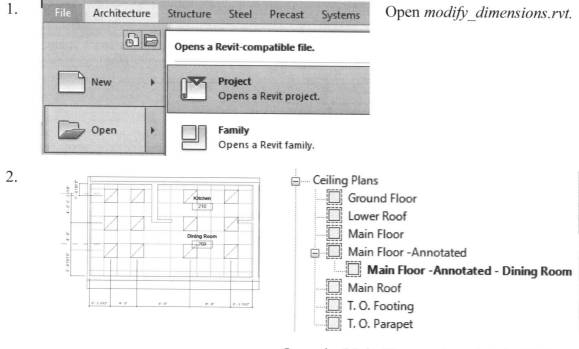

   Open *modify_dimensions.rvt.*

2.

   Open the **Main Floor – Annotated - Dining Room** ceiling plan.

3.

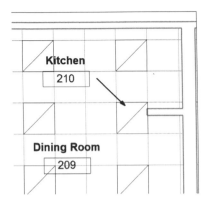

Locate the lighting fixture on the middle right of the room.

It is interfering with the wall and should be moved.

4.

Aligned

Select the **ALIGNED** dimension tool on the Annotate ribbon.

5.

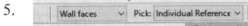

On the Options bar: Select **Wall faces.**

6.

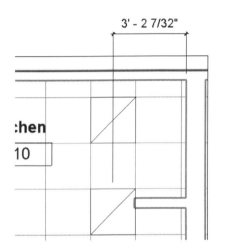

Place a horizontal dimension between the center of the lighting fixture and the inside face of the east wall.

7.

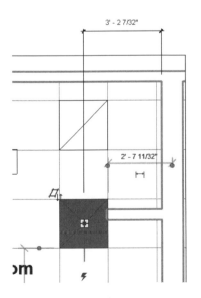

Select the lighting fixture.

Notice that the temporary dimension appears.

8.

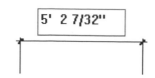

Change the value of the temporary dimension to **5' 2 7/32"**.

Click **ENTER** to update the dimension.

Left click in the window to release the selection.

9.

Notice that the permanent dimension updates.

Save as *ex9-3.rvt*.

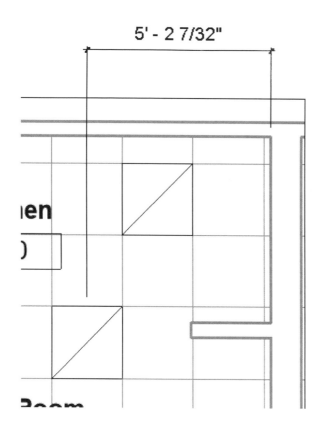

Exercise 9-4:

Create Ordinate Dimensions

Drawing Name: detail_view.rvt
Estimated Time: 30 minutes

This exercise reinforces the following skills:
- ❑ Detail Components
- ❑ Detail Views

1. 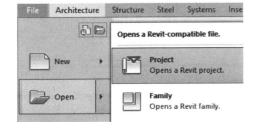 Go to **File**.

   Select **Open→Project**.

   Open *detail_view.rvt*.

2. Select **Drafting View** from the View ribbon.

3. 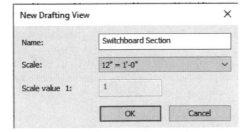 Type **Switchboard Section** in the Name field.

   Set the Scale to **12" = 1'-0"**.

   Click **OK**.

4. 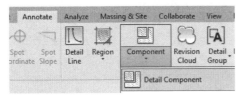 Switch to the Annotate ribbon.

   Select **Detail Component**.

5.  Select **Load Family**.

6.

| File name: | switchboard-section.rfa |
| Files of type: | All Supported Files (*.rfa, *.adsk) |

Locate the *switchboard-section* family.

This is in the downloaded files.
Click **Open**.

7.

Place three switchboard section families as shown.

8.

Select **Symbol** from the Annotate ribbon.

Symbol

9.

Revit   ×

No Generic Annotations family is loaded in the project. Would you like to load one now?

Yes   No

If this dialog appears:

Click **Yes**.

10.

Load Family Mode

Select **Load Family**.

11.

| File name: | Centerline_I.rfa |
| Files of type: | All Supported Files (*.rfa, *.adsk) |

Select *Centerline_I.rfa* from the downloaded files.

The centerline family provided with Revit uses metric units and is too small for this view.

12.

Place the symbol below the center mark in the first two squares.

13.

A
Text

Select **Text** from the Annotate ribbon.

14.

Type BUSWAY CONNECTION for the text value.

Use the grips to set the text to be one line and centered over the first square.

15.

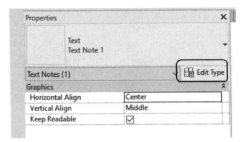

Select the text.

On the Properties palette:

Select **Edit Type**.

16.

Change the Text Size to **2"**.

Click **OK**.

17.

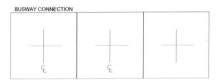

Adjust the position of the text.

18.

Copy the text so it is over the second square.

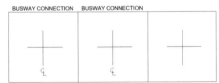

19.

Select **Aligned** dimension.

20.

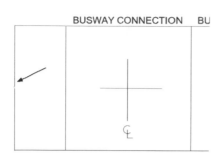

BUSWAY CONNECTION   BU

Place a dimension to the left of the first square.

21.

Select the dimension that was just placed.

Select **Edit Type** on the Properties palette.

22.

Select **Duplicate**.

23.

Name: Ordinate - Arial 2"

Type **Ordinate – Arial 2"** in the Name field.

Click **OK**.

24.

Type Parameters

Parameter	V:
Graphics	
Dimension String Type	Ordinate
Tick Mark	Diagonal 1/8"
Line Weight	1

Set the Dimension String Type to **Ordinate**.

25.

Flipped Dimension Line Extensio	3/32"
Witness Line Control	Gap to Element
Witness Line Length	3/32"
Witness Line Gap to Element	1"
Witness Line Extension	3/32"
Witness Line Tick Mark	None

Set the Witness Line Gap to Element to **1"**.

26.

Dimension Line Snap Distance	1/4
Text	
Width Factor	1.000000
Underline	☐
Italic	☐
Bold	☐
Text Size	2"
Text Offset	1/16"
Read Convention	Up, then Left
Text Font	Arial

Set the Text Size to **2"**.

27.

Set the Alternate Units to **Below**.

Set the alternate units to **mm**.

In the Alternate Units Prefix, type **[**.

In the Alternate Units Suffix, type **]**.

This adds brackets around the alternate units.

28.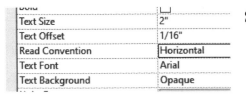

Set the Read Convention to **Horizontal**.

29.

Select **Edit** next to Ordinate Dimension Settings.

30.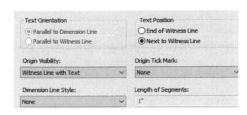

Set Text Position to **Next to Witness Line**.

Set Dimension Line Style to **None**.

Set Origin Tick Mark to **None**.

Click **OK**.

31.

Click **OK** to close the dialog box.

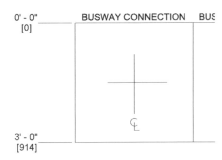

The first line selected when you placed your dimension will be the 0' 0" dimension. If the top line is not designated as the origin, delete the dimension and place again using the Ordinate dimension type.

32.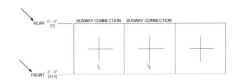

Add text to designate the front and rear of the switchboards.

33.

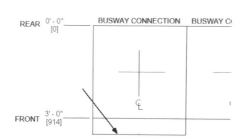

Place an aligned dimension below the first square.

34.

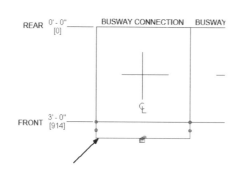

Select the dimension.

Select **Edit Type** on the Properties palette.

35.

Duplicate...   Select **Duplicate**.

36.

Name: Linear - Dual Arial 2"

Type Linear – **Dual Arial 2"** in the Name field.

Click **OK**.

37.

Parameter	Value	=
Graphics		☆
Dimension String Type	Continuous	
Leader Type	Arc	
Leader Tick Mark	Arrow Filled 30 Degree	
Show Leader When Text Moves	Away From Origin	
Tick Mark	Arrow Filled 30 Degree	
Line Weight	1	
Tick Mark Line Weight	5	

Set the Leader Tick Mark to **Arrow Filled 30 Degree**.

Set the Tick Mark to **Arrow Filled 30 Degree**.

38.

Flipped Dimension Line Extensio	3/32"
Witness Line Control	Gap to Element
Witness Line Length	3/32"
Witness Line Gap to Element	1"
Witness Line Extension	3/32"
Witness Line Tick Mark	None

Set the Witness Line Gap to Element to **1"**.

39.

Dimension Line Snap Distance	1/4
Text	
Width Factor	1.000000
Underline	☐
Italic	☐
Bold	☐
Text Size	2"
Text Offset	1/16"
Read Convention	Up, then Left
Text Font	Arial

Set the Text Size to **2"**.

Click **OK**.

40.

Text Font	Arial
Text Background	Opaque
Units Format	1234.57 ["]
Alternate Units	Right
Alternate Units Format	1235 [mm]
Alternate Units Prefix	[
Alternate Units Suffix	]

Click on the **Units Format**.

41.

Format

☐ Use project settings

Units: Inches

Rounding: Rounding increment:

2 decimal places ⌄   0.01

Unit symbol:

None ⌄

☐ Suppress trailing 0's

☐ Suppress 0 feet

☐ Show + for positive values

☐ Use digit grouping

☐ Suppress spaces

Disable **Use project settings**.

Set the Units to **Inches**.

Set the Rounding to **2 decimal places.**

Click **OK**.

42.

Text Background	Opaque
Units Format	1234.57 ["]
Alternate Units	Right
Alternate Units Format	1235 [mm]
Alternate Units Prefix	[
Alternate Units Suffix	]

Set the Alternate Units to **Right**.

Set the alternate units to **mm**.

In the Alternate Units Prefix, type **[**.

In the Alternate Units Suffix, type **]**.

This adds brackets around the alternate units.

43.

Bold	☐
Text Size	2"
Text Offset	1/16"
Read Convention	Horizontal
Text Font	Arial
Text Background	Opaque

Set the Read Convention to **Horizontal**.

Click **OK**.

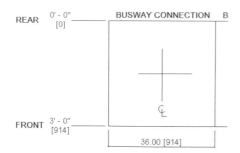

The arrowheads are too small in this dimension.

44.

Switch to the Manage ribbon.

Select **Arrowheads** under Additional Settings→Annotations.

45.

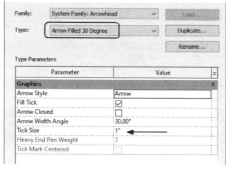

Select the **Arrow Filled 30 Degree** from the drop-down list.

Change the Tick Size to **1"**.

Click **OK**.

46.

The new dimension family should look similar to the image.

47.

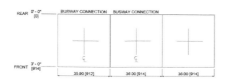

Place additional dimensions.

48.   Save as *ex9-4.rvt.*

Text Notes

Text notes can be added to a view with or without leaders. The appearance of text is controlled using Text Styles. Text is the information added to a view to label or provide a description about the building elements. It is view-specific and automatically updates when the view scale changes.

You can use the default test types which load with a template or create custom text types. Text can have leaders automatically added to designate specific elements in a view.

When you add text, the text and leaders will automatically snap into alignment with existing text and leaders. Text can be modified after it is placed to assign it to a new text style or to change the size, font, justification, width, leader, etc. You can import text notes from Word.

Guidelines for Working with Text and Tags

- Add text and tags after a view is created and the elements have been placed.
- Create copies of main model views and annotate the copied view. This way you have one view with annotations and one without. This makes it easier to make changes to the model in the non-annotated view.
- Plan and crop the documentation views to add text notes easily. Text notes and their associated leaders should not obscure the display of building elements.
- Align text notes to make it easier for reading.
- Create different text types for specific situations.

Exercise 9-5:

Adding a Text Note

Drawing Name: *text.rvt*
Estimated Time: 5 minutes

This exercise reinforces the following skills:

❑ Adding Text to a View

1.  Open *text.rvt*.

2.
   Floor Plans
   Ground Floor
   Ground Floor w Dimensions
   Lower Roof
   Main Floor
   Main Floor - Dining Room
   Main Floor- Power
   Main Roof
   Main Roof - Building 1

   Open the **Main Floor – Power** floor plan.

3.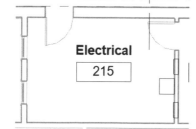
   Electrical
   215

   Zoom into **Room 215-Electrical**.

4. Annotate   Activate the **Annotate** ribbon.

5. **A** Text   Select the **Text** tool from the ribbon.

6. 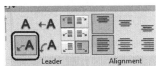 Enable the two segment leader option on the ribbon.

7. Verify that the **3/32" Arial** text style is selected on the Properties palette.

8.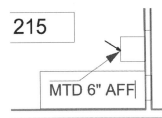

Left click to start the leader at the transformer.

Place a second left click to the left to start the text.

Type **MTD 6" AFF**.

Click **Close** on the ribbon to complete the text.

9.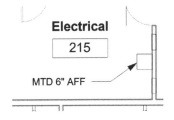

Save as *ex9-5.rvt*.

Exercise 9-6:

Create a Text Type

Drawing Name: *text_style.rvt*
Estimated Time: 5 minutes

This exercise reinforces the following skills:

❑ Defining a new Text Type

1.  Open *text_style.rvt*.

2. 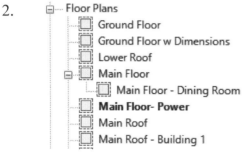 Open the **Main Floor – Power** floor plan.

3. 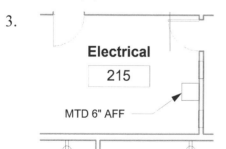 Zoom into **Room 215-Electrical**.

   Select the text note with the leader.

4.  Select **Edit Type** on the Properties palette.

5.  Select **Duplicate**.

6.

Name: 3/32" Romans

Change the Name to **3/32" Romans**.

Click **OK**.

7.

Parameter	Value
Graphics	
Color	Black
Line Weight	1
Background	Transparent
Show Border	☐
Leader/Border Offset	5/64"
Leader Arrowhead	Arrow Filled 15 Degree
Text	
Text Font	RomanS
Text Size	3/32"
Tab Size	1/2"
Bold	☐
Italic	☐
Underline	☐
Width Factor	1.000000

Family: System Family: Text
Type: 3/32" Romans
Type Parameters

Change the Background to **Transparent**.

Change the Leader Arrowhead to **Arrow Filled 15 Degree**.

Change the Text Font to **RomanS**.

Click **OK**.

8.

215

MTD 6" AFF

Left click anywhere in the display window to release the selection.

Review the changes to the text.

9.

Electrical

MTD 2·15 AFF

Notice that if you move the text box over the room tag, it is transparent.

Save as *ex9-6.rvt*.

Keynotes

Keynotes are used to tag elements or materials with standard CSI keynotes or custom keynotes.

A keynote parameter is available for all model elements (including detail components) and materials. You can tag each of these elements using a keynote tag family. The keynote value is derived from a separate text file that contains a list of keynotes.

Many companies develop their own keynotes to be used on their projects.

To assign a custom keynote file or verify the location of the keynote file, go to the Annotate ribbon and access Keynoting Settings under the Keynote tool.

The keynote text file is located under the Libraries folder:

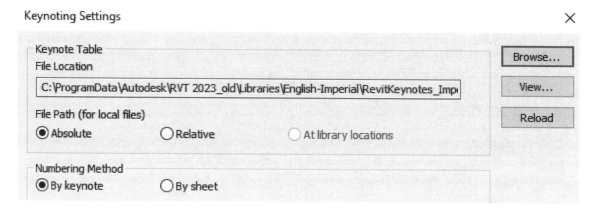

You will not be able to place a keynote unless you have the file location set correctly.

Exercise 9-7:

Using Keynotes

Drawing Name: *keynotes.rvt*
Estimated Time: 20 minutes

This exercise reinforces the following skills:

❑   Add tags to lighting fixtures using keynotes

1.   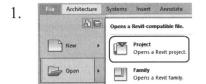   Open *keynotes.rvt*.

2.   Open the **Pendant Lighting** section under Lighting.

    Electrical
      Lighting
        Ceiling Plans
        Sections (Building Section)
          Pendant Lighting
          Recessed Lighting Fixture
          Sconce Lighting Fixture
          Sconce Lighting Fixture - Callout 1
          Section 3

3.   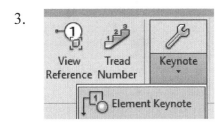   Activate the **Annotate** ribbon.

    Select **Element Keynote** under Keynote.

4.   On the Type Selector:

    Search
    Keynote Tag
      Keynote Number
      Keynote Number - Boxed - Large
      Keynote Number - Boxed - Small
    Keynote Text

    Select **Keynote Number**.

5.   Enable **Leader** on the ribbon.

6.

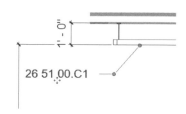

Select the pendant light fixture and place the keynote text.

7.

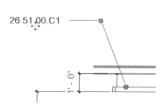

Place the keynote.

Select **Edit Type** on the Properties palette with the keynote selected.

8.

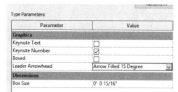

Set the Leader Arrowhead to **Arrow Filled 15 Degree**.

Click **OK**.

9.

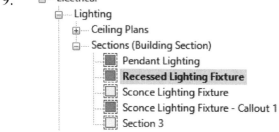

Open the **Recessed Lighting Fixture** section under Lighting.

10.

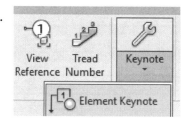

Activate the **Annotate** ribbon.

Select **Element Keynote** under Keynote.

11.

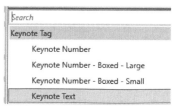

On the Type Selector:

Select **Keynote Number**.

12.

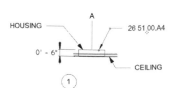

Place the keynote.

13.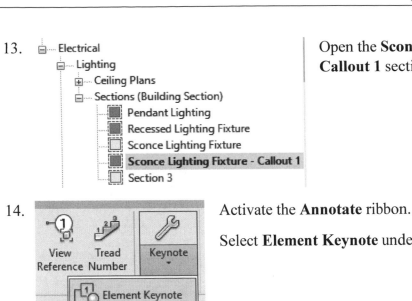

Open the **Sconce Lighting Fixture – Callout 1** section under Lighting.

14.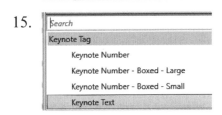

Activate the **Annotate** ribbon.

Select **Element Keynote** under Keynote.

15.

On the Type Selector:

Select **Keynote Number**.

16.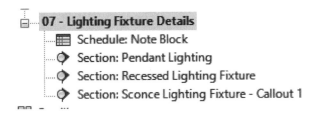

Place the keynote.

17.

Open **07- Lighting Fixture Details** under Sheets.

Review the three lighting fixture views.

18. Save as *ex9-7.rvt*.

Keynote Legends

Keynote Legends automatically create legends from the Keynotes that you have used in your project. By default, it only has two columns: the key number and the related key text. Place the legend on any sheet with views using keynote tags.

Exercise 9-8:

Create a Keynote Legend

Drawing Name: *keynote_legend.rvt*
Estimated Time: 10 minutes

This exercise reinforces the following skills:

- ❑ Legends
- ❑ Schedules
- ❑ Sheets
- ❑ Keynotes

1.  Open *keynote_legend.rvt*.

2. 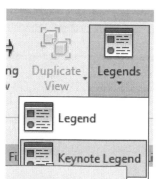 Activate the **View** ribbon.

   Select **Keynote Legend** under Legend.

3.

Name:

Keynotes

OK

Type **Keynotes** in the Name field.

Click **OK**.

4.

Scheduled fields (in order):

Key Value
Keynote Text

The available fields are already pre-selected.

5.

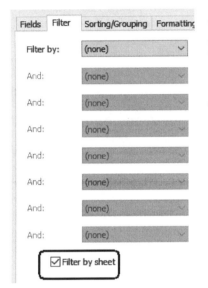

Fields | Filter | Sorting/Grouping | Formatting

Filter by: (none)

And: (none)

And: (none)

And: (none)

And: (none)

And: (none)

And: (none)

And: (none)

☑ Filter by sheet

Activate the Filter tab.

Enable **Filter by sheet.**

This means that only keynotes appearing in views on the sheet will be listed in the legend.

6.

Fields | Filter | Sorting/Grouping | Formatting | Appearance

Graphics

Build schedule: ⦿ Top-down
○ Bottom-up

Grid lines: ☑ Wide Lines   ☐ Grid in headers/footers/spacers

Outline: ☑ Medium Lines

Height: Variable   ☑ Blank row before data

Text

☐ Show Title

☐ Show Headers

Title text: Schedule Default

Header text: Schedule Default

Body text: 3/32" Arial

Activate the Appearance tab.

Enable **Grid lines**.

Enable **Outline**.

Set the Grid lines to use **Wide Lines**.

Set the Outline to use **Medium Lines**.

Disable **Show Title**.

Disable **Show Headers**.

Set the Body Text to **3/32" Arial**.

Click **OK**.

7. 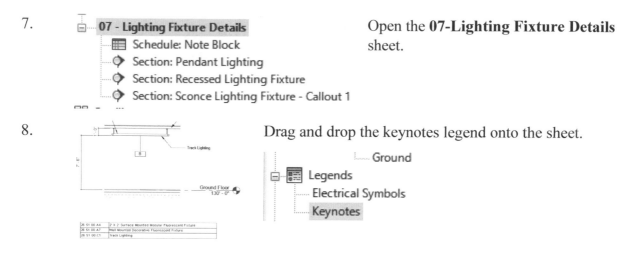 Open the **07-Lighting Fixture Details** sheet.

8. Drag and drop the keynotes legend onto the sheet.

9. Save as *ex9-8.rvt*.

Tags

A tag is an annotation for identifying elements in a drawing.

Every category in the family library has a tag. Some tags automatically load with the default Revit template, while others you need to load. If desired, you can make your own tag in the Family Editor by creating an annotation symbol family. In addition, you can load multiple tags for a family.

You can tag elements in a view of a host model, linked model and nested model. Tags are view-specific. This means tags will only appear in the views where they are placed. Tags will automatically scale when the view scale is changed.

Tags display the parametric information about the associated/selected elements.

You can create custom tags by editing predefined tags and saving them with a new name.

Tags can be added with or without a leader. Tags can be oriented horizontally or vertically. Once tags are placed, they can be modified using the Type Selector or by changing their properties.

You can Tag by Category, Tag All, or do a Multi-Category Tag.

Tag by Category places tags based on family. For example, tagging all light fixtures or tagging all electrical panels. When you tag building elements by category, Revit identifies the element type and automatically provides the appropriate tag (if the tag has been loaded into the project).

Tag All places tags on any element where that tag has been loaded into the project. Tags are placed on any elements that don't have an existing tag. This function can be useful when you place and tag rooms in a floor plan view and you want to see the same tags in the reflected ceiling plan. Load all the desired tag families into the project prior to selecting Tag All. You can opt to tag all elements or selected elements only.

Tag Multiple allows you to tag all untagged elements in more than one category in one operation.

Guidelines for Using Tags

- Add tags after a view has been created and the model is fairly well along.
- Duplicate the view and place tags on one version of the view and no annotations on the other view. Use the non-annotated view to make changes to the model.
- Load more than one tag type for building elements so they can easily be tagged.

Exercise 9-9:

Tag Light Fixtures

Drawing Name: *tags_1.rvt*
Estimated Time: 5 minutes

This exercise reinforces the following skills:

- ❑ Add tags to lighting fixtures

1.  Open *tags_1.rvt*.

2. 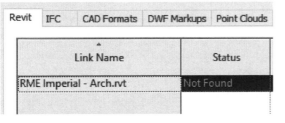 Go to the Insert ribbon.

Click **Manage Links**.

Reload the missing file from the downloaded exercise folder.

Click **OK**.

3.   Annotate   Activate the **Annotate** ribbon.

4. 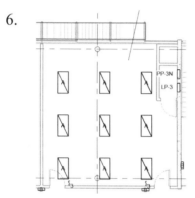 Select **Tag by Category** from the ribbon.

5. On the Options bar:

   Enable Horizontal.

   Disable Leader.

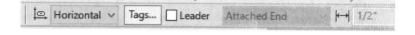

6.   Left click on each lighting fixture in Classroom 5.

   Note that the tags are centered on each light fixture selected.

   Right click and Click **Cancel** or Click **ESC** to exit the command.

7. Save as *ex9-9.rvt*.

Exercise 9-10:

Tag Devices

Drawing Name: *tags_3.rvt*
Estimated Time: 5 minutes

This exercise reinforces the following skills:

- ❑ Add tags to electrical equipment
- ❑ Tag All

1.   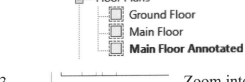   Open *tags_3.rvt*.

2.   Open the **Main Floor Annotated Power** floor plan.

   - Electrical
       - Lighting
       - Power
           - Floor Plans
               - Ground Floor
               - Main Floor
               - **Main Floor Annotated**

3.   Zoom into the electrical equipment in the middle of the view.

4.      Select the Switchboard on the left.

   Note that there is no Panel Name assigned in the Properties palette.

   Repeat for the two devices next to the switchboard. None of the devices have been assigned names.

5.   Annotate   Activate the **Annotate** ribbon

6. Select **Tag All** from the ribbon.

7. 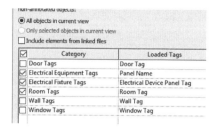 Enable

- Electrical Equipment Tags

- Electrical Fixture Tags

- Room Tags

Disable **Leader** at the bottom of the dialog.

Click **OK**.

 *? symbols appear on all the unnamed devices.*

8. Select the Metering Switchboard tag on the left.

Type **SB1** for the panel name.

Click **ENTER**.

Click ESC to release the selection.

9.  Select the Metering Switchboard device: SB1.

In the Properties palette, note that the Panel Name has updated.

Click ESC to release the selection.

10.  Select the device tag in the middle.

You can use the TAB key to cycle through selections.

Type **SB2** for the panel name.

Click **ENTER**.

Click ESC to release the selection.

11.  Select the device tag on the right.

You can use the TAB key to cycle through selections.

Type **T1** for the panel name.

Click **ENTER**.

Click ESC to release the selection.

12. Save as *ex9-10.rvt*.

Symbols

A symbol is a 2D graphical representation of an annotation element or other object.

Exercise 9-11:

Define a Ground Symbol

Drawing Name: *none*
Estimated Time: 5 minutes

This exercise reinforces the following skills:

- ❑ Revit families
- ❑ Symbols
- ❑ Detail Lines

1. 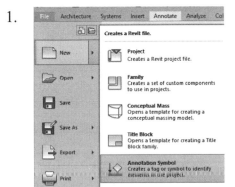 Go to **File→New→Annotation Symbol**.

2.   File name: Generic Annotation.rft

     Files of type: Family Template Files (*.rft)

     Select the *Generic Annotation* template.

     Click **Open**.

3.   *The intersection point of the two green dashed lines is the insertion point of the symbol.*

     Left click on the text note.

     Right click and select **Delete**.

4.   Line

     Activate the **Create** ribbon.

     Select the **Line** tool.

5.   Draw a 2" line at the intersection point, so that it is 1" on each side.

6.   Draw a 1.25" line ½" below the first line, so that it is .625" on each side.

7.   Draw a 1/2" line ½" below the first line, so that it is .25" on each side.

8.   File name: ground symbol

     Files of type: Family Files (*.rfa)

     Save as *ground symbol.rfa* in your work folder.

Exercise 9-12:

Place a Symbol

Drawing Name: *symbol.rvt*
Estimated Time: 5 minutes

This exercise reinforces the following skills:

- ❏ Symbols
- ❏ Load Family
- ❏ Views
- ❏ Properties palette

1. 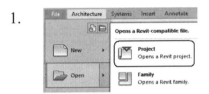 Go to **File→Open→Project**.

   Open *symbol.rvt*.

2. Open the **Equipment** elevation under Power.

3. 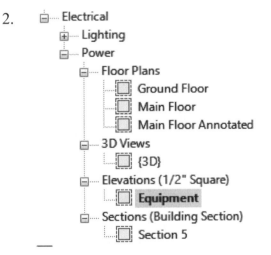 Select the **Symbol** tool from the Annotate ribbon.

4.

The symbol we want is not available in the Type Selector.

Select **Load Family** on the ribbon.

5.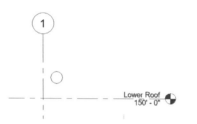

Open the *Note Block.rfa* file located in the downloaded exercises.

6. Left click to place.

Cancel out of the command.

7. Select the note block.

In the Properties palette:

Type **2** in the Note Number field.

Type **Verify Placement with Subcontractor** in the Note Text field.

8. Save as *ex9-12.rvt.*

Exercise 9-13:

Creating Arrowhead Styles

Drawing Name: *arrow_heads.rvt*
Estimated Time: 10 minutes

This exercise reinforces the following skills:

- ❑ Revit families
- ❑ Arrowheads
- ❑ Text Notes

1. 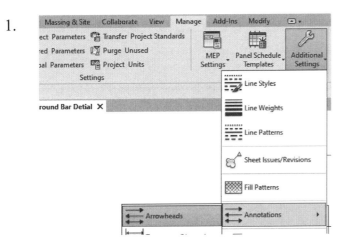 Activate the Manage ribbon.

   Go to **Additional Settings→Annotations→ Arrowheads**.

2. Select the *Arrow Filled 15 Degree* type.

   Click **Duplicate**.

3. Type **Arrow Filled 15 Degree – Small**.

   Click **OK**.

4. Set the Tick Size to **5/256"**.

   Left click in the Width Angle box to get the new value accepted.

   Click **OK**.

Type Parameters	
Parameter	Value
Graphics	
Arrow Style	Arrow
Fill Tick	☑
Arrow Closed	☐
Arrow Width Angle	15.00°
Tick Size	5/256"
Heavy End Pen Weight	7

5.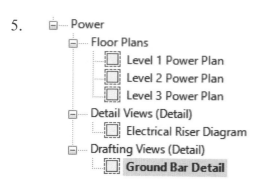

Open the **Ground Bar Detail** view.

6.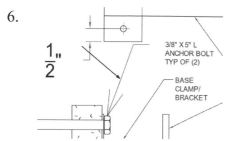

Select the Anchor Bolt text note.

7.

Select **Edit Type**.

8.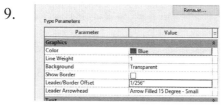

In the Leader Arrowhead field:

Select **Arrow Filled 15 Degree – Small**.

9.

Change the Leader/Border Offset to **1/256"**.

Click **OK**.

10.

The note updates and appears similar to the existing notes.

Save as *ex9-13.rvt*.

Global Parameters

Use global parameters to drive or report values.

You can use global parameters in a project to:

- drive the value of a dimension or constraint.

- associate to an element instance or type property to drive its value.

- associate to an instance or type project parameter.

- report the value of a dimension, so this value can be used in equations of other global parameters.

A global parameter is a parameter that you create inside a Project that can be used to assign or report a value across the entire project. Global parameters are useful to ensure your project meets local building codes. We can use global parameters to ensure that minimum clearance distances are met.

Exercise 9-14:

Using Global Parameters

Drawing Name: *global parameters.rvt*
Estimated Time: 10 minutes

This exercise reinforces the following skills:

- ❑ Global Parameters
- ❑ Dimensions

The OSHA standard (29 CFR 1910.303 (g)) requires sufficient access and working space around all equipment serving 600 volts or less. For equipment serving between 120 volts and 250 volts, the regulations require a minimum of three feet of clearance. The width of the working space in front shall be 30 inches minimum or width of the equipment.

1.  Go to **File→Open→Project**.

    Open *global parameters.rvt*.

2.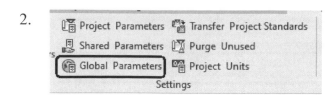

Switch to the Manage ribbon.

Select **Global Parameters**.

3.

Select **New** (located at the bottom of the dialog).

4.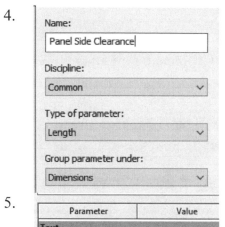

Type **Panel Side Clearance** in the Name field.

Set the Discipline to **Common**.

Set the Type of parameter to **Length**.

Group parameter under **Dimensions**.

Click **OK**.

5.

Parameter	Value
Text	
Panel Side Clearance	3 0

Set the Value to **3' 0"**.

6.

Select **New** (located at the bottom of the dialog).

7.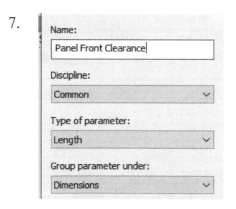

Type **Panel Front Clearance** in the Name field.

Set the Discipline to **Common**.

Set the Type of parameter to **Length**.

Group parameter under **Dimensions**.

Click **OK**.

8.

Dimensions	
Panel Front Clearance	2' 6"
Panel Side Clearance	3' 0"

Set the Value to **2' 6"**.

Click **OK**.

9.   Zoom into the area next to Storage Rm 215.

10.   Select the **Aligned** tool from the Annotate ribbon.

11.   Place two dimensions between the panels.

12.   Place an aligned dimension between T1 and the wall face.

       Hint: Select Wall Faces on the Options bar.

13.   Select the dimension placed between SB1 and SB2.

14.   On the ribbon, select **Panel Side Clearance** below the Label: drop-down list.

15.   The panel's position adjusts.

16. Select the dimension placed between SB2 and T1.

17. On the ribbon, select **Panel Side Clearance** below the Label: drop-down list.

18. The panel's position adjusts.

19. Select the dimension placed between the wall and T1.

20. On the ribbon, select **Panel Side Clearance** below the Label: drop-down list.

21. The panel's position adjusts.

Move the conduits back on top of SB2.

22. 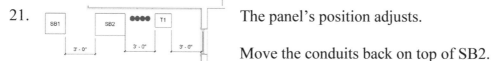 Select the **Aligned** tool from the Annotate ribbon.

23.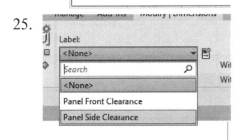

Place dimensions for the panels located above the panels that were just dimensioned.

You may need to move the panel tags in order to see the panels.

24.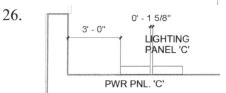

Select the dimension between the wall face and PWR PNL "C".

25.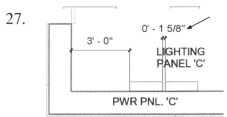

On the ribbon, select **Panel Side Clearance** below the Label: drop-down list.

26.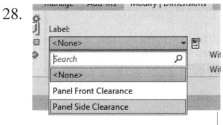

The panel's position adjusts.

27.

Select the dimension between the PWR PNL "C" and LIGHTING PANEL "C".

28.

On the ribbon, select **Panel Side Clearance** below the Label: drop-down list.

29.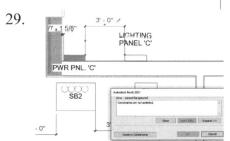

The PWR Panel C is positioned inside a wall which causes an error.

Click Cancel on the dialog.

30. Select Pwr Panel "C".

31. Select the **Pin** tool on the Modify ribbon.

This fixes the panel into position so it doesn't move.

Left click anywhere in the window to release the selection.

32. Select the dimension between the PWR PNL "C" and LIGHTING PANEL "C".

33. On the ribbon, select **Panel Side Clearance** below the Label: drop-down list.

34. This time Lighting Panel C shifts to the right.

35. Reposition the dimensions so you have a clear space in front of the panels.

36. Switch to the Annotate ribbon.

Select **Filled Region**.

37. Use the Type Selector to set the filled region type to **Diagonal Up** on the Properties panel.

38. 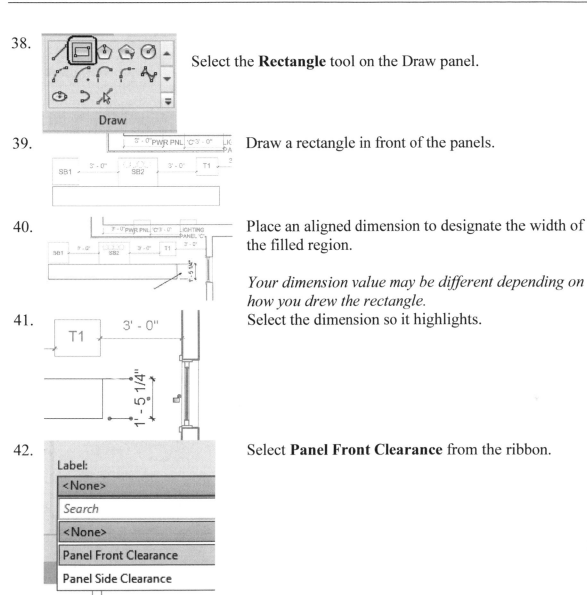 Select the **Rectangle** tool on the Draw panel.

39. Draw a rectangle in front of the panels.

40. Place an aligned dimension to designate the width of the filled region.

Your dimension value may be different depending on how you drew the rectangle.

41. Select the dimension so it highlights.

42. Select **Panel Front Clearance** from the ribbon.

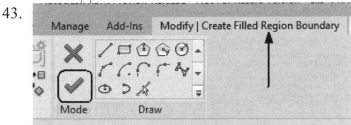

43. 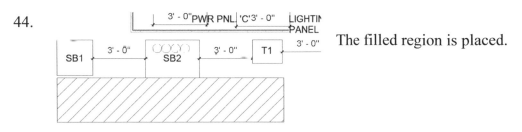 Select the ribbon tab labeled **Create Filled Region Boundary**.

Select **Green Check** to complete the filled region.

44. The filled region is placed.

45. Save as *ex9-14.rvt*.

Extra: Place a filled region in front of Pwr Panel C and Lighting Panel C and assign the front clearance global parameter to control the size of the region.

Lab Exercises

Open *lesson_9_lab.rvt.*

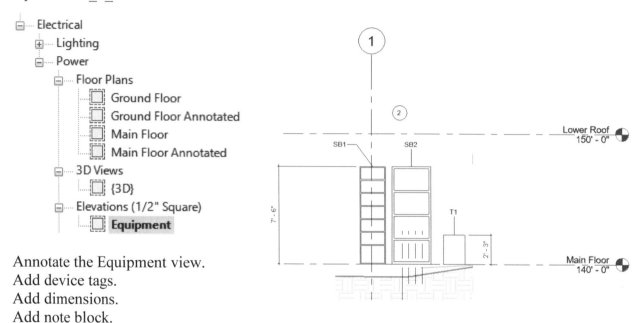

- Electrical
  - Lighting
  - Power
    - Floor Plans
      - Ground Floor
      - Ground Floor Annotated
      - Main Floor
      - Main Floor Annotated
    - 3D Views
      - {3D}
    - Elevations (1/2" Square)
      - Equipment

Annotate the Equipment view.
Add device tags.
Add dimensions.
Add note block.

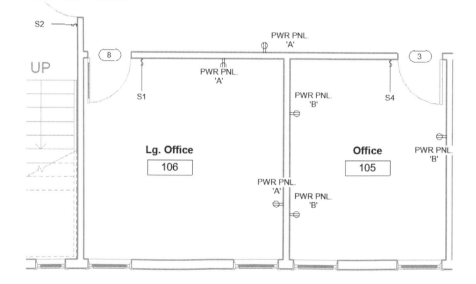

Create a view template for Power plans.

Disable visibility of sections, reference planes, wires, furniture, furniture systems.

Create a power view of the Ground Floor and Main Floor.

Apply the template.

Tag all the devices. Assign names to any unnamed devices.

Verify that all the devices have been assigned a circuit.

Sheets and Titleblocks

In Revit, sheets are included in the project file. You add sheets as needed, then place views on a sheet. Views can only be "consumed" once by placing on a sheet. If you need to use the same view on more than one sheet, you need to duplicate the view. Since schedules are also considered views, they also can only be placed once on a sheet. The only view that can be placed multiple times is a legend.

Sheets are the basis for construction documentation sets. A sheet allows you to place different views side by side on a page with titleblock information. You can print sheets to paper or save them to an electronic file, like a pdf.

You can activate a view on a sheet and modify the model elements. The building model will automatically update.

A viewport is a rectangular boundary around each view placed on a sheet. Each viewport has an identifying title below the boundary that displays the view name, view scale, and an identifier bubble. You can control the display of viewport titles and change the appearance of viewport titles. You can also create your own viewport title families.

Some architects use a cartoon set to plan the document requirements for a project. A cartoon set is a rough plan for the sheets that you want to include in the construction document set, and the drawings, schedules, or other information to show on each sheet.

By creating a cartoon set, you can ensure that the final construction document set includes all desired information. You can also use this method to ensure that the construction document set meets standards established by your organization.

With Revit, you can create a digital cartoon set. First add the required views (drawings and schedules) to the project and sketch the basic design of the building model. Add the desired sheets to the project and give them appropriate names and numbers. Then add the views to the appropriate sheets. If desired, you can set the view scales, titles, and other attributes now, so the resulting sheets use the desired settings.

Even though these views and sheets do not yet show the completed design, they provide an overall structure for the project. As you develop the building model in the project views, the schedules update accordingly, and the sheets display the desired information. This technique streamlines the project documentation process.

When you create a digital cartoon set that reflects corporate standards or a typical project setup, you can use the project to create a project template.

You can include information that is external to a project on the sheets that Revit generates. You can use external text, spreadsheets, and images on sheets.

Guidelines for Working with Sheets

- Create and name several copies of views for different design and documentation purposes.
- A viewport name on a sheet can be different from the view name in the Project Browser.
- Create viewport types that do not display the title or extension line.
- Create sheets using your organization's titleblocks and place views on the sheet at an early stage of your project. The views will update as the model progresses and you can use the sheets to provide design reviews to team members during the design phase.
- Create and save different print setups as part of the project template. Name the print setups to coincide with different project stages, so it is easy for you to create documentation sets for design reviews at different points in the project.

Exercise 10-1:

Add a Sheet

Drawing Name: *sheets.rvt*
Estimated Time: 25 minutes

This exercise reinforces the following skills:

- ❏ Sheets
- ❏ Duplicate Views
- ❏ Add Views to a Sheet
- ❏ View Properties
- ❏ Activate View
- ❏ Deactivate View
- ❏ View Scale
- ❏ Visibility/Graphics Overrides
- ❏ Edit Crop

1. 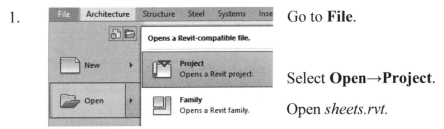   Go to **File**.

    Select **Open→Project**.

    Open *sheets.rvt*.

2. Activate the **View** ribbon.

3. Select the **New Sheet** tool on the ribbon.

4. Highlight the E size titleblock.

    Click **OK**.

5.   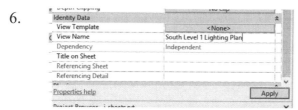   Locate the **Level 1 Lighting Plan** in the Project Browser.

Right click and Select **Duplicate View→Duplicate with Detailing**.

6.   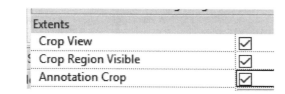   Under Identity Data in the Properties palette:

Change the View Name to **South Level 1 Lighting Plan**.

7.

```
⊟─── Electrical
   ⊟─── Lighting
      ⊟─── Floor Plans
            ▢ Level 1 Lighting Plan
            ▣ Level 2 Lighting Plan
            ▢ Level 3 Light Plan
            ▣ North Level 1 Lighting Plan
            ▢ South Level 1 Lighting Plan
```

Notice that the name updates in the Project Browser.

8.

Extents	
Crop View	☑
Crop Region Visible	☑
Annotation Crop	☑

Under Extents in the Properties palette:

Enable **Crop View**.

Enable **Crop Region Visible**.

Enable **Annotation Crop**.

These can also be toggled in the Display bar at the bottom of the screen.

9.

```
⊟─ 🔲 Sheets (all)
      ─── E101 - 1st Floor Lighting Plan
      ⊞─── E101.1 - NORTH LEVEL 1 LIGHTING PLAN
      ⊞─── E101.2 - Unnamed
      ─── E102 - 1st Floor Lighting Plan
      ⊞─── E201 - 2nd Floor Lighting Plan
      ─── E202 - 2nd Floor Power Plan
      ─── E601 - Panel Schedules
      M101 - Unnamed
```

In the Project Browser under Sheets:

Open the **M101-Unnamed** sheet.

10.   Select the **View** tool on the ribbon to add a view to the sheet.

11.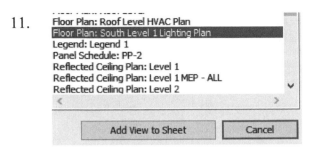

Locate the **South Level 1 Lighting Plan**.

Click **Add View to Sheet**.

12.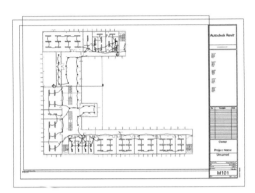

Left click to place the view on the sheet.

13.

Select the view.

Select **Activate View** on the ribbon.

This allows you to modify the view and the elements in the view.

14.

Under Graphics in the Properties palette:

Change the View Scale to ¼" = 1'-0".

Left click on **Edit** next to Visibility/Graphics Overrides.

15.

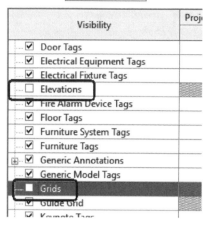

Activate the Annotation Categories tab.

Disable **Elevations** and **Grids** to hide those elements.

Click **OK** to close the dialog.

16.

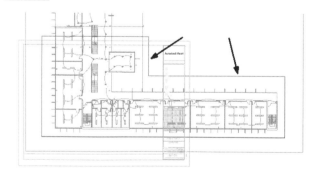

Click on the viewport to activate the grips.

17.

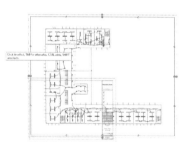

Select **Edit Crop** from the ribbon.

18.

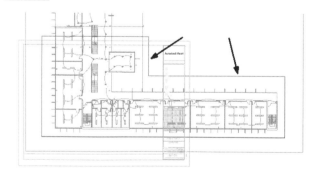

Add two lines.

Use the Trim tool to eliminate the extending lines.

Reposition the lines to enclose only the lower half of the floor plan.

Check to make sure you have created a closed polygon.

19.

Click **Green Check** to complete the Edit Crop operation.

20.

Change the View Scale back to **1/8" = 1'-0"**.

21.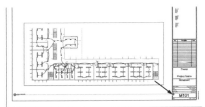

Toggle **Hide Crop Region** on the Display bar to hide the crop region on the view.

22.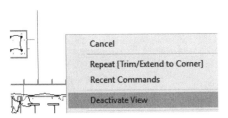

Right click on the view.

Select **Deactivate View.**

23.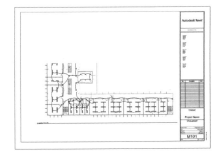

Select the view to activate the title line.

Use the grip to shorten the extension line on the title.

24.

Click outside the view to be able to select the title line to reposition it.

25. Save as *ex10-1.rvt*.

Exercise 10-2:

Add Views to a Sheet

Drawing Name: *views 2.rvt*
Estimated Time: 20 minutes

This exercise reinforces the following skills:
- Views
- Sheets

1. Go to **File**.

   Select **Open→Project**.

   Open *views 2.rvt*.

2. Open the **Ground Floor Lighting** Sheet.

3. 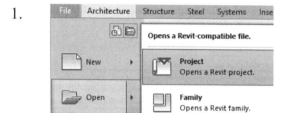 Select the **Insert View** tool from the View ribbon.

4. Select the **Reflected Ceiling Plan: Ground Floor Lighting Fixtures & Switches**.

   Click **Add View to Sheet**.

   Left click in the display window to place the view.

5. 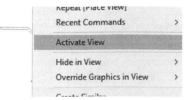 Right click and select **Activate View**.

6.

Toggle on the Crop Region.

7.

Use the grips to crop the view.

8.

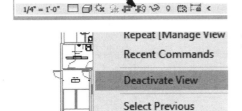

Toggle off the Crop Region.

9.

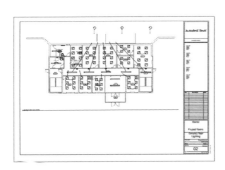

Right click and select **Deactivate View**.

10.

E1 30 x 42 Horizontal
E1 30x42 Horizontal

Select the titleblock.

Use the Type Selector to change to an **E 30x42 Horizontal** size titleblock.

11.

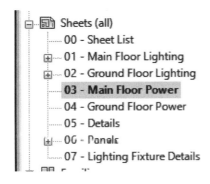

Reposition the view on the sheet.

You can't re-scale the view because a view template has been applied.

12.

Sheets (all)
00 - Sheet List
01 - Main Floor Lighting
02 - Ground Floor Lighting
03 - Main Floor Power
04 - Ground Floor Power
05 - Details
06 - Panels
07 - Lighting Fixture Details

Open the **Main Floor Power** Sheet.

13. Select the **Insert View** tool from the View ribbon.

14. Floor Plan: Ground Floor -Power
Floor Plan: Ground Floor Annotated
Floor Plan: Lower Roof
Floor Plan: Main Floor
Floor Plan: Main Floor Annotated
Floor Plan: Main Floor Power
Floor Plan: Main Roof
Floor Plan: Site
Floor Plan: T. O. Footing

Select the **Floor Plan: Main Floor Power**.

Click **Add View to Sheet**.

Left click in the display window to place the view.

15. E1 30 x 42 Horizontal
E1 30x42 Horizontal

Select the titleblock.

Use the Type Selector to change to an **E 30x42 Horizontal** size titleblock.

16.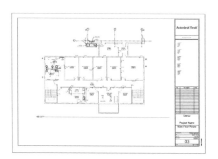

Reposition the view on the sheet.

You can't re-scale the view because a view template has been applied.

To turn off the visibility of the tag above the view, activate the view, window around the tag, right click and select Hide in View→Elements.

Deactivate the view when you are done.

17. Sheet List
Sheets (all)
  00 - Sheet List
  01 - Main Floor Lighting
  02 - Ground Floor Lighting
  03 - Main Floor Power
  04 - Ground Floor Power
  05 - Details
  06 - Panels
  07 - Lighting Fixture Details

Open the **00 Sheet List** sheet.

18. Select the **Insert View** tool from the View ribbon.

19. Schedule: PANEL A INFORMATION
Schedule: PANEL B INFORMATION
Schedule: PANEL C INFORMATION
Schedule: Room Schedule
Schedule: Sheet List
Section: Callout (2) of Section 1
Section: Callout of Section 1

Select the **Schedule: Sheet List**.

Click **Add View to Sheet**.

Left click in the sheet to place the view.

20.

Sheet List			
Sheet Number	Sheet Name	Current Revision	Current Revision Date
01	Main Floor Lighting		
02	Ground Floor Lighting		
03	Main Floor Power		
04	Ground Floor Power		
05	Details		
06	Panels		
07	Lighting Fixture Details		

Adjust the column widths on the schedule.

21.   Save as *ex10-2.rvt*.

Guide Grid

Guide grids help arrange views so that they appear in the same location from sheet to sheet or align views on the same sheet.

You can display the same guide grid in different sheet views. Guide grids can be shared between sheets.

When new guide grids are created, they become available in the instance properties of sheets and can be applied to sheets. It is recommended to create only a few guide grids and then apply them to sheets. When you change the guide grid's properties/extents in one sheet, all the sheets which use that grid are updated accordingly.

You can change the appearance of Guide Grids using Object Styles on the Manage ribbon.

Exercise 10-3:

Align Views on a Sheet

Drawing Name: *align_views.rvt*
Estimated Time: 15 minutes

This exercise reinforces the following skills:
- ❑ Guide Grid
- ❑ Insert View
- ❑ Sheets
- ❑ Visibility/Graphics
- ❑ Move

1. 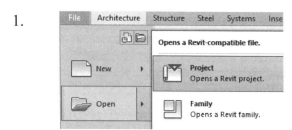 Go to **File**.

     Select **Open→Project**.

     Open *align_views.rvt*.

2. Sheets (all)
    - 00 - Sheet List
    - 01 - Main Floor Lighting
    - 02 - Ground Floor Lighting
    - 03 - Main Floor Power
    - 04 - Ground Floor Power
    - 05 - Details
    - 06 - Panels
    - **07 - Lighting Fixture Details**

    Open **07- Lighting Fixture Details** sheet.

3. ʝs Guide
    Grid

    Select the **Guide Grid** tool from the View ribbon.

4.

Assign Guide Grid ✕

○ Choose existing:

◉ Create new:
Name: Guide Grid 1

OK   Cancel

Enable **Create new**.

Click **OK**.

5. Use the MOVE tool to adjust the position of the grid so it aligns with the titleblock outline. Use the grips to adjust the size of the grid.

Use the ALIGN tool to align the outside of the grid to the title block border.

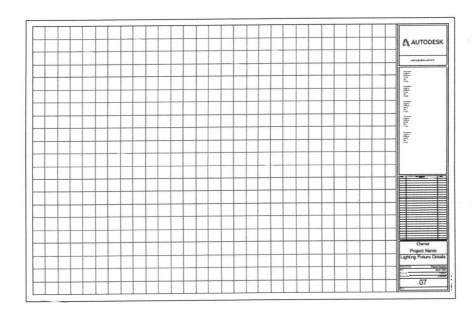

6.

View

Select the **Insert View** tool from the View ribbon.

7.

Section: Callout of Section 2
Section: Conduit
Section: Pendant Lighting
Section: Recessed Lighting Fixture
Section: Sconce Lighting Fixture
Section: Sconce Lighting Fixture - Callout 1
Section: Section 1
Section: Section 2
Section: Section 3

Highlight **Section: Pendant Lighting**.

Select **Add View to Sheet**.

Left click in the display window to place the view.

8.

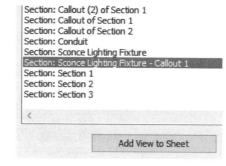

Use the MOVE tool to align the Ground Floor level with the grid.

9.

Select the **Insert View** tool from the View ribbon.

10.

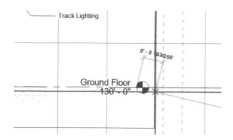

Highlight **Section: Sconce Lighting Fixture – Callout 1**.

Select **Add View to Sheet**.

Left click in the display window to place the view.

Notice that the Pendant Lighting section view is no longer available in the list because it was placed on a sheet.

11.

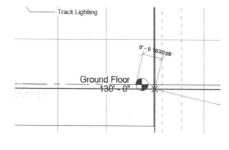

Use the MOVE tool to align the Ground Floor level with the grid.

Note that the views are aligned using the guide grid.

12.

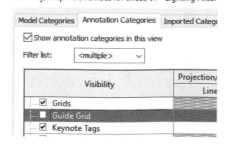

Type **VV** to launch the Visibility/Graphics Overrides dialog.

On the Annotation Categories tab:

Disable **Guide Grid**.

Click **OK**.

13. Save as *ex10-3.rvt*.

Titleblocks

Titleblocks are Revit families. You can load standard titleblocks into a project or create a custom titleblock using the Family Editor. Most companies have their own custom titleblock designed.

The title block you define includes the sheet size. If you delete the title block, the sheet of paper is also deleted. This means your title block is linked to the paper size you define.

You need to define a title block for each paper size you use.

Guidelines for Working with Titleblocks

- You should create a titleblock for each different sheet/paper size used when plotting. Keep in mind that the industry is moving more and more to pdf digital documents and less towards plotting.
- Create a titleblock to represent different phases of construction on a project.
- Load titleblocks into your project templates so they are easily accessible.
- Create custom labels to make it easier to fill in the data on titleblocks.

Exercise 10-4:

Update a Titleblock

Drawing Name: *titleblock_1.rvt*
Estimated Time: 10 minutes

This exercise reinforces the following skills:
□   Edit a Titleblock

1. 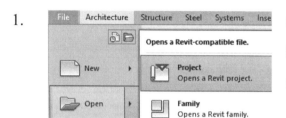   Go to **File**.

    Select **Open→Project**.

    Open *titleblock_1.rvt*.

2. 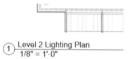   In the Project Browser:

    Locate the Sheets category.

    Double left click on **E201-Unnamed**.

    This opens the view.

3.   Zoom into the view title.

    You see that Level 2 Lighting Plan has been placed on the sheet.

4.    Zoom into the lower right corner of the titleblock.

5.    Left click on the text that reads **Unnamed**.

    Modify the text to **2nd Floor Lighting Plan**.

6.  Change Owner to **River City**.

7.  Change Project Name to **Office Building**.

8.  Double left click on the **E301 – NORTH LEVEL 1 LIGHTING PLAN SHEET** in the Project Browser.

9. Zoom into the lower right corner of the titleblock.

Note that the titleblock on this sheet has updated to the new owner and project names.

10. Save as *ex10-4.rvt*.

Exercise 10-5:

Load a Titleblock

Drawing Name: *load_titleblocks.rvt*
Estimated Time: 10 minutes

This exercise reinforces the following skills:
- ❑ Load a Titleblock
- ❑ Place a View
- ❑ Modify View Scale for a view
- ❑ Modify a title line
- ❑ Position a view on a sheet

1. 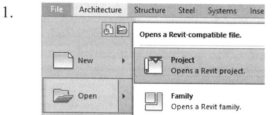 Go to **File**.

   Select **Open→Project**.

   Open *load_titleblocks.rvt*.

2. 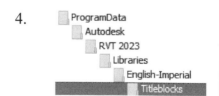 Right click on Sheets in the Project Browser.

   Select **New Sheet**.

3. Load... Select the **Load** button at the top of the dialog.

4. ProgramData
   Autodesk
   RVT 2023
   Libraries
   English-Imperial
   Titleblocks

   Browse to the Titleblocks folder under *Libraries\US Imperial*.

   Imperial Lib...   *You have a shortcut for the Imperial Libraries folder on the left pane of the dialog.*

5.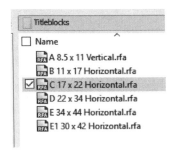

Select the **C 17 x 22 Horizontal** titleblock.

Click **Open**.

6.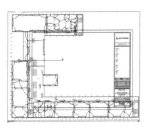

Highlight the **C 17 x 22 Horizontal** titleblock.

Click **OK**.

7.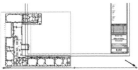

Locate the **Level 1 Power Plan** in the Project Browser.

Hold down your left mouse button to drag and drop the view onto the sheet.

8.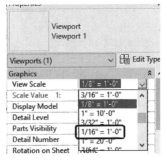

The view is quite a bit bigger than the titleblock.

9.

On the Properties palette:

Change the View Scale to **1/16" = 1'-0"**.

10.

The view size adjusts.

Use the small grip at the end of the title line to adjust the title size.

11.

Reposition the view on the sheet.

Note that the project information is updated on the sheet to match the existing sheets.

Save as *ex10-5.rvt*.

Exercise 10-6:

Adding Project Information to a Titleblock

Drawing Name: *project_information.rvt*
Estimated Time: 10 minutes

This exercise reinforces the following skills:
- ❑ Project Information
- ❑ Titleblocks
- ❑ Titleblock properties

1. Go to **File**.

   Select **Open→Project**.

   Open *project_information.rvt.*

2. Select the **Manage** ribbon.

3. Select **Project Information** from the ribbon.

4. Type **IBEW** for Organization Name.

   Type **Local 595** for Organization Description.

5. 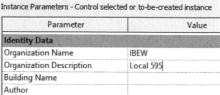 Left click in the Project Address field.

   Select the ... button located on the right.

6. 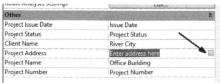 Type in an address.

   Click **OK**.

7.

For Project Issue Date: Type today's date.

For Project Status: Type **Client Review**.

For Project Number: Type **20110-67**.

Click **OK**.

8.

In the Properties palette:

In Checked By: Type **M. Teacher**.

In Drawn By: Type your first initial and last name.

In Sheet Name: Type **1st Floor Power Plan**.

9.

Note that the titleblock updates with the information entered in the Properties palette.

10. Save as *ex10-6.rvt*.

Exercise 10-7:

Creating a Custom Titleblock

Drawing Name: *none*
Estimated Time: 45 minutes

This exercise reinforces the following skills:

- ❑ Titleblock
- ❑ Import CAD
- ❑ Labels
- ❑ Text
- ❑ Family Properties

1.  Go to **File → New → Titleblock**.

2. 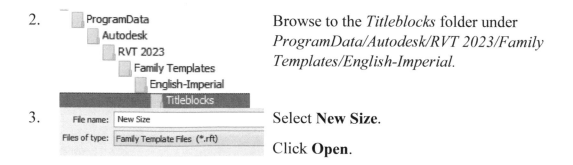 Browse to the *Titleblocks* folder under *ProgramData/Autodesk/RVT 2023/Family Templates/English-Imperial*.

3. File name: New Size
    Files of type: Family Template Files (*.rft)

    Select **New Size**.

    Click **Open**.

4.

Pick the top horizontal line.

Select the dimension and change it to **30″**.

Pick the right vertical line.

Select the dimension and change it to **42″**.

5.

Zoom In Region
Zoom Out (2x)
Zoom To Fit

Right click in the graphics window and select **Zoom to Fit**.

You can also double click on the mouse wheel.

6.

Import CAD

Activate the Insert ribbon.

Select **Import→Import CAD**.

7.

File name:	Architectural Title Block
Files of type:	DWG Files (*.dwg)

Locate the *Architectural Title Block* in the exercise files directory.

8.

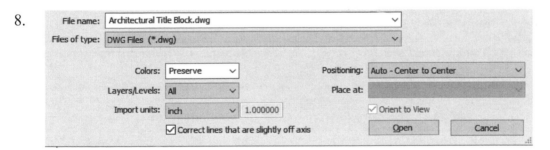

Set Colors to **Preserve**.
Set Layers to **All**.
Set Import Units to **Inch**.
Set Positioning to: **Auto - Center to Center**.
Click **Open**.

9.

Import detected no valid elements in the file's Paper space. Do you want to Import from the Model space?

Yes	No

Click **Yes**.

10. 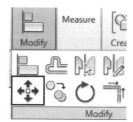 Select the title block.

Use the **Move** tool on the Modify panel to reposition the titleblock so it is aligned with the existing Revit sheet.

11. Use the RESIZE tool to scale the title block to fit on the sheet.

12.

Select the imported title block so it is highlighted.

Select **Explode → Full Explode** from the Import Instance panel on the ribbon.

You may need to adjust the position of the text after the explode operation.

We will place an image in the rectangle labeled Firm Name and Address.

13. Activate the **Insert** ribbon.

Select the **Import→Image** tool.

14.

Open the *ibewLogo250.png* file from the downloaded Class Files.

15.

Place the logo on the right side of the box.

Scale and move it into position.

To scale, just select one of the corners and drag it to the correct size.

16. **A**

Text

Select the **Text** tool from the **Create** Ribbon.

17.

Select **Edit Type**.

18.

Change the Text Size to **1/8″**.

Click **OK**.

Revit can use any font that is available in your Windows font folder.

19.

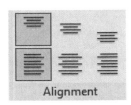

Select **Left** on the Alignment panel.

20.

Type in the name of your school or company next to the logo.

Position the text and logo so they look correct in the rectangular outline.

21. Select the **Label** tool from the Create ribbon.

Labels are similar to attributes. They are linked to project properties.

22.  Left pick in the Project Name and Address box.

23. Click **OK**.
Left click to release the selection.
Pick to place when the dashed line appears.

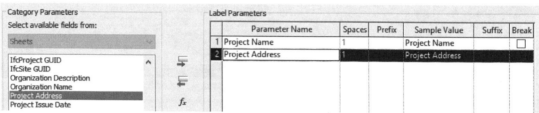

Select **Project Name**.
Use the **Add** button to move it into the Label Parameters list.
Add **Project Address**.
Click **OK**.

24.  Cancel out of the Label command.

Position the label.
Use the grips to expand the label.

25. 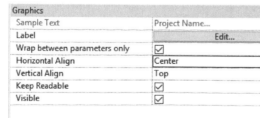 Select the label.

In the Properties panel:

Enable **Wrap between parameters**.

Verify that the Horizontal Align is set to Center.

26.

Project Name
Project Address

Use Modify→Move to adjust the position of the project name and address label.

27.

☑ 🔠 Edit Type

Select **Edit Type**.

28.

Duplicate...

Select **Duplicate**.

29.

Name: Tag 1/8" Arial

Change the Name to **Tag 1/8" Arial**.

30.

Change the Text Size to **1/8"**.

Click **OK**.

Text	
Text Font	Arial
Text Size	1/8"
Tab Size	1/2"
Bold	☐
Italic	☐
Underline	☐
Width Factor	1.000000

31.

Project Name and Address

Project Name
Project Address

Adjust the size and position of the label.

32.

A
Label

Select the **Label** tool from the Create ribbon.
Use the Type Selector to set the label type to **Tag 1/8" Arial**.

33.

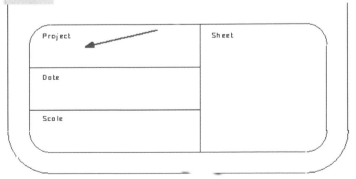

Left click in the Project box.

34. Locate **Project Number** in the Parameter list.

Move it to the right pane.

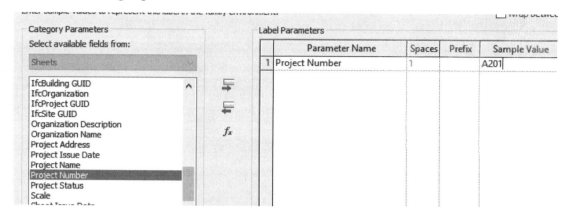

Add a sample value, if you like.

Click **OK**.

35.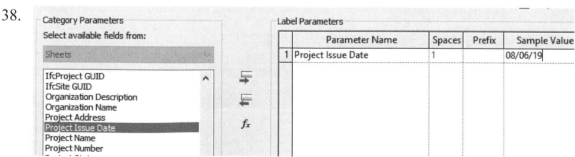

Left click to complete placing the label.

Select and reposition as needed.

36. ![Label] Select the **Label** tool from the Create ribbon.

Label

37. Date

Left click in the Date box.

Scale

38.

Parameter Name	Spaces	Prefix	Sample Value
1 Project Issue Date	1		08/06/19

Highlight **Project Issue Date**.
Click the **Add** button.
In the Sample Value field, enter the default date to use.
Click **OK**.

39.  Position the date in the date field.

40.  Left click in the Scale box.

41.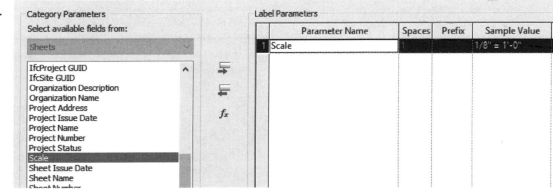

Highlight **Scale**.

Click the **Add** button.

Click **OK**.

42. 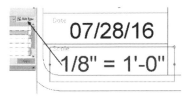 Highlight the scale label that was just placed and select **Edit Type** on the Properties panel.

43.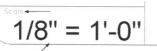

Set the Background to **Transparent**.

Click **OK**.

44.

Cancel out of the label command.

Note that the border outline and the block text are no longer hidden by the label background.

45.

Adjust the lines, text and labels so that everything looks clean.

46.

Select the **Label** tool.

47.

Left click in the Sheet box.

48.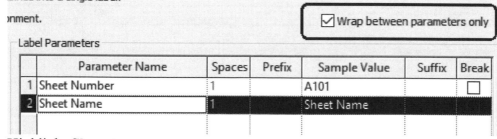

Highlight **Sheet Number**.
Click the **Add** button.
Highlight **Sheet Name**.
Click the **Add** button.
Enable **Wrap between parameters only.**
Click **OK**.

49.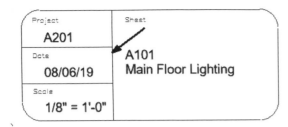

Shift the vertical divider line to the left to allow more space for the sheet name.

Position the Sheet label.

50. Save the file as *Titleblock 30 x 42.rfa.*

Exercise 10-8:

Using a Custom Title Block

Drawing Name: *titleblock_2.rvt*
Estimated Time: 10 minutes

This exercise reinforces the following skills:

- ❑ Title block
- ❑ Import CAD
- ❑ Labels
- ❑ Text
- ❑ Family Properties

1. Open *titleblock.rvt*.

2. Activate the Insert ribbon.

   Select **Load Family** in the Load from Library panel.

3.

   Browse to the exercise folder.

   Locate *Titleblock 30 x 42.rfa.*

   Select it and Click **Open**.

4. Activate the **01 -Main Floor lighting** sheet.

Sheet List
Sheets (all)
  00 - Sheet List
  01 - Main Floor Lighting
  02 - Ground Floor Lighting
  03 - Main Floor Power

5.

C 17 x 22 Horizontal

Search
B 11 x 17 Horizontal
  B 11 x 17 Horizontal
C 17 x 22 Horizontal
  C 17 x 22 Horizontal
D 22 x 34 Horizontal
  D 22 x 34 Horizontal
E1 30 x 42 Horizontal
  E1 30x42 Horizontal
Titleblock 22 x 34
  Titleblock 22 x 34
Titleblock 30 x 42
  Titleblock 30 x 42

Select the title block so it is highlighted.

You will see the name of the title block in the Properties pane.

Select the **Title Block 30 x 42** using the Type Selector.

6.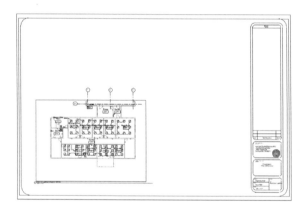

You may need to reposition the view on the title block.

7.

Zoom into the title block.

Note that the parameter values have all copied over.

8.

Activate the Manage ribbon.

Select **Project Information**.

9.

Fill in the dialog.

Organization Name: **IBEW**

Author: *Your initial and last name*

Project Issue Date: *Today's date*

Project Status: **In Design**

Client Name: **B. Brown**

Project Address: **123 Main St Anytown, USA**

Project Name: **ANYTOWN FIRE STATION 35**

Project Number **FS12**

Click **OK**.

10.

The titleblock updates with the project information.

ANYTOWN FIRE STATION 35
123 Main St
Anytown, USA

Project	Sheet
FS12	
Date	01
08/06/19	Main Floor Lighting
Scale	
1/8" = 1'-0"	

11. Save the file as *ex10-8.rvt*.

Revisions

Every project requires revisions. Revisions are tracked using a revision schedule. The revision schedule is a "nested" family that is hosted by the titleblock. When you make a change to the model and you want to issue a revision, you can draw a revision cloud in the view, or the cloud can be drawn on the sheet displaying the view. Revision clouds are "view-specific". This means they are only visible in the view where they are placed. Revision clouds are "tagged" using a revision tag which is linked to the parameters in the revision schedule.

Exercise 10-9:

Defining a Revision Schedule

Drawing Name: *revisions.rvt*
Estimated Time: 10 minutes

This exercise reinforces the following skills:

- Setting up Revision Control in a project.

1.    Go to **File**.

Select **Open→Project**.

Open *revisions.rvt.*

2.    Activate the **View** ribbon.
Select **Revisions** on the **Sheet Composition** panel.

This dialog manages revision control settings and history.

Numbering can be controlled per project or per sheet. The setting used depends on your company's standards.

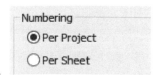

Enable **Per Project**.
One Revision is available by default. Additional revisions are added using the **Add** button.

3. 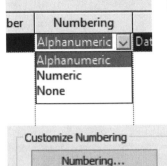 The visibility of revisions can be set to **None**, **Tag** or **Cloud and Tag**. Use the **None** setting for older revisions which are no longer applicable so as not to confuse the contractors.
Set the revision to show **Cloud and Tag**.

4. Select **Alphanumeric** under the Numbering column.

5. Click **Numbering** under Customize Numbering.

6. Highlight **Alphanumeric**.

    Click **New**.

7. Type **Alpha-1** for the Name.

    Enable **Alphanumeric**.

8. Change the sequence to remove the letters **I** and **O**.

    Click **OK**.

9.

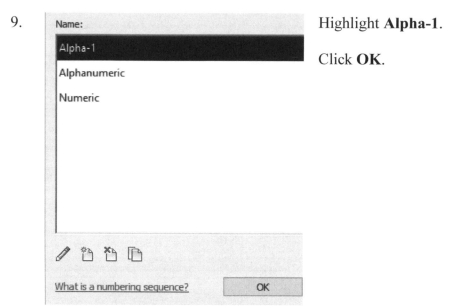

Highlight **Alpha-1**.

Click **OK**.

10. Enter the two revision changes shown.

Revision	Numbering	Date	Description	Issued	Issued	Issued by	
A	Alphanumeric	08.06	Change 96" Pendant Light to M125 Recessed Flange Light	☐	Joe	Sam	Cloud and Tag
B	Alphanumeric	08.07	Change Sconce Model to Sconce Light-Sphere	☐	Joe	Sam	Cloud and Tag

11. Click **OK** to close the dialog.
You can delete revisions if you make a mistake. Just highlight the row and Click
Delete.

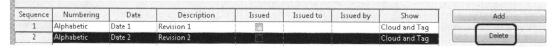

Sequence	Numbering	Date	Description	Issued	Issued to	Issued by	Show		
1	Alphabetic	Date 1	Revision 1	☐			Cloud and Tag		Add
2	Alphabetic	Date 2	Revision 2	☐			Cloud and Tag		Delete

12. Save the project as *ex10-9.rvt*.

Exercise 10-10:

Modify a Revision Schedule in a Title Block

Drawing Name: *revision_schedule.rvt*
Estimated Time: 20 minutes

This exercise reinforces the following skills:

- Titleblock
- Revision Schedules

1.  Go to **File**.

   Select **Open→Project**.

   Open *revision_schedule.rvt*.

2.  Open the **01 – Main Floor Lighting** sheet.

3.  Zoom in to the **Revision Block** area on the sheet.

   Note that the title block includes a revision schedule by default.

4.

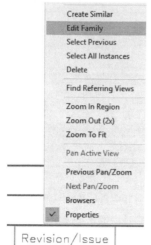

Select the titleblock.

Right click and select **Edit Family**.

5.

Window around the lines and text in the revision block to select.

Right click and select **Delete**.

6.

Go to the View ribbon.

Select **Revision Schedule**.

7.

Scheduled fields (in order):

Revision Sequence
Revision Number
Revision Description
Revision Date

The Scheduled fields should be:

- Revision Sequence

- Revision Number

- Revision Description

- Revision Date

*Do **NOT** remove the Revision Sequence field. This is a hidden field.*

8.

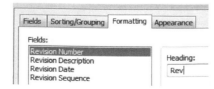

Select the Formatting tab.

Change the Revision Number Heading to **Rev**.

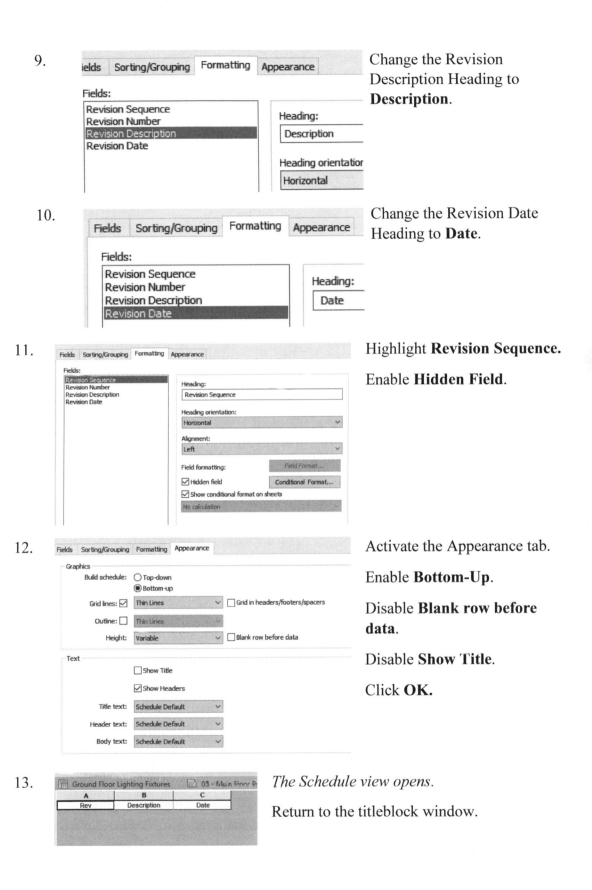

9. Change the Revision Description Heading to **Description**.

10. Change the Revision Date Heading to **Date**.

11. Highlight **Revision Sequence.**

   Enable **Hidden Field.**

12. Activate the Appearance tab.

   Enable **Bottom-Up.**

   Disable **Blank row before data**.

   Disable **Show Title.**

   Click **OK.**

13. *The Schedule view opens.*

   Return to the titleblock window.

14.

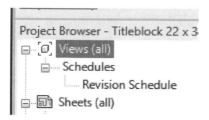

The Revision Schedule is now listed in the browser.

Drag and drop it into the correct location on the sheet above the firm name and address.

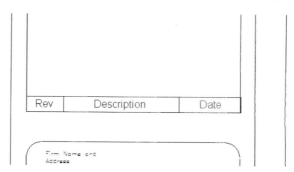

15.

Adjust the column width of the schedule using the grips so it fits properly in the title block.

16.

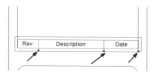

Save the file.

When prompted to replace the existing file, select **Yes**.

17.

Select **Load into Project and Close** on the Family Editor panel on the ribbon.

18. If you have more than one project open:

Place a check next to *revision_schedule.rvt.*

Click **OK**.

19.

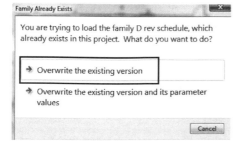

Select **Overwrite the existing version** if this dialog appears.

You will only see this dialog if you have already loaded the new title block in the project.

20. Save as *ex10-10.rvt.*

Exercise 10-11:

Add Revisions in a Title Block

Drawing Name: *revisions_2.rvt*
Estimated Time to Completion: 30 Minutes

This exercise reinforces the following skills:

- Add revision clouds to a view.
- Tag revision clouds.

1.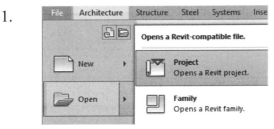

   Go to File.

   Select Open→Project.

   Open *revisions_2.rvt*.

2. Open the **01-Main Floor Lighting** sheet.

   Sheets (all)
   - 00 - Sheet List
   - **01 - Main Floor Lighting**
   - 02 - Ground Floor Lighting
   - 03 - Main Floor Power
   - 04 - Ground Floor Power

3.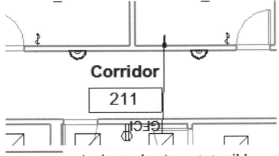

   Zoom into **Corridor 211**.

4.

   Activate the **Annotate** ribbon.

   Select **Revision Cloud** from the Detail panel.

5.

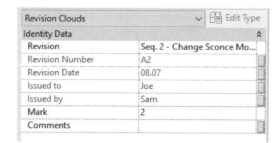

On the Properties pane:

Select the Revision that is tied to the revision cloud – **Seq 2 – Change Sconce.**

Type **2** in the Mark field.

Under Comments: Enter **Check with buyer on schedule**.

6.

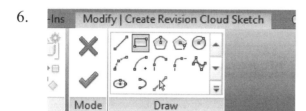

You can use any of the available Draw tools to create your revision cloud.

Select the **Rectangle** tool.

7.

Draw the Revision Cloud on the sconce indicated.

Note that a cloud is placed even though you selected the rectangle tool.

8.

Select the **Green Check** on the Mode panel to **Finish Cloud**.

9.

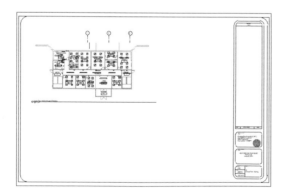

Sheets (all)
- 00 - Sheet List
- 01 - Main Floor Lighting
- **02 - Ground Floor Lighting**
- 03 - Main Floor Power

Open the **02-Ground Floor Lighting** sheet.

10. 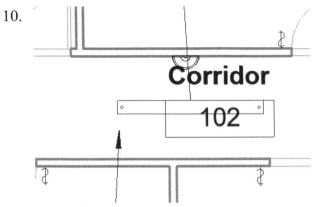 Zoom into **Corridor 102**.

11. 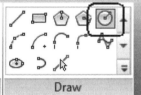 Activate the **Annotate** ribbon.

Select **Revision Cloud**.

12.

Revision Clouds		∨	Edit Type
Identity Data			
Revision	Seq. 1 - Change 96" Pendant Light t...		
Revision Number	A1		
Revision Date	08.06		
Issued to	Joe		
Issued by	Sam		
Mark	1		
Comments	Check with buyer on schedule		

Select the Revision that is tied to the revision cloud – **Seq 1 – Change 106" Pendant Light...**

Type **1** in the Mark field.

Under Comments: Enter **Check with buyer on schedule**.

13. 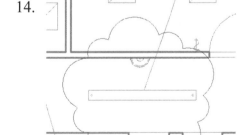 Select the **Circle** tool.

14. Draw the Revision Cloud on the lighting fixture to the left of the room tag.

Note that a cloud is placed even though you selected the circle tool. The first mouse click places the center of the cloud, so try to pick a point centered on the pendant light.

15. 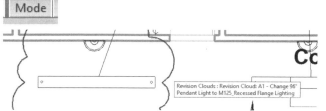 Select the **Green Check** on the Mode panel to **Finish Cloud**.

If you mouse over the revision cloud, you will see a tooltip to indicate what the revision is.

Revision Clouds : Revision Cloud: A1 – Change 96" Pendant Light to M125_Recessed Flange Lighting

16.

No.	Description	Date
A1	Change 96" Pendant Light to M125_Recessed Flange Lighting	08.06

Zoom into the title block and note that the revision block has updated with the revision that is marked on the sheet.

17.

A2	Change Sconce Model to Sconce Light-Sphere	08.07
Rev	Description	Date

Open 01-Main Floor Lighting sheet.

Zoom into the title block and note that the revision block has updated with the revision that is marked on the sheet.

18. Zoom into **Corridor 211**.

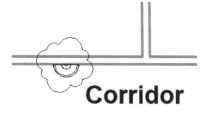

Corridor

211

19. 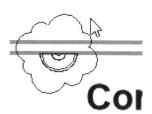 Select the **Tag by Category** tool from the Tag panel on the Annotate ribbon.

20. Left pick on the revision cloud to identify the category to be tagged.

Coi

21. There is no tag loaded for Revision Clouds. Do you want to load one now?

If you see this message:

Click **Yes**.

22.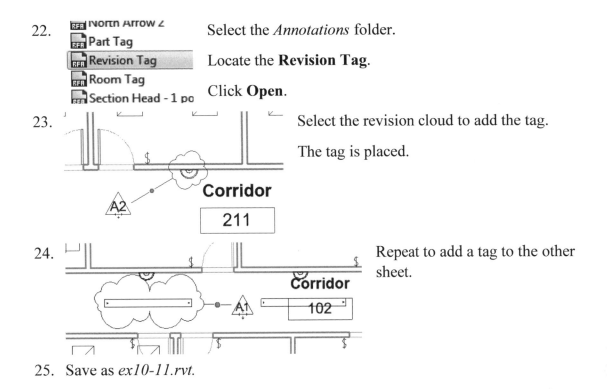

Select the *Annotations* folder.

Locate the **Revision Tag**.

Click **Open**.

23. Select the revision cloud to add the tag.

The tag is placed.

24. Repeat to add a tag to the other sheet.

25. Save as *ex10-11.rvt*.

View Lists

You can use a view list to see and modify parameters for multiple views at once. For example, suppose you include the Detail Level and Scale parameters in a view list. From the view list, you can change the detail levels of selected views to coarse, medium, or fine, or change view scales to use consistent settings. You can also use a view list to apply specific view templates. You can change the view name or view title that displays on sheets. By using a view list in this way, you can identify and correct inconsistent view settings using a schedule.

A lot of users use the view list to determine which views have been placed on which sheets and which views have not been consumed (placed on a sheet).

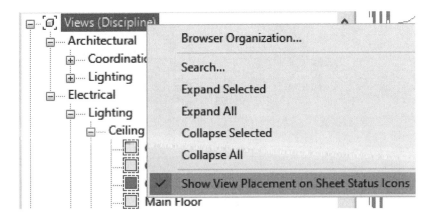

The project browser now displays whether a view has been placed on a sheet.

Exercise 10-12:

Using a View List to Check Sheets

Drawing Name: *view_list_2.rvt*
Estimated Time: 10 minutes

This exercise reinforces the following skills:

- ❑ View List
- ❑ Schedules
- ❑ Sheets

1.
    Go to **File**.

    Select **Open→Project**.

    Open *view_list_2.rvt*.

2. Activate the View ribbon.

    Select the **View List** tool under Schedules.

3.

Scheduled fields (in order):

Sheet Number
Sheet Name
View Name
Title on Sheet
View Template

Select the following fields:

- Sheet Number

- Sheet Name

- View Name

- Title on Sheet

- View Template

Click **OK**.

4.

The schedule opens.

Determine which views should be placed on sheets.

Review the view templates and view titles to determine if any changes need to be made.

5. Save as *ex10-12.rvt*.

Sheet Organization

Establish the organization of your drawing sheets in order to make sheet management consistent from project to project. Setting up sheet organization is similar to organizing the views in your Project Browser. You can set up the Browser – Sheets system family with different types to organize your sheets in the desired manner. To access the system family, right click on Sheets in the Project Browser and select Browser Organization.

Parameters can be applied to sheets and those parameters can be used to filter and sort the sheets. The parameters can be included as project parameters in your template so that when a new sheet is created, the parameters are automatically added to the sheet. You can also use the parameters to organize your drawing views.

Exercise 10-13:

Defining Sheet Organization

Drawing Name: *sheet_organization.rvt*
Estimated Time: 20 minutes

This exercise reinforces the following skills:

   ❑ Project Browser
   ❑ Project Parameters

1. 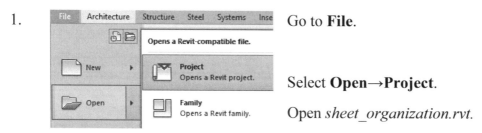 Go to **File**.

            Select **Open→Project**.

            Open *sheet_organization.rvt*.

2. Manage Select the **Manage** ribbon.

3. Project Parameters Select the **Project Parameters** tool.
  Shared Parameters
  Global Parameters

4. Select **New Parameter**.

  Closer
  Frame Silencer
  Hardware type
  Hinges
  Kickplate
  Lockset
  Occupant
  Sub-Discipline

5. Enable **Project parameter**.

6. In the Name field: Type **Sheet Type**.

   Enable **Instance**.

   Set the Type of Parameter to **Text**.

   Group Parameter under **Identity Data**.

7. Enable **Sheets** in the Categories pane.

   Click **OK**.

8. The new parameter is now listed.

   Click **OK** to close the dialog box

9. Highlight the **00-Sheet List** sheet in the Project Browser.

10. Type **Cover** in the Sheet Type field in the Properties palette.

    This is the new parameter you just created.

11.

Sheet Name	Main Floor Lighting
Sheet Issue Date	08/04/19
Sheet Type	Ceiling Plan

Sheets (all)
  00 - Sheet List
  01 - Main Floor Lighting

Highlight the **01-Main Floor Lighting** sheet in the Project Browser.

Type **Ceiling Plan** in the Sheet Type field in the Properties palette.

12.

Sheets (all)
  00 - Sheet List
  01 - Main Floor Lighting
    Reflected Ceiling Plan: Main Floor Lighting F
  02 - Ground Floor Lighting
    Reflected Ceiling Plan: Ground Floor Lighting

Highlight the **02-Ground Floor Lighting** sheet in the Project Browser.

13.

Sheet Issue Date	08/04/19
Sheet Type	
Appears In Sheet ...	Ceiling Plan
Revisions on Sheet	Cover
Other	

Select **Ceiling Plan** in the Sheet Type field in the Properties palette.

14.

Sheets (all)
  00 - Sheet List
  01 - Main Floor Lighting
    Reflected Ceiling Plan:
  02 - Ground Floor Lighting
    Reflected Ceiling Plan:
  03 - Main Floor Power

Highlight the **03-Main Floor Power** sheet in the Project Browser.

15.

Sheet Issue Date	08/04/19
Sheet Type	Floor Plan
Appears In Sheet ...	☑

Type **Floor Plan** in the Sheet Type field in the Properties palette.

16.

Sheets (all)
  00 - Sheet List
  01 - Main Floor Lighting
    Reflected Ceiling Plan: I
  02 - Ground Floor Lighting
    Reflected Ceiling Plan: (
  03 - Main Floor Power
  04 - Ground Floor Power

Highlight the **04-Ground Floor Power** sheet in the Project Browser.

17.

Sheet Issue Date	08/04/19
Sheet Type	
Appears In Sheet ...	Ceiling Plan
Revisions on Sheet	Cover
Other	Floor Plan

Select **Floor Plan** in the Sheet Type field in the Properties palette.

18. 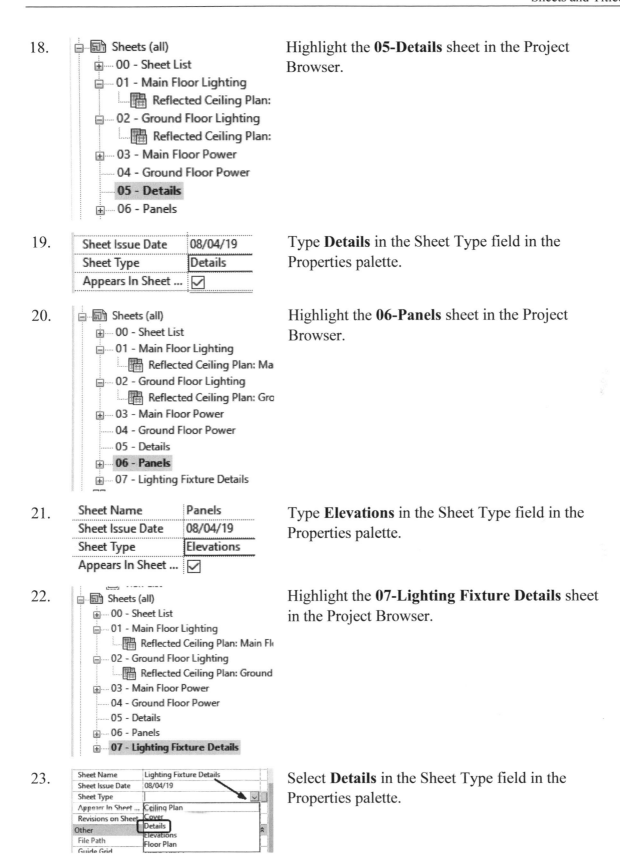 Highlight the **05-Details** sheet in the Project Browser.

19. Type **Details** in the Sheet Type field in the Properties palette.

20. Highlight the **06-Panels** sheet in the Project Browser.

21. Type **Elevations** in the Sheet Type field in the Properties palette.

22. Highlight the **07-Lighting Fixture Details** sheet in the Project Browser.

23. Select **Details** in the Sheet Type field in the Properties palette.

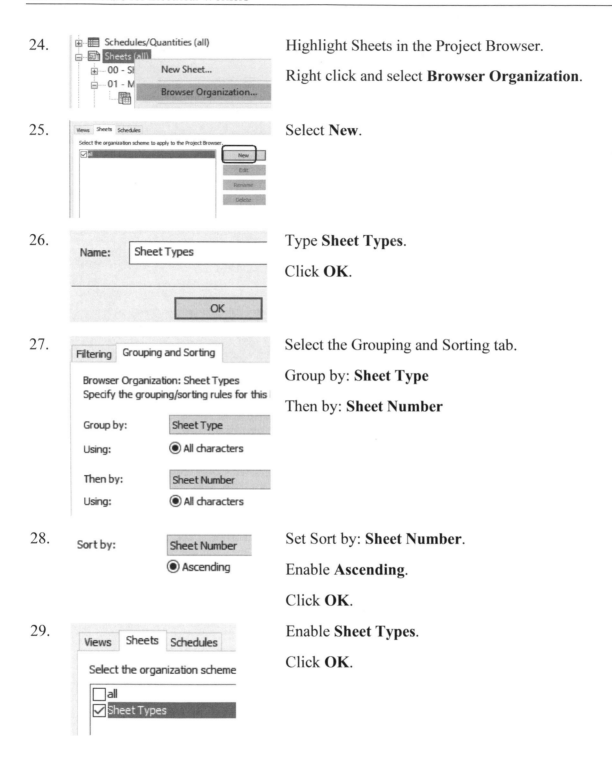

24. Highlight Sheets in the Project Browser.

Right click and select **Browser Organization**.

25. Select **New**.

26. Type **Sheet Types**.

Click **OK**.

27. Select the Grouping and Sorting tab.

Group by: **Sheet Type**

Then by: **Sheet Number**

28. Set Sort by: **Sheet Number**.

Enable **Ascending**.

Click **OK**.

29. Enable **Sheet Types**.

Click **OK**.

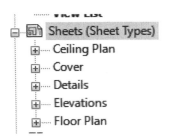

Sheets are now placed under the defined Sheet Types to make it easier to locate them.

30. Save as *ex10-13.rvt*.

Printing Documentation

You can print views and sheets to PDF (Portable Document Format) files.

The resulting PDF files can be shared with other team members, viewed online, or printed. When printing multiple views and sheets to PDF, you can specify whether each view or sheet is saved in a separate PDF file, or one PDF file contains all selected views and sheets. If you decide to save multiple views and sheets to individual files, you cannot cancel the print job once it starts.

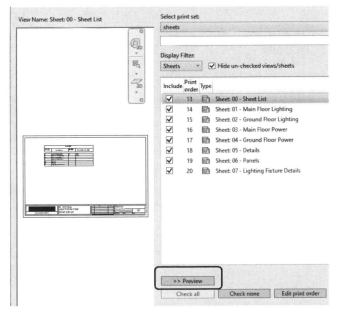

You can preview the views and sheets to be printed prior to committing to the print operation.

Exercise 10-14:

Printing a Documentation Set to PDF

Drawing Name: *plotting.rvt*
Estimated Time: 10 minutes

This exercise reinforces the following skills:
- ❏ Plot
- ❏ Sheets

1.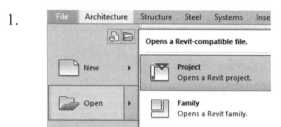

   Go to **File**.

   Select **Open→Project**.

   Open *plotting.rvt.*

2.

   Go to **File→Print→Print**.

3.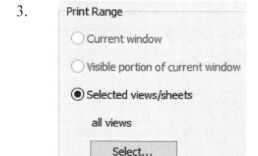

   Under Print Range:

   Enable **Selected Views/Sheets**.

   Click **Select**.

4. 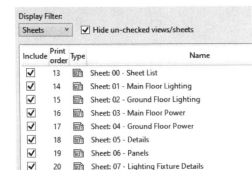   Enable **Sheets** under the Display Filter.

5.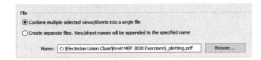

Verify all the sheets are checked.

Click **Select**.

6. 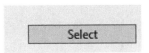   Select **Adobe PDF** as the printer.

7.   Enable **Combine multiple selected views/sheets into a single file**.

Use the Browse button to select the folder to place the PDF file.

If you don't combine into a single file, every sheet will be a unique file.

Click **OK**.

8.   Browse to the folder where the file is saved to review.

Close the project file without saving.

Lab Exercises

Open plotting.rvt.

Inspect the sheets, add missing views as needed, clean up the documentation by adding tags, dimensions, and assigning view templates.

About the Author

Elise Moss has worked for the past thirty years as a mechanical designer in Silicon Valley, primarily creating sheet metal designs. She has written articles for Autodesk's Toplines magazine, SolidProfessor, engineering.com, AUGI's PaperSpace, DigitalCAD.com and Tenlinks.com. She is President of Moss Designs, creating custom applications and designs for corporate clients. She has taught CAD classes at Laney College, DeAnza College, Silicon Valley College, and for Autodesk training centers. Autodesk has named her as a Faculty of Distinction for the curriculum she has developed for Autodesk products, and she is a Certified Autodesk Instructor. She holds a baccalaureate degree in mechanical engineering from San Jose State.

She is married with two sons. Her older son, Benjamin, is an electrical engineer. Her younger son, Daniel, works with AutoCAD Architecture in the construction industry. Her husband, Ari, has a distinguished career in software development.

Elise is a third-generation engineer. Her father, Robert Moss, was a metallurgical engineer in the aerospace industry. Her grandfather, Solomon Kupperman, was a civil engineer for the City of Chicago.

She can be contacted via email at elise_moss@mossdesigns.com.

More information about the author and her work can be found on her website at www.mossdesigns.com.

Other books by Elise Moss

AutoCAD 2023 Fundamentals
Autodesk Revit 2023 Architecture Certification Exam Study Guide